Praise for *Our Country*

"'History is written by the victors,' but not this book. Cook provides a comprehensive, meticulously detailed and accurate account of the events and forces that shaped the U.S. from its creation, through its rise and current decline. The plight of the native inhabitants at the hands of the colonists and the supremacy of money powers and the deep states they control in shaping U.S. and global policies are covered in depth. This is a fascinating must read for understanding the past and the post-neoliberal, civilizational world emerging from the current tectonic geopolitical change."

DR. FADI LAMA, International Adviser for
the European Bank of Reconstruction and Development (EBRD)

"Richard Cook takes you through an incredible journey of America's true history with a sensitivity to the evil that was never fully extracted from the USA after the revolution of 1776 while also shedding light on the courage and goodness which gave rise to the best of the republic over its young life. If you want a look towards true hope for the future centered on solutions while also acknowledging the darkness and traps facing us in these precarious times, then read *Our Country, Then and Now.*"

MATTHEW EHRET, Editor-in-Chief *Canadian Patriot Review*,
author of *The Clash of the Two Americas*

Our Country, Then and Now

RICHARD C. COOK

Clarity Press, Inc.

©2024 Richard C. Cook

ISBN: 978-1-949762-85-3
Ebook ISBN: 978-1-949762-86-0

In-house editor: Diana G. Collier
Book design: Becky Luening

Photo credit (upper): The reconstructed colonial Capitol building in Williamsburg, Virginia. On May 15, 1776, in this building, the Fifth Virginia Revolutionary Convention resolved to instruct the Virginia delegates to Congress in Philadelphia to enter a motion for American independence. In accordance with these instructions, Richard Henry Lee of Virginia presented a motion to Congress on June 7 that stated: "Resolved, that these United Colonies are, and of right ought to be, free and independent States, that they are absolved from all allegiance to the British Crown, and that all political connection between them and the State of Great Britain is, and ought to be, totally dissolved." (Wikimedia Commons, Photo/Smash the Iron Cage)

Photo credit (lower): The U.S. Capitol is seen through a fence with barbed wire during the second impeachment trial of former President Donald Trump in Washington, Friday, February 12, 2021. (AP Photo/Jose Luis Magana)

ALL RIGHTS RESERVED: Except for purposes of review, this book may not be copied, or stored in any information retrieval system, in whole or in part, without permission in writing from the publishers.

Library of Congress Control Number: 2023946989

Clarity Press, Inc.
2625 Piedmont Rd. NE, Ste. 56
Atlanta, GA 30324, USA
https://www.claritypress.com

This book is dedicated to:

Adele, Frederick, Melissa, Nathaniel, Timothy;
Amy, Julie, Marissa, Rose

Table of Contents

Chapter 1: Beginnings. 1
Chapter 2: Money and Banking 20
Chapter 3: Varied Paths. 39
Chapter 4: "Manifest Destiny" 63
Chapter 5: Civil War 84
Chapter 6: "The Crime of 1873". 106
Chapter 7: The Gilded Age 125
Chapter 8: "Rule Britannia" 151
Chapter 9: The Money Trust. 172
Chapter 10: "The War to End All Wars". 191
Chapter 11: The Roaring Twenties and the Depression 206
Chapter 12: The National Security State. 230
Chapter 13: Nightmare. 251
Chapter 14: The Rockefeller Republic 276
Chapter 15: "Springtime in America" Faces Early Frost . . . 309
Chapter 16: The Bush/Cheney Catastrophe 336
Chapter 17: U.S. Governance Hits the Wall 365
Chapter 18: Betrayals and Challenges 405
Appendix: Monetary Reform 429
Acknowledgements. 432
Index 435

CHAPTER 1

Beginnings

Introduction

In late 2023, the question arising from the failed U.S. proxy war against Russia in Ukraine is: Are we witnessing the collapse of U.S. foreign policy? When nations fail, it's often a slow-motion train wreck. But at some point, the locomotive falls off the track. Can that day be arriving? If so, we will then be faced with the task, first, of picking up the pieces, and then of finding a way towards recovery, possibly in an entirely new direction. My question is: *What should that direction be?*

This topic is glossed over in the wasteland of the mainstream media which is overseen by the "Deep State"[1] and follows its usual practice of dumbing down the masses and pandering to those in power. Yet the U.S. decline is being treated exhaustively by many independent writers both in the U.S. and abroad. Still, it's easy to mistake the trees for the forest. My aim is to offer a bird's eye view of that forest. Naturally, this will take some time; hence this book, which is more a personal testimony than an attempt at a rigorous scholarly defense of the issues addressed herein.

A Personal Witness

I am a personal witness to the forest, as have been my American ancestors. I intend to enliven the story with first-person accounts from my own involvement in events, and I also provide anecdotes and narratives from my forebears, going back to the days of the first European settlements.

I was born in Montana, grew up in Michigan and Virginia, spent my federal government career in Washington, DC, and now live in Maryland. The firsthand portions of the narrative may be somewhat skewed to these places. But I believe they are representative of the U.S. as a whole.

My first ancestor from Europe, Thomas Bliss, arrived in Massachusetts from England in 1636 as a religious dissenter escaping persecution. He was a Puritan. I also have Native American ancestry through my French-Canadian grandfather. My male forebears fought in many of the American wars through

WWII, including the Revolutionary War, the War of 1812, the Mexican War, the Civil War, and the 20th century world wars.

A great-great-grandfather on my mother's side arrived from Ireland during the Great Famine and fought for the Union in the American Civil War. A great-grandfather on my father's side acquired land in an Oklahoma land rush, was a friend to the Shawnees and spoke their language.[2]

My grandfather on my father's side was a gambler and card sharp. My grandfather on my mother's side was a scaler in the Anaconda logging camps near Seeley Lake, Montana. When young, my father spent time in and around Montana's Flathead Reservation. All these people and places have a part in my story.

While my mother grew up in Montana, she later became a tour guide for Colonial Williamsburg. My father was the first in his family to attend college and became a chemist for Dow Chemical. When we moved to Virginia when I was thirteen, we lived a mile from the CIA's training center at Camp Peary, not far from the largest U.S. conglomeration of military bases at Hampton Roads. This area was the location of three of America's most historic sites—Jamestown, Williamsburg, and Yorktown—as well as one of the centers of contemporary moral darkness.

I graduated from the College of William and Mary, the *alma mater* of Thomas Jefferson, and worked as an analyst for the federal government for thirty-two years at several civilian agencies, including the Jimmy Carter White House. I also taught history for two years and worked a farm in West Virginia for a year.

I became a whistleblower during the Challenger disaster then spent the rest of my career at the U.S. Treasury Department. One of my major focuses there was monetary and financial history. I learned that an overwhelming portion of American's present dysfunction has financial roots.

I have five children and seven grandchildren. Since I retired to become a writer on public policy over the last fifteen years, I have watched U.S. foreign affairs slam into the wall time and again. The Biden administration and its apocalyptic trajectory may be taking this country close to the end of the line. Matters are that serious.

America Becomes Home to a Distinct Non-European Culture

The Europeans who emigrated to the Americas came due to a variety of motives. Some came for religious or political freedom, some for physical survival by leaving a Europe that was suffocating to the working and farming classes, some for adventure, some hoping for wealth, and some as fugitives fleeing prosecution. Immigration continues to this day and is an urgent political issue.

My earliest direct ancestor was a member of the Bliss family, which originally crossed the English Channel to England from France after the Norman Conquest in 1066. They came from Blois in northern France, hence the family name. The Protestant Reformation came five centuries later, leading to some members of the Bliss family becoming religious dissenters and suffering persecution at the hands of the Church of England.

Today the Bliss family is prolific both in Britain and the U.S. My direct ancestor was Thomas Bliss of Devonshire, England. We start with his father:

> Thomas Bliss, the progenitor, lived in Belstone Parish, Devonshire... He was a wealthy landowner and a Puritan, persecuted on account of his faith by both civil and religious authorities under the direction of the infamous Archbishop Laud, [and] was maltreated, impoverished, and imprisoned.
>
> He was reduced to poverty and his health ruined by the persecution of the Church of England. He is supposed to have been born about 1550 or 1560. The year of his death was 1635. When the parliament of 1628 assembled, Puritans, or Roundheads as the Cavaliers called them, accompanied the members to London.
>
> Two of the sons of Thomas Bliss, Jonathan and Thomas, rode from Devonshire on iron grey horses and remained for some time in the city, long enough at least for the king's officers and spies to learn their names and condition, and whence they came; and from that time forth, with others who had gone to London on the same errand, they were marked for destruction.
>
> They were soon fined a thousand pounds for non-conformity and thrown into prison, where they remained many weeks. Even old Mr. Thomas Bliss, their father, was dragged through the streets with the greatest indignity. On another occasion the officers of the high commission seized all their horses and sheep, except one poor ewe that in its fright ran into the house and took refuge under a bed.
>
> At another time the three brothers, with twelve other Puritans, were led through the marketplace in Okehampton with ropes around their necks, and fined heavily, and Jonathan and his father were thrown into prison, where the sufferings of the son eventually caused his death. At another time, the king's officers seized the cattle of the Bliss family and most of their household

goods, some of the articles of furniture being highly valued for their beauty and age, since they had been in the family for centuries.

In fact, the family was so reduced in circumstance that it was unable to secure the release of both Jonathan and his father, so the younger man had to remain in prison, and at Exeter he suffered thirty-five lashes with a three-corded whip, which tore his back in a cruel manner. Before Jonathan was released the estate had to be sold.[3]

After Thomas Bliss, Sr. and his son, Jonathan, died from maltreatment by crown officials, his other two sons, one also named Thomas, my ancestor, emigrated to America. There are several extant accounts of his life.

Name: Thomas Bliss. Sex: M. Birth: in Daventry, England. Death: 7 Jun 1649 in Rehoboth, Bristol County, MA…. Thomas Bliss was born circa 1590 in Daventry, England, and married there Nov. 22, 1614, (1) Dorothy Wheatley, daughter of John Wheatly. She died in England, mother of four children. Thomas married (2) the widow of Nicholas Ide and mother of [the younger] Nicholas Ide who was born in England, circa 1624.

Thomas brought his family to New England in 1636 and took up residence with or near his uncle [also named] Thomas Bliss, who lived on the south shore of Boston Bay. His other uncle, George Bliss, was then living in Lynn, Mass. on the north side of the Bay. They arrived too late in the year to build new homes, and they had to buy houses where they were available…. In 1643 the family left Weymouth with the first group of settlers for Seekonk, which in time was renamed Rehoboth. Thomas, a blacksmith, held Commonage Rights…there he participated in the early land divisions. He died there in 1649, his will proved at Plymouth, June 8, 1649.

I would characterize Thomas Bliss, Jr., as an early American, not simply an Englishman or European. Unlike the French and Spanish, the English settled America through private initiatives, not government-directed expeditions. As a blacksmith, Thomas Bliss was an artisan of the local economy and his own boss.

Not a wealthy man, he yet owned his home, ran his own business, took part in local government, and was able to bequeath his children some items of value. I believe he was as well off in things that count as most people in our own day, despite our gadgets and complexities. He appears to have had no debt.

Native Americans

The New World which the Europeans began to settle after Columbus's 1492 voyage was occupied by millions of Native Americans, termed "Indians," who had a complex history with diverse cultures and languages going back at least 25,000 years and possibly longer.

Modern genetic research has confirmed that they first migrated to North America from Siberia, though some Native Americans contend they have always been here and point to creation legends whereby their people emerged from the earth or came down from the stars. Many ancient peoples worldwide have legends of origin from the Pleiades. Native Americans have inhabited the Western Hemisphere for at least fifty times as long as people of European origin.

It has been said that the "discovery of America" by Columbus initiated the greatest demographic disaster in known history.

My friend Calvin D. Trowbridge, Jr., has written about the Native American "holocaust." Native American resistance to European settlers began to collapse by 1607–1620, taking place through what Trowbridge called *Three Bloody, Diseased, Deadly Decades*.[4] Around the year 1600, on the eve of English settlement, the present continental U.S. had a native population of at least fifteen million. The population of England was then only about four million.

By 1900, three centuries later, there were only about 250,000 full-blooded Native Americans remaining in the U.S., though, as reflected in census demographics, that number climbed steadily to about 9.7 million by 2020, bolstered by those identifying as mixed race.

Disease and the collapse of living standards due to white aggression, poverty, and war caused the huge decline in Indian populations. The recurring pattern was, first, the arrival of devastating pandemics, the biggest killers being smallpox, plague, influenza, malaria, and measles. Next, the Europeans themselves, who had transmitted disease through contact with the natives, appeared in larger numbers.

Massachusetts and the rest of New England began to be settled after 1620. At first, both Indians and the white settlers tried to live together on the basis of informal agreements or written treaties, but hostilities soon broke out when the whites began to covet Indian lands and tried to play off disparate bands of Indians against each other.

Violence between the whites and Indians became endemic, despite the efforts of a few uncondescending settlers like Roger Williams of Rhode Island. There have always been some among the whites who respected the Indians and tried to find ways to live in peace. But these were exceptions. The worst of the New England violence was the Pequot War of 1636–1637, followed by King Phillip's War of 1675–1678, when the decisive battles for control of New England were fought.

The English had what to them seemed good intentions. It was only six years before the start of the Pequot War that Puritan leader John Winthrop had given his sermon on *A Model of Christian Charity* on March 21, 1630, at Holyrood Church in Southampton, England. This was before the first group of Massachusetts Bay colonists embarked on the ship *Arbella*. Winthrop quoted the *Gospel of Mathew* (5:14) where Jesus warns, "A city on a hill cannot be hid."

Winthrop told his fellow Puritans that their new community would be "as a city upon a hill, the eyes of all people are upon us." He said:

> If we shall deal falsely with our God in this work we have undertaken and so cause him to withdraw his present help from us, we shall be made a story and a byword through the world.

I would submit that the jury is still out on John Winthrop's warning.[5]

In contrast to the maltreatment of the Indians by the English, the French who settled what is now Canada were more tolerant, the men often marrying native women. Meriwether Lewis of the Lewis and Clark expedition wrote in 1804 of the French Canadians he encountered: "Not an inconsiderable portion of them can boast a small dash of the pure blood of the aborigines of America."[6]

One possible reason for the differing treatment of Indians by the English and French was that the English desired land, while the French wanted trade. Though the French established the European-style cities of Quebec and Montreal, the northerly climate of New France did not particularly favor farming, while the region of the St. Lawrence River and the Maritime Provinces offered vast numbers of fish to be harvested to go along with a seemingly unlimited supply of beaver and deerskins further inland.

So, the French trappers and traders roamed the expanses of North America to collect the furs, purchasing them from the Indians across an enormous territory. They paid the Indians with trade goods that included beads which the Indians treated as money, cooking utensils and other brass and iron products, and eventually knives and muskets.

Jesuit priests followed the fur traders on missions of conversion, sometimes succeeding and sometimes not. One of the nations they encountered was the Iroquois. The Iroquois of what became Ontario and New York were the most vociferous in brokering the fur trade with, first, the French, and later the English, leading to the internecine Beaver Wars among the Iroquois and rival tribes. These tribes had been decimated by disease and were sometimes exterminated altogether.

A huge number of Native Americans died from white man's diseases, especially smallpox. Though the Indians often fought back against white encroachment, there were also tribal members willing to sell the land and relocate to more remote locations or, in the West, to the shadows of frontier Army forts.

Eventually, most of the surviving Indian tribes east of the Mississippi were pushed out, leaving behind only individuals who blended into the woods and wastelands, with their descendants often remaining there today. Examples of still-intact eastern Indian communities are the Iroquois in New York State and southern Canada, a substantial Cherokee community in North Carolina, the Lumbee Indians of North Carolina, and the Seminoles in Florida.

The American West is another story, with its large number of federally recognized reservations, some of them of great physical expanse but often arid and poor in resources. While many western Indians live as individual or family residents in cities and towns, Indian country in general has growing political power through its multitude of surviving, though often struggling, tribal communities.

The Indians have also begun to attain a measure of financial independence through monetary settlements with the government for past abuses and acquiring the authority to operate casinos. Among the most successful in adapting to modern conditions while retaining Indian identity have been the Salish, Pend d'Oreilles, and Kootenais of the Flathead Nation in western Montana. More about these tribes later.

Speaking too briefly of Central and South America, the Spanish and Portuguese enslaved the Indians, at least the ones that would accept Christianity, or drove them to the bottom of their caste system. These were often peoples who had built rich and advanced civilizations, such as the Aztecs, Mayans, and Incas.

Today in Latin America, some countries have a much more prominent native heritage and culture than others. Compare, for instance, Mexico with Argentina. The times my wife Karen and I have spent in the native areas of

Yucatan in Mexico at the Mayan ruins in places like Cobá and Tulum, as well as in living Mayan settlements, have been precious to us.

On the treatment of Native Americans by their conquerors, Stephen E. Ambrose in his book *Crazy Horse and Custer: The Parallel Lives of Two American Warriors* writes:

> From the time of the first landings at Jamestown, the game went something like this: you push them, you shove them, you ruin their hunting grounds, you demand more of their territory, until finally they strike back, often without an immediate provocation so that you can say, "They started it." Then you send in the Army to beat a few of them down as an example to the rest. It was regrettable that blood had to be shed, but what could you do with a bunch of savages?[7]

African Americans

Native Americans often kept members of rival tribes as slaves and were enslaved in turn by the Spanish in New Mexico and California. But starting in 1619, with arrival of the first slave ship at Jamestown, the English colonists began to import black Africans to work the plantations in the South.

The ruins of old Jamestown can still be seen along the James River in the Colonial National Historical Park. The story of Jamestown is a tapestry of white, black, and Indian cultures that lasted nearly a century before the capital of the Virginia colony moved to Williamsburg in 1693. Descendants of the Africans who arrived here in the 17th century have a much longer family history in this country than most white Americans.

The slave trade earned fortunes for European and American ship owners, sea captains, merchants, and financiers. Some modern American family fortunes originated with slave transport and trading. Baltimore was an early center for the slave trade. A couple of miles from our home in Maryland, a slave auction block still stood on a street corner until it was removed a couple of years ago.

Mixed Race Ancestry

We should note that a large number of U.S. residents today are of mixed-race ancestry. This includes not only the 3.2 percent of those who identify through the U.S. Census as mixed race, but also many European, African, Hispanic, Native American, or Asian people who have ancestors of different races in their genetic make-up.

There are many whites in the American South who have a degree of African ancestry and many in the West who have some Native American. What might have been an invitation to tolerance has often been a cause of shame, though less so today than in the past. Today, DNA research has made our genetic background more accessible. An interesting study on the subject is "The Genetic Ancestry of African Americans, Latinos, and European Americans Across the United States," published by the *American Journal of Human Genetics*.[8]

The Melting Pot

This short account can't begin to give due credit to all the people from other civilizations and cultures from around the world who have come to the U.S. over the centuries that give some degree of truth to the phrase, "Melting Pot"—if one excludes the dispossessed natives confined to reservations and the Africans who were first enslaved, then their descendants segregated. This includes the countless number of refugees from the wars, many American-instigated, that have devastated the world in the past century, bringing people from China, Korea, the Philippines, Southeast Asia, India, the Middle East, Africa, the Pacific Islands, and elsewhere, including those fleeing the current conflict in Ukraine.

In the chapters that follow, many of these conflicts will be discussed. We might also mention the large number of immigrants to the U.S. from Africa, India and elsewhere to address domestic labor shortages, especially in skilled professions like medicine. There are also the many immigrants gaining residential or citizenship rights through marriage. Often immigrants are brought in to work at lower rates of pay than their American counterparts.

Independence and the Constitution

By the mid-18th century, the white inhabitants of the British colonies on the Eastern Seaboard had grown in population and prosperity to the point of achieving political independence from the British Crown. The British had alienated the Americans beyond recall through the economic imperialism directed by their Parliament.

In 1775, British North America had about 2.5 million residents. About 500,000 of these were enslaved Africans, roughly 20 percent of the total. During the Revolution that began that year, what became Canada remained a British colony and became home to the Loyalists who fled the thirteen lower colonies. Quebec had been a colony of Great Britain since the Treaty of Paris in 1763 which ended the Seven Years War, known in America as the French and Indian War.

Taking ideas from the 1215 Magna Carta, along with English common law, and elaborated by concepts from the European Enlightenment, including that of a social contract and the existence of natural human rights, the U.S. colonies formed a republic whose constituent states retained considerable power. This was foreshadowed by a number of republics emerging in Europe on a smaller scale during the Renaissance, mainly Italian city-states, most notably Venice and Florence. The Dutch Republic began in 1579–1588. England's own attempt to establish a republic failed after the Civil War in the mid-1600s.

The U.S. republic became an overnight international sensation. The guiding genius of the Revolution was Philadelphia printer, author, and American statesman, Benjamin Franklin. The philosophers of the Enlightenment who contributed to the concept of human rights and freedom were among the influences brought to bear on U.S. independence.

But English philosopher Thomas Hobbes (1588–1679) had also concluded that humanity's natural condition is a state of perpetual war, fear, and amorality. Thus, only a balanced government could hold a society together.

The U.S. Constitution vested considerable power with the landowning oligarchs and town merchants. Individuals of a more democratic bent were able to secure a Bill of Rights guaranteeing certain individual freedoms. The enslaved Africans were not included in these freedoms, nor were the Indians. Over time, it was the local financiers in alliance with those of Europe who gained ascendancy.

Another Constitutional innovation was to establish a large and growing free trade area among the component states, potentially almost unlimited in territorial extent. Already, through freedom of trade and a robust and growing population, the U.S. was on its way to becoming a power on a par with many European states. Most of those who provided leadership were Protestants. The only colony where Catholicism had gained a foothold was Maryland. So true freedom of religion was not yet realized.

At a time when the British had been moving toward greater accommodation with the Indians, the Revolution freed the U.S. government and the constituent states to step up their extermination/displacement campaigns to the west along the frontier and into the Ohio territory. During the war, General George Washington waged a devastating campaign against the Iroquois, most of whom were allied with Britain. The Iroquois had established farms and orchards on their land in New York. The U.S. Army destroyed their cornfields and cut down the fruit trees.

The Iroquois also engaged in atrocities against the Americans, as at the Wyoming Massacre on July 3, 1778, in Luzerene, Pennsylvania where 360 American settlers were killed by a force of 1,000 Loyalists and Iroquois under the command of British Colonel John Butler. One of my direct ancestors, forty-nine-year-old Joseph Ogden, was killed there and buried with the others in a mass grave.

When the Revolutionary War ended in 1783, with the U.S. acquiring land from the Atlantic Ocean to the Mississippi River and from Canada to Florida, the question was how to deal with the Indians who had lived there from ancient times. The treatment meted out was harsher than the British had ever contemplated.

The difference in interests between the slaveholding South and the free-labor North immediately asserted itself. In the South, the demands of plantation agriculture caused the planters to push west from the Carolinas and Georgia into Alabama and Mississippi, eventually ending in formal action by the federal government to banish the southern Indian tribes to the land beyond the Mississippi River. Hence, the Cherokee Trail of Tears.

The Seminole wars against the southeastern tribes went on for generations, driving the remaining Indians and escaped Africans deep into Florida. These never surrendered but were effectively out of reach until recently. Now, having eventually started a successful bingo parlor and casino, the Seminoles are owners of the Hard Rock International business empire.

In the Ohio country—the Old Northwest—free farmers began to move in by the thousands to cultivate the fertile farmlands, first in the western Appalachian foothills, then the level expanses of Ohio, Indiana, Illinois, and northwards into Michigan and Wisconsin. The Indians of this huge region resisted vigorously. While the Indian wars with the Plains tribes are the most famous, the battles between the U.S. Army and the tribes of the Ohio Valley were the bloodiest.

"Right of Discovery" and "Right of Conquest"

In 1823, Chief Justice John Marshall defined the doctrine of the "right of discovery" and how it related to the "right of conquest" under which the whites had been acting toward the Indians since first contact:

> The United States...have unequivocally acceded to that great and broad rule by which its civilized inhabitants now hold this country. They hold, and assert in themselves, the title by which it was acquired. They maintain, as all others have maintained, that

discovery gave an exclusive right to extinguish the Indian title of occupancy, either by purchase or by conquest.[9]

In an 1831 decision, Marshall's court held that Indian tribes were "domestic dependent nations," without an inherent right to sovereignty and self-determination. This doctrine ruled U.S. Indian policy until limited sovereignty was acknowledged for federally recognized tribes in 1935.

Marshall traced the right "to extinguish the Indian title of occupancy" to the authority asserted by the British crown that had passed to the U.S. through its achievement of independence. This was different from the idea expressed by some, especially New Englanders, that *God had given the land* to the whites or at least designated them as lawful successors to Indian habitation.

Under the law as interpreted by Marshall, no presumption of "divine right" was asserted. Such an assertion was neither possible nor necessary to a republic. Under a republican form of government, such a right was simply proclaimed. The concept of "right of conquest" dared anyone to come along and take that "right" away.

A word for this attitude is *hubris,* which was viewed by the Greeks as a fatal flaw. That flaw, in my opinion, was what Chief Justice John Marshall, was already expressing in 1823 and that the U.S. has been acting out in the modern world, an attitude that is well-known across the globe. Such arrogance cannot go unpunished by this principle that finds expression across all civilizations—"Do not be deceived: God is not mocked, for whatever one sows, that will he also reap."[10]

Thomas Jefferson and the Louisiana Purchase

Living as I did in Williamsburg, Virginia, from the age of thirteen to twenty-three, and attending the College of William and Mary, I experienced Thomas Jefferson's memory and legacy close-up. Contributing to this, I watched Colonial Williamsburg's introductory film, *The Story of a Patriot*, many times, which featured Jefferson as a colonial-era member of the House of Burgesses.

As primary author of the *Declaration of Independence* Jefferson stated his belief in the "inalienable rights" of "life, liberty, and the pursuit of happiness." Ambiguous words to be sure. We still ask today how much thought he gave to women, enslaved Africans, or Indians. Regardless, Jefferson was one of the great men of the age with a profound influence on world history. As U.S. president, Jefferson carried out the Louisiana Purchase.

After France was pushed out of North America through the Seven Years War, Spain took over the territory from west of the Mississippi River to the Continental Divide that separates the eastern and western slopes of the Rocky Mountains. From that point, a traveler might cross the remaining mountains to the Pacific Ocean, passing where California's northern border lies today. From there, Spain ruled over California, New Mexico, and the area later called Texas. The regions in the Northwest, the Oregon Country, were claimed by both the U.S. and Britain.

To the south, the capital of the French-held Louisiana Territory was New Orleans at the Mississippi's mouth. New Orleans had been founded by French explorers in 1718, who controlled the trade moving down the vast expanse of the Mississippi-Ohio-Missouri basin to the Gulf of Mexico. France ceded Louisiana to Spain in 1769. But with the defeat of Spain by Napoleon, his brother Joseph Bonaparte, now the King of Spain, returned Louisiana to France. For U.S. President Thomas Jefferson, who had long coveted the entire western region for eventual U.S. expansion, the fact that the great Napoleon, instead of the decrepit Spanish Empire, now controlled the Louisiana Territory, was appalling.

Britain and Napoleon's France had been locked in a death struggle that would not be resolved until the Battle of Waterloo in 1815. Jefferson spent much of his eight-year presidency trying to maneuver between the two, with limited success. When he had tried to abolish America's trade with either in order to maintain American neutrality, the U.S. economy was devastated.

But once Napoleon gained control of the port of New Orleans, Jefferson realized that U.S. independence from Britain could be jeopardized. He wrote a letter that would ring presciently over a century later, when the U.S. tried to maintain neutrality during World War I—yet another war involving Great Britain and a continental rival—"The day that France takes possession of New Orleans fixes the sentence which is to restrain her forever within her low water mark. From that moment we must marry ourselves to the British fleet and nation."[11]

To avoid the fate of perpetual dependence on Britain, Jefferson now asked France whether it would like to avoid any future possibility of war with the U.S. over Louisiana by ceding the territory in its entirety. When France refused, Jefferson offered $2 million for New Orleans alone, but France refused again. Then, as suddenly as news traveled overseas by ship in those days, Jefferson learned that France was offering to sell the Louisiana Territory in its entirely for $15 million. It appeared that by now, Napoleon had given up any hope of outflanking Britain by reestablishing a French North American empire.

It was an offer Jefferson could not refuse. It is popularly believed that Jefferson had qualms about purchasing the Louisiana territory on constitutional grounds. Actually, he had no such reservations. Those complaints came from the Federalists who opposed anything and everything Jefferson tried to accomplish. So, the constitutional questions were a speed bump. The real issue was money.

When Jefferson became president in 1801, the government of the U.S. was worse than penniless. After George Washington became president in 1793, Secretary of the Treasury Hamilton got Congress to enact legislation whereby the federal government would pay all the debt from the Revolutionary War, both federal bonds owed to the public and those issued by the states. Repayment on the debt would be at full face value of the bonds and notes presented for redemption.

During Washington's two terms, the government budget was small, with ninety percent going to pay the Army, Navy, and interest on debt. Most of the spending on the Army was to fight the Indians in the Old Northwest. By 1800, the government's debt stood at $83 million.

With Jefferson now in the White House, Swiss-born Secretary of the Treasury Albert Gallatin was trying to bring government finances under control. Despite this, in 1803 he strongly supported the purchase of the Louisiana Territory for the designated $15 million. The cash portion was $2 million in gold, plus $11.25 million in 15-year bonds at six percent and $1.75 million in short-term promissory notes ("IOUs"). Most of this was paid directly to France, with $3.75 million being held back as assumption of debts owed by the French government to U.S. citizens.

Napoleon was a practical man who saw no merit in sitting on a pile of bonds issued by a barely reputable republic, so he sold them at a rate of $87.50 on the hundred to Alexander Baring of the British Barings Bank. *It is not generally acknowledged in the history books that the U.S. effectively bought the Louisiana Purchase from Barings Bank, not from the French government.*

This transaction was part of a recurring theme of financial sleight-of-hand involving the U.S. government and private bankers, especially British ones. We will hear more of Alexander Baring later in this history.

Any type of government borrowing, of course, is a mortgage against the nation's future. But Jefferson, for decades, had his eyes on the enormous territory west of the Mississippi as a necessary acquisition for the future growth of the North American empire that he was sure the U.S. would someday become. He had no doubt that the future rate of return on his $15 million investment would be high, and he was right.

Jefferson also owned the world's largest collection of books and maps about western North America. Still, no white man had ever penetrated into its deepest heart. Jefferson now unleashed the expedition he had been hoping would take place.

On July 4, 1803, in the third year of Jefferson's presidency, word arrived from Europe that Napoleon had in fact agreed to sell the Louisiana Territory to the U.S. That same day, Captain Meriwether Lewis of the U.S. Army set out down the Ohio River from Pittsburg on the first leg of Jefferson's most important project, the Lewis and Clark expedition. The co-leader of the "Corps of Discovery," William Clark, joined the party near Louisville, Kentucky. It was thus a military project.

The Lewis and Clark Expedition

Stephen E. Ambrose's *Undaunted Courage: Meriwether Lewis, Thomas Jefferson, and the Opening of the American West* is an outstanding treatise on the Lewis and Clark expedition. A point Ambrose makes is that Jefferson was frugal in the extreme in spending public funds. The initial budget for the Lewis and Clark expedition was only $2,500.

Here are some excerpts from Ambrose's book that have mainly to do with Jefferson's attitude toward the Indians. Ambrose writes, for instance, that "[Jefferson] thought the only difference between Indians and white men was religion and the savage behavior of the Indians, which was caused by the environment in which the Indian lived."[12]

Jefferson told Meriwether Lewis to befriend the Indians they encountered, to avoid hostilities, to explore the establishment of trade relations, to collect information about their habits and ways of life, and to let them know that ownership of their lands had changed. They would now learn that President Thomas Jefferson was their new "father," who would protect and look after them if they behaved well.

Part of Jefferson's program, which included giving a few trinkets and articles of clothing as gifts, was to wean the Indians away from the English traders coming down from Canada and to prevent any similar mercantile incursions from the Spanish in Mexico. It was not until 1872, however, that the last of the British mercantile ventures, the Hudson's Bay Company, ceased operations in U.S.-controlled territory.

What did the Indians think of all this? Scottish writer and future diplomat Charles MacKenzie visited Lewis and Clark at Fort Mandan, in what is today North Dakota. The fort was actually a rude collection of huts built for a layover during the winter of 1804–1805. Commenting on his witnessing an exchange between the Mandan Indians and the two American captains, where the latter were carefully writing down everything the Indians had to say—through interpreters, of course—MacKenzie later wrote, "The Indians... concluded that the Americans had a wicked design on their country."[13]

MacKenzie's perception was not far wrong. In *Undaunted Courage,* Stephen E. Ambrose wrote of a visit paid by a party of Osage Indians to Thomas Jefferson in 1804, having been sent to Washington, DC, by Captain Lewis:

> [Jefferson]...intended to win their loyalty through a combination of bribes and threats, the traditional American Indian policy. "We shall endeavor to impress them strongly not only with our justice and liberality," he wrote, "but with our power."[14]

Jefferson's intention was gradually to reorient the Indians of the West to become farmers and ranchers once they had outgrown their "wild" lifestyle on the Plains, where they hunted and made war on each other. (Indeed, which nations in Europe did not?) Once settled down, they would spread out and populate the region west of the Mississippi. After this was accomplished, white settlers would be allowed to settle until the entire expanse was populated. Notably, part of Jefferson's plan was to expel the whites who had already made homesteads in the territory.

Jefferson's plan was pie in the sky. For one thing, by that time, the Indians had been so reduced in numbers by disease that they were far from being a viable population in relation to the spaciousness of the land. Second, very few wanted to become farmers and ranchers in the American style. Those that did farm their land were already doing so on a small, household scale, combined with a nomadic lifestyle. Most significantly, they wanted to live in their own traditional communities and did not want to be blended with (i.e. assimilated into) the dominant white culture.

Jefferson assumed the mountains of the West were no higher than the Appalachians. He had no concept of the vast expanse and variety of the rugged landscape the Lewis and Clark expedition would encounter. And most importantly, there was no way that the government would be able to expel the existing white settlers, whose numbers were growing by the day, or keep new settlers out, given the enormous pressure of westward migration and the land hunger of the growing domestic U.S. population, to say nothing of the Europeans still being enticed to emigrate.

That said, the Lewis and Clark expedition was one of the great exploratory voyages in history, of a land of often pristine wilderness that would rapidly be overrun during coming decades.

Despite Jefferson's admonitions, the Lewis and Clark expedition did not enjoy perfectly harmonious relations with the Indians they encountered. We have heard of the help given to the expedition by the young Indian woman Sacagawea, who did help ease some difficult situations, and the party did spend the winter camped nearby the more or less friendly Mandan tribe. But there were also rough moments and a couple of violent confrontations. One of the tribes with whom relations were less than perfect was the Hidatsas. Ambrose wrote:

> Lewis had other difficulties with the Hidatsas. They must have believed his protestations that the Americans meant them no harm, but they resented the lack of presents, and resented even more what one of them called "the high-sounding language the American captains bestowed upon themselves and their nation, wishing to impress the Indians with an idea that they were great warriors, and a powerful people, who, if exasperated, could crush all the nations of the earth."[15]

Again, we are looking at *arrogance*. In fact, American *arrogance* never ceased, and I would argue that it has a lock on the exercise of U.S. foreign policy to this day.

Despite Thomas Jefferson's power tactics against the Indians, he made a speech to the Osage delegation during their meeting in Washington in 1804 that displayed understanding of the ideals that should guide any brand of American statesmanship, then or now. He said:

> It is so long since our forefathers came from beyond the great water, that we have lost the memory of it, and seem to have grown out of this land, as you have done....We are all now of one family, born in the same land & bound to live as brothers; & the strangers from beyond the great water are gone from among us. The great Spirit has given you strength, and has given us strength; not that we might hurt one another, but to do each other all the good in our power.... No wrong will ever be done you by our nation.[16]

Despite the obvious hypocrisy with regard to brotherhood with native nations, Jefferson understood that the U.S. was a new kind of nation made up of many peoples, not just an extension of Europe.

By the 1830s, the Indian tribes had been banished from almost all of the lands east of the Mississippi River. In the West, they were herded onto reservations, and by the late 19th century, the government was trying to abolish even those refuges of historic native culture. A place where the Indians escaped the worst of U.S. oppression was western Montana, where the large Flathead and Blackfeet Reservations are now located.

There were also the Hopi, Pueblo, and Navaho lands of the Four Corners. Indian Territory in Oklahoma was also a kind of haven, though much of the land was later opened to white homesteaders. Today there are 574 Indian tribes recognized by the federal government and more by the states. There are 325 separate federally recognized Indian reservations. The world of the American Indian remains vast, complex, and rich in spirit, however straitened the circumstances.

Meanwhile, from the 1790s to the Civil War, the U.S. developed as three largely distinct regions—the South, with its slaveholding states, the free states of the North, and the West, where a nascent culture was just beginning to be formed. The North was on its way to becoming an industrial powerhouse, while the slaveholding South lived off plantation agriculture, particularly tobacco and cotton. Jefferson's philosophy of one nation living in brotherhood was soon forgotten, and indeed had never been widely entertained.

ENDNOTES FOR CHAPTER 1

1 I define "Deep State" as all those federal agencies—civilian and military—that conceal their actions behind a wall of secrecy which enables them to pursue a largely self-determined course of action without accountability.

2 Carol Clark Johnson, *Fullers, Sissons, and Scotts, our Yeoman Ancestors: 46 New England and New York families* (Mobile, Ala.: American International, 1976), p. 17.

3 William Richard Cutter, *New England Families*, Volume 3 (1911), 1824.

4 Calvin D. Trowbridge, Jr., *Three Bloody, Diseased, Deadly Decades* (Outskirts Press, 2019).

5 Winthrop intended his speech to serve as a warning to the community to live up to the Biblical text in their actions in the New World. More than three-and-a-half centuries later, President Ronald Reagan used the phrase "city on a hill" in a speech to indicate that this was what the U.S. actually represented to the world. So Reagan presumed that the U.S. had already arrived at that ideal state. I would argue that this was a statement of hubris by Reagan that distorted both the Biblical text and Winthrop's intentions.

6 Stephen E. Ambrose, *Undaunted Courage: Meriwether Lewis, Thomas Jefferson, and the Opening of the American West* (Simon & Schuster, 1997), 138.

7 Stephen E. Ambrose, *Crazy Horse and Custer: The Parallel Lives of Two American Warriors* (Anchor, 1996), 322. This template by which the U.S. military deals with its foes has persisted to the present day. The way the Biden administration

baited Russia into attacking Ukraine in 2022 followed the same pattern. Once Russia launched its attack, Biden and his officials immediately blamed Russia for its "unprovoked aggression."

8 "The Genetic Ancestry of African Americans, Latinos, and European Americans Across the United States," *American Journal of Human Genetics,* January 8, 2015.

9 *Johnson v. M'Intosh* 21 U.S. 523, cited in Trowbridge, ii.

10 Gal. 6:7 ESV.

11 Alexander DeConde, *This Affair of Louisiana* (Scribner, 1976), 113–14.

12 Ambrose, *Undaunted Courage,* 55.

13 Ibid, 204.

14 Ibid, 342. This approach seems to differ little from the way the U.S. treats many foreign nations today: i.e., "bribes and threats."

15 Ibid, 189.

16 Ibid, 343.

CHAPTER 2

Money and Banking

"Hard Money" and Paper Notes

For a meaningful discussion of US history, we must consider money. The problem of the U.S. money supply was of increasing concern, if not an obsession, during the formative years of the nation, and has not been resolved to this day. In fact, with the current assault on the U.S. dollar internationally, dubbed "de-dollarization," we are seeing yet another financial crisis.

In the earliest years of the nation, coinage of foreign nations was the most reliable currency, especially the Spanish gold dollar, which was used both before and after U.S. independence. Barter was also widespread, mainly in land and agricultural commodities.

"Real" money was always viewed as authenticated by gold and silver, which was why English privateers preyed on the early Spanish galleons returning with plunder from the New World. Enough made it to Spain to fuel a vast European monetary expansion in Western Europe. But whatever limited amount arrived in British North America through trade quickly disappeared back to London under the prevailing mercantilist policies of the day.

In the days of Indian trading, beaver skins served as money. This continued into the 19th century. In dealing with the Indians, both sides dealt in wampum, which was the reason colored beads became an important part of Indian trade goods—especially the highly-prized blue beads.

But as the "best" money remained gold and silver coins which were known as "hard money," or "specie," that could readily be weighed, measured, transported—or stolen, one of the main purposes of early "counting houses" and later banks, was to keep coinage or bullion safely under lock and key. Trade was also carried on in gold dust kept in leather pouches attached to one's belt, requiring a buyer and seller to have a reliable set of scales.

Then there was paper money. But the paper money expedients adopted by colonial governments could be attacked, as happened with the Currency Act of 1767. This prohibition of colonial paper money by the British

Parliament plunged the colonies into a deflation that became a leading cause of the American Revolution.

Before, during, and after the Revolution, a multitude of methods were used to create, print, and circulate paper notes. This included issuances of bond certificates by land banks in Massachusetts and Pennsylvania, where borrowers mortgaged their real estate; the printing of Continental Currency during the Revolution by the Continental Congress; the occasional issuance of bills of credit by individual colonies to pay their bills; and a couple of primitive banks with lending prerogatives.

The use of paper money led to price inflation, depending on the degree of confidence individuals receiving payment had in the issuing entity. Often confidence was close to zero. But without a solid and widely acknowledged medium of exchange, a society becomes desperate. Even if money is counterfeited, as the British did in their attack on the U.S.'s Continental Currency, the bogus notes themselves may still have trade value.

Constitutional Confusion

The U.S. Constitution ratified in 1788 was ambiguous on the subject of money in terms of creating a consistent, dependable supply of a "medium of exchange" or "store of value," which are the dual purposes money is supposed to achieve.

Article I, Section 8, Clause 5 of the U.S. Constitution states that "Congress shall have the power to coin money, regulate the value thereof, and of foreign coin, and fix the standard of weights and measures." There is no mention of how this "coined" money is to be entered into commerce.

Paper money is not mentioned at all, except that the states are *prohibited* from issuing "bills of credit." This meant that the states could not "print money" and spend it into circulation.

But neither was the federal government itself specifically authorized to issue bills of credit, even though the *original* draft of the Constitution did in fact allow it. But the clause was removed. Alexander Hamilton had argued:

> To emit an unfunded paper as the sign of value ought not to constitute a formal part of the Constitution, nor even hereafter to be employed; being, in its nature, pregnant with abuses, and liable to be made the engine of imposition and fraud; holding out temptations equally pernicious to the integrity of government and to the morals of the people.

Nevertheless, the U.S. government *has* asserted the prerogative of issuing bills of credit, particularly during the Civil War with the issuance of Greenbacks. Such practice was found constitutional by the Supreme Court, which ruled that bills of credit were inherent in the government's right to borrow, as bills of credit are fundamentally a debt instrument.

Government bills of credit receive their force by being acceptable in payment of taxes. They can then become legal tender in the economy at large.

The Constitution also gave Congress the power to regulate interstate commerce, which has been interpreted to provide for oversight with respect to national banks and creation of a *national* banking system, but not to *state* banks. What has been called "the Commerce Clause" refers to Article I, Section 8, Clause 3, which gives Congress the power "to regulate commerce with foreign nations, and among the several states, and with the Indian tribes."

Note that Indian tribes, as a distinct category of social and economic organization, here receive a Constitutional recognition that has never been rescinded.

Congress is also authorized to collect money through taxes and other means. Article I, Section 8, Clause 1 states that:

> The Congress shall have Power To lay and collect Taxes, Duties, Imposts and Excises, to pay the Debts and provide for the common Defence [sic] and general Welfare of the United States; but all Duties, Imposts and Excises shall be uniform throughout the United States.

This clause implies that the only two allowable sources of government revenue are 1) taxes and 2) borrowing, a restriction that has been a perpetual source of controversy and conflict, causing the federal government, as well as state and local governments, to be constantly on the cusp of default through potential inability to pay their legally mandated obligations.

The Constitution does not specify the nature of the medium of exchange that can lawfully be used by taxpayers or lenders. So what is "legal tender"? If Congress wanted to, it could designate the lids of Mason jars to be legal tender, as in a recent novel on what life in the U.S. would be like under a currency collapse.

The provision for incurring debt also implies some kind of market or mechanism for the buying and selling of government bonds, particularly because bonds themselves are assets that may be marketed commercially and that may circulate at varying rates of return. Today, the market for government debt is run by the Federal Reserve as "fiscal agent" to the U.S. Department of the Treasury.

The other major provision in the Constitution that has strongly influenced the circulation of money is Article IV, Section 3, Clause 2, the "Property Clause," that states, "the Congress shall have power to dispose of and make all needful rules and regulations respecting the territory or other property belonging to the U.S." Obviously, this applies to land that is bought and sold by the government in the marketplace or where the government is obligated to pay full market value for land confiscated for public purposes.

The Property Clause also applies to land acquired from the Indians under the Treaty Clause of the Constitution. Article II, Section 2, states that "the President shall have Power, by and with the Advice and Consent of the Senate to make Treaties, provided two-thirds of the Senators present concur."

No one knows how much land the Indians have ceded to the U.S. through treaty or how much has simply been seized through confiscation or occupation. No one even knows for sure how many Indian treaties there have been. Estimates range from 374 to 500.

The right of Indians to make treaties with the government was cancelled by Congress in the 1870s, meaning that since then, Congress has tried to rule over the tribes by decree. The federal court system has held, however, that past treaty obligations are enforceable, and the tribes have been granted limited rights to self-government by Congress since 1935.

This does not mean that the treaties were fair. In the 1809 Treaty of Fort Wayne, for instance, the Indians ceded to the U.S. 2.5 million acres of land in the Old Northwest for a reimbursement of two cents per acre. From 1814–1824, the southeastern tribes gave up fifty million acres of land, with then-commissioner Andrew Jackson writing seventy Indian removal treaties. Later, as president, Jackson negotiated treaties that opened another twenty-five million acres of land. This allowed the lands to be sold by the federal government to whites, often involving sale on credit, at a time when government at all levels had few other sources of income.

The vagueness of these Constitutional provisions, the lack of a coherent system of government financial management, the absence of any provision or definition for commercial banking, and the reliance on inference as a guide to action—"implied powers"—has resulted in the fact that throughout history, the creation and utilization of money in the U.S. has been chaotic.

This chaos left the door wide open for the entrance of financial predators and eventually allowed the complete takeover of the nation's financial system by the financier class through the Federal Reserve Act of 1913. Today, a majority of the U.S. population live from paycheck to paycheck; i.e., most people are "broke" and, moreover, helpless against an "inflation" that it seems no one can ever explain or do anything about.

The U.S. has been in an inflationary crisis since the 1970s—for a half-century. This translates to an ongoing devaluation of the U.S. dollar and progressive impoverishment of the working population. The government feeds this inflation by compounding annual cost-of-living allowances—COLAs—for entitlements, government salaries, etc.

Maybe all this chaos was why the Constitution left so many loopholes in the first place. The Constitution was drawn up by the top politically connected figures of the Revolution. They in turn had close ties to the merchants, investors, and speculators, including wealthy Europeans, who cashed in on the bonds that had kept Washington's army in the field.

In fact, a scandal erupted when speculators bought up the promissory notes being held by former soldiers who had been paid with Continental Currency or bonds. The new U.S. Treasury, under the first Secretary of the Treasury, Alexander Hamilton, allowed the redemption of these notes at face value, leading to charges of favoritism and corruption.

Gold and Silver

Again, money in the U.S. *has always been a problem*. The scarcity of gold and silver obviously limited the quantity of coins in circulation and made mining for gold and silver a lucrative occupation, where fortunes could be made overnight.

With the discovery of gold in the West, particularly in California in 1848, the amount of circulating coinage skyrocketed. The same thing happened with the mining of gold in Montana in the 1850s and 60s, in the South Dakota Black Hills in the 1870s, and in the Yukon and elsewhere in the world during the late19th and early 20th centuries, when the mining of precious metals became more industrialized.

The richest bonanzas came with the exploitation of the South African gold mines by the British, after they defeated the Dutch settlers in the Boer Wars from 1880–1902. It was South African gold and diamonds, exploited mainly by Cecil Rhodes and the Rothschilds, that propelled Great Britain to world dominance by the end of the 19th century.

Gold and silver must be minted into coinage to become an effective medium of exchange. The first U.S. Mint was established in Philadelphia in 1792. Individuals possessing gold or silver could take their holdings to the Mint to be stamped into coins. Congress established a value ratio between gold and silver of 15:1.

But this coinage would be far from adequate. Obviously, a scarcity of money in a developing economy limits commerce to the point of economic crisis as populations grow and industry evolves. With the California Gold Rush, a second U.S. Mint was established in San Francisco in 1854. Still, it was not enough.

During an age that viewed the only real backing for money to be specie, or precious metals, the national currency continued to stand on shaky ground. This is where banking began to creep in. As stated earlier, an apparent solution to the shortage of ready money has always been printed paper. During colonial times, paper money was issued by brokers for hogsheads of tobacco and by government agencies, such as the Massachusetts and Pennsylvania Land Banks, where loans were made using real estate as collateral. But the paper money we know today is bank-issued, i.e., Federal Reserve Notes.

There was no bank-issued paper money in America until the Revolutionary War, when the Bank of North America was set up in Philadelphia. Yet paper notes of credit had long been in use in the western world to support trade. Throughout history, including at times in our own day, merchants or manufacturers also paid their obligations in "scrip," a type of paper note redeemable in goods or services at the issuing business location.

Scrip could circulate within a community and be bought and sold as a commodity, either at face value or a discount. Use of scrip becomes common during times of economic hardship, such as the Great Depression, and in modern times, is reflected in the "local currency" movement. "Cryptocurrency" like Bitcoin is a kind of "scrip." But it's an inferior form of money because it is not legal tender and cannot be used to pay taxes unless purchased with authorized currency. It can also be attacked at will by the government.

Fractional Reserve Banking

It's "fractional reserve banking" that can make the problems with paper money explosively worse. The practice of fractional reserve banking had been institutionalized in Europe going back to the Middle Ages by allowing a broker, a bank, or a gold merchant or goldsmith to lend paper promissory notes in *excess* of the amount of gold or silver deposited by the business's customers.

The money supply could thereby be multiplied many times over the original value of gold on deposit. Such multiplication can easily lead to inflation and default by whomever is unlucky enough to hold the notes when a crash comes—as it always does.

Despite its obvious disadvantages, fractional reserve banking became predominant in the western world. Some even say it allowed Europe to conquer the globe, given its capacity to expand Europe's purchasing power far beyond its tangible backing, i.e., its real value in specie. Obviously, this power to issue paper money not only had the potential of making those so doing the richest people in a nation but also the controllers of princes and kings, who depended on such loans in furtherance of their internecine wars or colonial expansion.

True, the money was usually redeemable in gold, or possibly silver by the issuing bank, which was fine, as long as all the holders of paper did not show up at the same time seeking redemption. Such a "run" on a bank would result in bankruptcy and collapse. Throughout history, this has often happened. When it does, individuals may lose their entire lifetime savings.

Back in the day, the bankers could even be prosecuted under the law, sometimes even be put to death. But today, if a bank defaults, it simply goes out of business. The bank's owners may then start a new one. Bank deposits now have limited protection from default through the Federal Deposit Insurance Corporation, established in 1933 as one of the key reforms of the New Deal. Prior to that, people like my grandfather kept their money under the siding inside the garage and similar hiding places. While the practice has often been mocked in cartoons and elsewhere, it wasn't without wisdom.

Paper money created "out of thin air" has been made "legal tender" by government decree. This, combined with charges against loans levied by a bank through interest has made it easy for sober-minded people to declare bank lending a thing of the devil.

Over time, governments began to require that a certain ratio be observed in the amount of money a bank could lend vs. the amount it had to hold back in reserve as customer deposits but formerly as gold kept in a vault, in an effort to prevent defaults. This "reserve ratio" has been pegged at about 16% in today's U.S. banks but has varied historically, running up to around 25% or even 50%. Incredibly, there have been times when the reserve ratio has been zero, including in the U.S. since 2020.[1]

With the coming of Electronic Funds Transfer in the late 20th century, the reserve ratio could easily be manipulated, but we'll come to that later. Today, the power of banks to issue interest-bearing loans is virtually unlimited. But it's where most of the circulating currency in the modern economy comes from. This is why it's said we have a "debt-based" financial system.

Eventually, the financial system based on fractional reserve banking became the basis of economic relationships in the U.S. and all other countries. This was because it was the easiest way for the monetary system to keep up with modern industrial growth while delivering profits to the system's controllers and to the politicians whose legislation the controllers need to enforce the obligations of their debtors. Accordingly, government today devotes most of its efforts to protecting the wealth of the creditor class.

Money "created out of thin air" could and would eventually be used for all manner of financial speculation, including purchasing entire businesses that are then stripped of assets and which shed debts by declaring bankruptcy. If you think such practices are morally criminal, you are right.

Moneylenders' methods of enforcing debtors' obligations have varied, but they are always severe. In Shakespeare's *Merchant of Venice*, the money-lender Shylock demanded his "pound of flesh." While this was metaphoric, at that time debtors could be cast into prison. Debtors' prisons became established institutions in England, Western Europe, and America by the seventeenth century. In England, a lender could have a debtor imprisoned without trial for a fee of one shilling. By 1628, 10,000 people were imprisoned in England for debt, many dying in prison before a friend or relative bailed them out.

Even though both the federal government and the states eventually outlawed imprisonment for debt, the practice has been revived in recent times. Today, people can go to jail for debt for many reasons, especially if they are too poor to pay court fees or fines. Fathers who fail to pay child support are routinely jailed for "contempt of court."

Usury

Banks make money through "usury," which is the charging of interest for a monetary loan. Often the argument is made that usury does not apply to all interest charges, and that the term should only be applied as a term of censure if the rate of interest being charged is "excessive."

Today, if the economy slows down, the Federal Reserve lowers its interest rate to stimulate borrowing. If the economy overheats, the Fed raises the interest rate to "reduce demand." We are as accustomed to this system as we are to the air we breathe. It's why financial advisers say that "the Fed" rules the economy.

But for a thousand years, the Church outlawed usury as immoral and destructive, especially when the failure to pay interest on a loan caused a person to be imprisoned or to have his property seized. It remains condemned today by Islam, giving rise to what is called Islamic banking.

The critique of usury has its roots even earlier, in the teachings of Aristotle (384–322 BCE), who noted that the natural purpose of money is as a means of exchange. Money in and of itself, barring the intrinsic value of the metal used in coinage, is an abstraction, without real existence. For this emblem of unreality to generate something else of real existence was, to Aristotle, an abomination.

Usury was wrong because it enabled the holders of money—whose natural use was as a means of exchange between goods with practical value in the real world—to get more money through charging for its use—money which they effectively had not earned but had created out of nothing, far exceeding the actual wealth that they held. Money thus became, systemically, a need and then an end in itself. This illegitimate growth of money through interest, by which the recipient could command even more tangible goods to appear, was viewed as magical, even demonic.

This did not mean that no one in the Middle Ages practiced usury. Of course they did, including church prelates. But that didn't make it right or any less ruinous. Obviously, usury was worse when compound interest was extracted, rather than simple interest. It's why the total payment of a mortgage exceeds by far the price of a house.

In fact, the roots of usury go back even deeper in history than Aristotle. Dr. Michael Hudson, an economist who has carried his research of the origins of money back to the ancient Middle East, tells a story that he came across as follows: The Devil was unhappy because the human race was developing in such a splendid fashion, with people obeying the injunctions of their spiritual leaders, and leading happy, productive lives.

So the Devil called a meeting of his little devils and told them that if anyone could figure out a means to ruin people's lives and reduce them to slavery, they would get a great reward. Well, after a few days one of the little devils came back and said he had the solution. He said that the solution was to lend people money and to charge them compound interest on the loans. A big smile broke across the Devil's face. "That's it!" he said. "Now they will all be ruined by debt and will be my slaves forever."

This story should make it perfectly clear that today, with the ubiquity of fractional reserve lending at compound interest, the Devil rules.

The First Bank of the United States

Banking appeared in America during the Revolutionary War but made its debut as a government-sponsored institution with the chartering by Congress of the Bank of the United States in 1791. The "First" Bank was the brainchild of Secretary of the Treasury Alexander Hamilton.

In his *Report on a National Bank,* Hamilton cited the need to increase the nation's money supply as the main purpose of the Bank. Opponents argued that the lack of a metallic currency would cause inflation and that the Bank would entail usury. Hamilton said that "the abundance of a country's precious metals was not so important as the quantity of the productions of labor and industry."

Both sides had a point in that specie had inherent economic value, but so did the nation's productivity. Hamilton saw the bank as a financial source for "internal improvements," which at that time meant mainly roads and canals. At the time, funding was being secured elsewhere. In 1806, for example, Congress authorized funds for a National Turnpike that would extend from Baltimore to the Mississippi River. Funding was provided by the sale of lands taken by treaty or confiscation from the Indians.

The Bank immediately became one of the most powerful institutions in the new nation, rivaling the Army and Navy. The Bank got its operating capital by the selling of shares to domestic and foreign investors, mainly Dutch at first, and by serving as a depository for customs duties and excise taxes collected by the new federal government.

Most of the Bank's initial capital consisted not of specie, but of IOUs from investors, protecting them from further loss if the bank failed. Eighty percent of the stock in the bank was owned by private investors. The rest was in U.S. government bonds.

The Bank from the start was a profit-making instrument for the financier class. Investor shares were expected to increase in value as time went on. The interest paid on the loans issued by the Bank would come out of the producing and trading capacity of the new nation, so the Bank immediately pressured the U.S. to pursue economic growth, a pressure that had political, social, economic, and military ramifications.

Any system based on fractional reserve banking increases the level of competition between nations for trade supremacy, causing import duties and taxes to rise, and pushing government negotiators to exact more advantages from treaties made with other parties, leading to a higher degree of international conflict.

In his *Report on Manufactures,* Hamilton developed the theory of protective tariffs for the nation's infant industries but also saw tariffs as a fund-raising measure. In fact, tariff duties became the main source of federal government income for decades. The most powerful government agencies were customs houses at ports of entry, especially New York.

Hamilton also anticipated selling land in the Northwest Territory for twenty cents an acre that the U.S. had acquired from the Indians for two cents. The actual sale price to settlers was ninety cents. In 1796, Congress adopted a permanent land ordinance providing for purchase of 640 acres, or a square mile, for $2 per acre cash with the remainder due in one year.

But these measures were only the first step in the pursuit of a sound economy. Bank lending was still viewed as needed, and the First Bank was Hamilton's chosen device.

It was how and to whom the Bank lent money that made it an immediate source of controversy and a potential source of corruption. The Bank could lend money for business and trade but could also lend it to the government to cover operational expenses. The government was a heavy borrower from the start—it had no other way to stay afloat when it undertook expenses beyond the existing revenues.

This marked the introduction of federal government deficit financing, though it had already been practiced by the Continental Congress through the sale of war bonds during the Revolution. Government debt would grow through the 1790s. Government deficit financing meant that in order to stay afloat, the government had to repay its debt either through economic growth in the tax base or through inflation. The growth imperative was fine as long as the nation's population, trade, and industry were also growing. But any slow-down, particularly at times of recession, depression, or war, could quickly become a crisis, or even a catastrophe.

If multiple nations ran their governments on the basis of deficit financing, competition would be inevitable, almost always leading to war. This was a root cause of the constant internecine warfare among the nations of Europe. They were all in debt to the financier class, and the financiers always demanded their "pound of flesh."

With the First Bank of the United States, despite its immediate advantages, the U.S. was now drawn into this system of international competition. The system also turned the more developed nations into predators with respect to less developed parts of the world. The race for colonies had been ongoing for over 250 years when the U.S. became independent. It would continue with a vengeance even until today with what is called *neo*colonialism.

Since World War II, the U.S. and UK in particular have been seeking economic colonies rather than outright ownership of whole nations. Even as former colonies began to seek political independence, they still were suppressed and fought over as economic dependencies.

Major international institutions, such as the World Bank and International Monetary Fund, also have, as one of their purposes, to maintain a neocolonial system based on lending and debt. Neocolonialism is also furthered through ownership or extraction of a nation's natural resources by banks, corporations, and investment firms drawing on their preferential access to bank-generated funds.

The U.S. Constitution did not contain a provision for the First Bank of the United States or for drawing the U.S. into the European system of banking and finance. But that didn't stop Hamilton from moving ahead, while Thomas Jefferson and his supporters saw the danger. The opposition of Jefferson to the Bank split the nation into rival political parties, a divide that continues to this day, though the origin of the split is forgotten.

Hamilton became the leader of the Federalists, with Jefferson the head of the Democratic-Republicans or, as they were sometimes called, "National Republicans." Hamilton now confided to the British ambassador that he saw the First Bank as the key to a future American "empire," much as the Bank of England had become for Great Britain.

Jefferson famously wrote:

I believe that banking institutions are more dangerous to our liberties than standing armies. Already they have raised up a money aristocracy that has set the government at defiance. This issuing power should be taken from the banks and restored to the people to whom it properly belongs. If the American people ever allow private banks to control the issue of currency, first by inflation, then by deflation, the banks and corporations that will grow up around them will deprive the people of all property until their children will wake up homeless on the continent their fathers conquered.

While Hamilton had the right idea in the need to generate spending power to finance public projects, the fears of Thomas Jefferson have come to pass. Today's U.S. political class is clueless about money and where it comes from, as are the American people, despite their experience-based aversion to banks.

Note, however, that Thomas Jefferson was as committed to an American empire as Hamilton. But he had a different concept. For Jefferson, who saw his purchase of the Louisiana Territory from France as a big step toward that empire, economic growth would be based on land, labor, and commerce, not bank-created money.

Note that the founding and growth of stock, and later, equity markets offer an alternative to usury. With the buying and selling of stock, the purchaser, or speculator if you will, assumed some of the risk of the enterprise, to the point of losing the investment if the activity caves in. The use of stock markets was similar to the idea of lending in Islamic countries, where a bank in effect purchases a share of the business that it is being asked to fund. This serves as a brake on unwise investment, as the lender, too, has something to lose, rather than continuing to gain irrespective of the success of the project by reaping interest from their unfortunate purported business partners.

With Jefferson's election in 1800, economists came to speak of the era of "Agrarian Republicanism." "It favored a society composed of small landholders and repudiated the Hamiltonian ideal of an alliance between the government and the capitalist group."[2]

It would be difficult to find two national leaders more different in background and orientation than Jefferson and Hamilton. Jefferson was a farmer to his bones. Hamilton was a capitalist. Each achieved greatness in his own way. And while Jefferson lived to a ripe old age, passing away in 1826 at the age of 83, Hamilton was shot dead in a duel with Vice-President Aaron Burr in 1804 before he had turned 50. Burr was a rival in the New York banking world, who was tried for treason for conspiring to create a competing nation to the U.S. out of western lands.

With deference to Hamilton, we should note that he was correct in the need for government to engage in infrastructure spending to support industrial growth. The government of France accomplished this during the 17th century by an effective system of taxation and state-owned enterprise. To this day, the U.S. government has failed to do the same, leading to gigantic shortfalls in our infrastructure development.

Skepticism About Banking

The U.S. population and its political representatives, at least among Jefferson's followers and successors, were deeply skeptical of the power of central banking. Too often people had seen the power of banks in putting merchants and farmers into debt, then seizing their property through foreclosure. Nevertheless, the First Bank of the United States lasted for twenty years, all the way through Jefferson's presidency, until 1811 and the presidency of James Madison, when the bank's charter was not renewed.

When Jefferson began his two terms of office in 1801, his allies within the National Republican movement took control of Congress. The government continued the use of customs duties on imports as its main source of revenue. But with the Republicans staunchly against any kind of direct taxation, the revenue system based on excise taxes that had produced the Whiskey Rebellion in 1794 was abolished.

Congress also abolished import duties on salt. Jefferson and his Secretary of the Treasury Albert Gallatin went along with these changes. But excise taxes were reinstated in 1804 for a "Mediterranean Fund" to allow the

Navy to fight the Barbary pirates in North Africa that were preying on U.S. shipping.

With the tax cuts being implemented, on top of the growth of national commerce and action by the Washington/Hamilton administration in the 1790s to pay down the national debt, the Jefferson administration was able to run a budget surplus every year except 1809. This also meant less need to borrow, which was fine with Treasury Secretary Gallatin, who understood public debt to be inflationary.

Gallatin showed real wisdom when we look at the devastating inflationary impacts of government borrowing in our own day. He also viewed government debt as being harmful to business investment because it absorbed capital that would ordinarily be used by the private sector to expand production. During Jefferson's two terms, the government was able to cut its total outstanding debt in half, so that interest payments as part of the federal budget also fell.

Fortunately, the operations of the First Bank were not part of the government's budget. When in 1802 it did not have the money to pay off its Dutch creditors, the Bank agreed to allow English banker Alexander Baring to buy all the Bank's outstanding equity—2,220 shares at a discount of forty-five percent—leading to a substantial profit.

This was the first entrance of British banking into U.S. national finance. Barings Bank had been founded in 1762 by Alexander Baring's father, a professor of theology in Bremen who emigrated from Germany. Its headquarters were in London. The Bank's rapid expansion as a major British financial power was financed by profits in the North American slave trade.[3]

Barings Bank started business in the U.S. in 1796, when it purchased one million acres of land in Maine when Maine was still part of Massachusetts. The tract consisted of almost 160 square miles of potentially valuable real estate. As noted previously, Barings was the seller of record of the Louisiana Purchase, having bought the Jefferson administration's promissory note from Napoleon.

Later, Barings helped to finance the U.S. government during the War of 1812. Rather odd, if you think about it, that the bank obviously had an interest in both sides of the conflict, but that's how banks operate. In case of war, get all parties in your debt—indicating the natural interest of banks in wars. Barings also became an agent in Britain for the U.S. government, a position it held until 1871.

By then, Barings was being called the "Sixth Great European Power," after England, France, Prussia, Austria, and Russia. But by the 1820s, its dominance was being challenged by the investment banking firm of Nathaniel

Rothschild, who had made a fortune helping to finance Britain's wars against Napoleon.

When the charter of the First Bank of the United States expired in 1811, the U.S. lost any semblance of a uniform national currency. With the cessation of the First Bank, which previously had been issuing bank notes when making loans that were declared legal tender, now notes could be issued only by state-chartered banks, which led to a hodgepodge of paper money. The Bank also had to export $7 million in specie to pay off Barings Bank and other foreign creditors.

As the War of 1812 approached—which many observers saw as a continuation of the original Revolution— the Madison administration refused to raise taxes. With the slowdown in commerce, customs duties were also in decline. So Treasury Secretary Gallatin fell back on borrowing from banks as a means of covering war costs. But with the First Bank gone, this meant borrowing from foreign banks, still chiefly Dutch.

Consequently, the U.S. government's debt skyrocketed, with total debt reaching $127.3 million by the end of 1815. This was eight times the nation's income for that year. In 1812, Congress also authorized the printing of $5 million in Treasury notes, paying 5.25 percent interest. Government creditors were forced to accept the notes, even though they were not legal tender, meaning there was no requirement for anyone to accept them in trade.

Treasury bonds were more reliable than notes, and by the end of 1812, the Treasury had sold $3.9 million to individual investors and $9.2 million to banks. Purchasers paid for bonds in installments, then used the equity to borrow money from state or foreign banks.

This cobbled-together system did increase the money supply, but the dubious nature of government-originated credit caused significant inflation. The inflation was not due to "too much money chasing too few goods," which is the standard economists' definition of inflation. The dubious nature of government-originated credit was reflected in the risk that people holding government or bank paper might lose their investment if the system crashed. The system did in fact crash at various inopportune moments during the years and decades to come.

Finally, in desperation, the government in 1813 authorized its biggest request for a loan to date—$16 million. This was bought by a syndicate consisting of David Parish, Stephen Girard, and John Jacob Astor. All three were foreign born—Parish and Astor were German, and Girard was French—and all three had made fortunes since coming to the U.S., though Parish later went bankrupt in Europe and drowned himself in the Danube River. Stephen Girard had become the richest man in Philadelphia, had helped liquidate the

First Bank of the U.S., and soon would become the lead financier for the Second Bank.

John Jacob Astor had made his fortune in the fur trade, setting up the first U.S. fur trading business in the Oregon Country. He pioneered trade with China and got even richer by taking part in the British-run opium business conducted in and against China. Opium beckoned numerous U.S. traders who carved out a niche by selling lower-quality Turkish opium to the growing numbers of addicted Chinese.

In 1814, the U.S. banking system collapsed when the state-chartered banks suspended specie payments in redemption of the smorgasbord of circulating paper notes whose value could not be reconciled with each other. With the First Bank of the United States now defunct, the government had placed deposits of its cash-on-hand in ninety-four separate state banks. Still, in 1815, President James Madison vetoed a bill for a new national bank. So the crisis continued.

The bill for the Second Bank of the United States was finally passed by Congress in 1816, giving the bank a twenty-year charter. This bank collapsed in 1833, when President Andrew Jackson withdrew federal deposits. Jackson said, "Either I will kill the bank, or the bank will kill me."

Around the same time, Martin Van Buren, Jackson's vice-president and successor as president, was warning against the U.S. becoming a "bank-ridden society." We shall tell more of the story of what came to be called the "Bank War" in the next chapter.

Meanwhile, here is a famous quote from Jackson:

> The bold efforts that the present bank has made to control the government and the distress it has wantonly caused, are but premonitions of the fate which awaits the American people should they be deluded into a perpetuation of this institution or the establishment of another like it.... If the people only understood the rank injustice of our money and banking system there would be a revolution before morning.[4]

The Growth of Private Banking

Private banking can operate without a central bank simply on the basis of business charters. The U.S. Constitution had nothing to say about such charters, except that if they were part of interstate commerce, they could be subject to Congressional regulation. But they weren't regulated during this early period, as they were considered stand-alone institutions within their individual states.

Such banking began to flourish during the period between the Revolutionary War and the Civil War through a vast array of individual banks chartered by state governments that one day came to be known collectively—and derisively—as "Wildcat Banks." But these banks were still essential to commerce.

Most of the banks lent money for ongoing business operations, for stocking of inventory, for meeting unusual demands on merchandise, or to farmers for hiring extra help for the fall harvest or spring planting. All these uses were viewed by borrowers as simply the cost of doing business, and interest rates were generally low. Rarely did the Wildcat Banks provide money for new business development, and, unlike modern times, almost never for speculation. Also, they were not used by state or local governments for infrastructure construction, a use viewed by Hamilton as an essential purpose of banking.

Some of these banks were actually owned and operated by state governments, though most were privately chartered to serve individual cities, towns, and farming areas. They were usually capitalized by local merchants and operated on a fractional reserve basis. They were required to redeem the paper money they issued in specie. Sometimes rival banks would organize runs in an attempt to put each other out of business by arranging for holders of paper notes to present them *en masse* for specie redemption.

Overall, what success the system achieved was encouraged by the fact that only by investing in *bona fide* productive enterprise could the banks stay in business. Even so, loose lending practices, financial uncertainties, runs, and even bad weather could lead to bank failures.

These banks printed and issued their own banknotes as loans. Obviously, this led to confusion, where there was little ability to compare the value of notes issued by different banks. Brokers bought and sold the notes, pocketing a commission. But the system served its intended purpose of enabling a circulating paper currency, though by the time of the Civil War, centralization and consolidation of banking was well underway.

Major banking centers grew up in locations where commerce and industry contributed to the capitalization of banks on a scale unknown in earlier times. The major banking centers were New York, Boston, and Philadelphia on the East Coast, and later, Chicago in the Midwest and San Francisco in the West. Over time, every American city that grew into a focal point of trade and manufacturing also became a banking center.

Social Consequences of Banking and Usury

Banking and usury can play out to baser motives, like greed and power. Putting other people into debt creates power relationships that degrade both parties. My wife Karen's great-grandmother came to America from Wales as a teenager, after her parents lost their farm when they co-signed a loan to a relative who then defaulted. Such cases, and many far worse, are legion in the history of the modern world.

Debt for any cause, but particularly for a victim of usury, preys on the mind, saps self-confidence, is a source of fear and anxiety, and may drive a person to self-destructive behavior, including alcohol or drug abuse, depression, or suicide. We see examples all around us of young people shackled with debt from student loans, a situation made worse by the fact that student loan debt has been excluded from being written off through bankruptcy. We can thank former U.S. Senator Joe Biden for that legislation.

Loan sharking is also one of the main abuses of organized crime, where debtors are threatened, hounded, or even murdered for inability to pay. Even in our day and age there are people who sell themselves into simulated or literal slavery to pay their debts or who prostitute themselves, their values, their time, or even their bodies for financial gain. Though often (but not always) greed is at their origin, usury causes or intensifies all these ill effects.

Even when debt is lawful, failure or inability to pay off a loan can be catastrophic, leading to default or bankruptcy. Individuals and families may lose their homes. If the debtor is a business, an institution, or even a nation, it may pass into receivership and possibly not even survive. Some believe the Roman Empire was driven to ruin by bad debts.

In every instance, usury creates an underlying tone whereby everything is seen only for its monetary value, where people lose touch with their real humanity. The family home becomes an investment, where the purchaser is waiting for an opportunity to cash in and move on when the market goes up. Every commodity that we purchase has built into its prices the interest required to repay money that has been borrowed at every stage of the manufacturing and distribution process. The entire system is infested with interest payments. It has been estimated that interest charges may account for up to fifty percent of retail purchase prices. And usury becomes the basis for creating a nation's circulating currency, as there are few other ways to generate the purchasing power required for people to pay for the necessities of life.

This leads to two distorting mechanisms that a society can choose from when seeking to ameliorate its burden of debt. One is to spend money faster. This is called the "velocity" of money, where the faster that money is spent, the more economic activity a given monetary base can support. Of course, velocity slows down and may even stop if the money is sucked out of the community by big outfits like Walmart that pay their employees minimum wage while taking all it can as profit accruing to far-off owners or stockholders and provisioning its stock not just from the U.S. but from global suppliers.

The other way to attack debt, as stated previously, is through inflation. If prices rise, debtors are at an advantage, as they can pay down their debts with money of lesser value compared to the original principle of the loan. Speaking again of homeowners, they are always in favor of inflating home values, though the situation backfires when local governments raise property assessments and rake in more on property taxes.

Of course, *local governments love inflation*. They know that the price of houses always goes up, almost never down. So does their tax haul.[5]

The same is the case with national governments. When governments go into debt, they do everything they can to generate inflation in order to pay down their own debts. When the government tells you they are "fighting inflation," they are lying.

Over the last half century, since the huge deficits of the Reagan era, federal government debt in the U.S. has skyrocketed. As night follows day, so has inflation followed debt, to the point where the value of the once sacrosanct U.S. dollar is being reappraised by most of the world.

ENDNOTES FOR CHAPTER 2

1 Board of Governors of the Federal Reserve System, Policy Tools: Reserve Rquirements. https://www.federalreserve.gov/monetarypolicy/reservereq.htm

2 Paul Studenski and Herman Edward Kroos, *Financial History of the United States* (McGraw-Hill, 1952), 65.

3 We should bear in mind this conjunction between banking and slavery. Today we speak of "debt slavery." It is not an idle phrase.

4 Occupy Wall Street tried to raise such a revolution after the financial crash of 2008–2009. Instead, President Barack Obama bailed out the banks with trillions of dollars of public funds.

5 I pointed this out recently to our town mayor. Of course he was indignant.

CHAPTER 3

Varied Paths

Settlement of the Old Northwest

Soon after the ratification of the Constitution, three new states joined the original thirteen: Vermont and Kentucky in 1791 and Tennessee in 1796. While "Vermont" refers to "green mountain" in French, the nation's Indian heritage is reflected in the names of many other states.

The name "Kentucky" is derived from the Iroquois "ken-tah-ten," "Land of Tomorrow." The name of Tennessee dates from 1567, when Spanish explorer Juan Pardo passed through a Cherokee village called Tamaqui. The territory north of the Ohio River, acquired from Great Britain through the Revolution and extending to the Great Lakes, was organized by Congress through the Northwest Ordinance of 1789. The territory comprised more than 260,000 square miles and was gradually divided to form the states of Ohio (1803—from the Iroquois meaning "Beautiful River"), Indiana (1816—"Indian Land"), Illinois (1818—named after the Illiniwek tribe), Michigan (1837—from the Ojibwa meaning "Michigamaa," or "Large Water"), Wisconsin (1848—"Mesconsing" or "Ouionsing," from the Algonquian meaning "Red Stone River"), and Minnesota (1858—from the Sioux for "Sky Tinted River").

A milestone passed after the War of 1812, when the "Old Northwest" was fully opened to white settlement. The War of 1812 ended with the 1814 Treaty of Ghent. The White House and other buildings in the capital had been burnt by the invading British. The most important outcome may have been that the U.S. failed in its attempt to conquer Canada and so would continue to share North America with a hostile British Empire.

The first U.S. president born west of the original thirteen states was Abraham Lincoln, born in Kentucky in 1809, later settling in Illinois. The state beyond the Appalachians that first acquired a significant measure of political prominence was Ohio. By 1865, Ohio was the birthplace of seven future U.S. presidents—Ulysses S. Grant, Rutherford B. Hayes, James

Garfield, Benjamin Harrison, William McKinley, William Howard Taft, and Warren Harding.

The Americans who populated the Old Northwest were mostly farmers who migrated from New England and New York State. There were also migrants from Virginia, where tobacco culture had played out the soil and the western mountainous areas were ill-suited to farming.

The soil of the vast region north of the Ohio River, much of it shaped by Ice Age glaciers, was perfect for agriculture. The climate, with four seasons and plenty of rain, was favorable, and land was cheap, with some granted free to war veterans. The network of rivers and lakes offered water power for mills, with good transportation at a time when roads were few and their condition poor.

The region offered valuable resources through the presence of minerals such as lead and iron, along with salt and abundant timber. The dearth of ready money was a constant problem, whether for everyday needs or to capitalize manufacturing or resource extraction. In the absence of large towns, there was little reason for investment by eastern banks, though local merchants and state-chartered banks extended modest amounts of credit. The economy depended on domesticated animals, especially horses and mules, hogs, and cattle. Debt and bankruptcy, as well as deadly disease, were regular parts of life. Infant mortality was common.

Though the Americans called their states after Indian names, they didn't want Indians in their presence or to hear of Indian claims of land for gardening or hunting. There were sporadic conflicts, but one culminating event: The Black Hawk War in Illinois, the last engagement fought between the Indians and the U.S. Army in the Old Northwest.

The Black Hawk War was big news in the state and nation for most of 1832. Much of the nation was now blanketed by newspaper coverage. It was a time when even minor skirmishes with the Indians were terrifying to the whites who, recalling their own egregious treatment of the Indians, fearfully rushed to stamp out anything looking like insurrection.

In 1820, the new state of Illinois had 55,000 white inhabitants. By 1830 the number had grown to 157,000. While there were still Indians living in scattered settlements, they had been decimated by illness and war, and what was left of the organized tribes had shifted west across the Mississippi.

My own Bliss family ancestors were among the early settlers of Illinois. As mentioned in Chapter One, they originated in Massachusetts with the arrival of Thomas Bliss in 1636. Subsequent generations moved from Massachusetts to Connecticut, then to New York State. From there they migrated to Illinois. Side branches came from New Jersey, Pennsylvania, and Delaware, the only slave state among the group.

In 1911, John F. Bliss of Princeville, Illinois, wrote a history of the Bliss family.[1] He writes of the Bliss migration:

> In Illinois, they ended up in a place called Princeville in northern Illinois near Peoria. Mr. Prince, after whom Princeville was named, was a frontiersman. Mr. Prince's log cabin stood on the ground now owned by our esteemed citizen, S. S. Slane…. Forty years ago or more, when as a boy I roamed the woods, this cabin stood. Mr. Prince had lived with the Indians for many years. He depended more on his hunting rifle for sustenance than upon tilling the soil.

He provides an anecdote about Prince:

> He must have had friendly relations with the Indians, for my mother told me that he was bitten by a rattlesnake. At that time he was the only white man in this part of the state. He used what remedies he had, but he grew much worse. Thinking he must die, he painfully drew himself up to the top of the roof of his cabin so that after death his body would not be eaten by wild beasts. In his extremity, some friendly Indians passed that way. They found him in this dying condition. They hurriedly held a consultation. Then they got busy. One hurried away out on the prairie. Soon he returned with an armful of herbs known later as "rattlesnake master." A kettle had been placed upon the fire, a poultice was soon made and applied to the bite, and the life of Prince was saved.

According to John F. Bliss, my great-great-grandfather Henry Bliss married Rebecca Smith. John F. Bliss writes:

> My grandfather, Henry Bliss, was born in East Town, Washington County, New York, Oct. 15, 1790. When he became a man he went west [The "West" at that time was western New York.], Chautauqua County, where he taught school during the winter and farmed during the rest of the year.
>
> At a social gathering one evening he met for the first time his future wife, Rebecca Smith, of Adams, Conn., who was visiting some other relation in that part of New York. The social function

turned into a dance in which all took part except my grandparents, who had religious scruples along that line. They were naturally thrown into each other's society for the evening, which proved to be very enjoyable to them.

This was the beginning of a courtship which ended in marriage on March 14, 1815. About this time, he was ordained as a minister in the Baptist church and held his relation to that church until he came to Illinois, when he united with the Christian Church sometime after. Their children were all born in New York, consisting of Hiram, Solomon, Esther, Nancy, Betsy, and Reuben.

There were a few tribes of Indians in western New York then. My father said they would often come to their house when he was a boy. They usually wanted salt. They always wanted to see the little white papoose. *He* was the white papoose. If they did not see him they would look for him, and many a time Indians... pulled him out from under the bed. He would kick and fight and they would laugh.

The early settlers were brave women as well as brave men, and my grandmother was one of them, as the following incident will show: Their home was in a clearing along the Chautauqua Lake. One day a deer took refuge from a pack of hounds, behind a large log near her home. A neighbor woman was sent to tell the men, who were chopping in the woods some distance away.

After the neighbor had gone, my grandmother heard the dogs coming. She was afraid they would frighten the deer away before the men came, so she took the butcher knife, quietly crawled up to the log, reached over and cut the throat of the deer. When the men arrived, she had it partly dressed.

Like all of the women of that time, she did the work of the house and made the clothing for the family, including the tailoring for the men. The song of the spinning wheel as my grandmother turned the wheel, with one hand holding the thread, I can hear yet, for fifty years ago the spinning wheel was in common use in all our homes. Economy was one of the virtues practiced in my grandmother's home. Pins were a valuable and scarce article. I have heard her say that a dozen pins were expected to last that many years and if one should be lost, diligent search was made for its recovery.

Next came the journey to Illinois:

And so my grandfather, the wood chopper, teacher, and preacher, with his wife and family of six children, Hiram, aged 19, Solomon, aged 17, Esther, aged 14, Nancy, aged 12, Betsy, aged 5, and Rheuben, aged 3, loaded their few household goods on a raft, said good-bye to their many relatives and friends of western New York, and set their faces toward the country of the setting sun.

The voyage had its dangers, for there were rapids which they must run and many a raft had gone to pieces. This was not the first time Hiram and Solomon had made this dangerous trip. They were possessed of great strength and physical endurance. They had spent their lives as woodsmen. They were expert swimmers, and they felt at home in water as well as on dry land.

They passed down the river into the Ohio and landed their raft safely at Cincinnati where they disposed of it. There they took passage on a boat for St. Louis, and from there to Peoria, the father and boys working for the support of the family. The next year they moved to near Southampton, a town at that time three or four miles west of Chillicothe.

Several towns in Ohio were called Chillicothe, which was an Indian name. This one was the center of the Indians' Hopewell mound-building culture, the last location of the main Shawnee settlement in Ohio, and a former capital of the state.

The trip further west by the Bliss family to Illinois took place in 1828, two years before Illinois became a state. They arrived at Princeville at the time of the Black Hawk War in 1832.

Betsy Elizabeth Bliss, mentioned above as the fourth child of Henry and Rebecca Bliss, was my great-great-grandmother. She married Clark Edward Hill, the grandson of Rev. John Hill, Sr., who had moved to Illinois from upper New York. So there were Protestant ministers on both sides of the family. Rev. Hill's mother was Mary Jackson from Northern Ireland. The father of Clark Edward Hill and the son of Rev. John Hill, Sr., was John Smith Hill, who died tragically in 1844 when he was caught in a blizzard while driving home from a business trip to a nearby town.

John F. Bliss writes of Betsy Hill:

Betsy Hill, daughter of Henry Bliss, was born in 1833. She is among our oldest and best-known citizens. She was five years old when she came to Peoria County. She has lived in this county seventy-three years. A man told me that she was the prettiest young lady in all the country. He said there were others who had the same opinion. This man was her husband, the late esteemed and respected Clark Hill of Monica.

The Hill family were more than early settlers. I think we could call them pioneers. They were a large family and of no small importance in the making of the history of Peoria County. Aunt Betsy has lived on the same farm since her marriage. She is the mother of seven children, three girls and four boys. The living are James, of Ohio; John, of Oklahoma; Clara Cook, of Wisconsin; and Milton, who lives on the old farm.

John F. Bliss's history captures something of the hardships and joys of life among the early settlers:

The history of one family of the early settlers of Illinois is largely the history of all. They had many things in common. They were largely descendants from the original colonists. They brought with them those sterling qualities which made them able to meet with an unyielding will the new problems and to successfully solve them with a courage which knew no defeat.

We of the present generation have a very limited conception of the suffering and deprivations our illustrious predecessors endured in settling a new country. We, their children and grandchildren, who sat at their knee on many a wintry night in the old family home, heard from their lips the stories which to us never lost interest, and which we rehearse to our children.

Culture and Values

John Bliss's history could be repeated thousands of times for the families that settled the Old Northwest, from Ohio to Minnesota. They were prolific, with many children, many of whom died of childhood diseases. The settlers were largely English in origin and Protestant in religion. They were farmers who worked at many sidelines, including blacksmithing and horse trading. Henry Bliss's brother, Zenas, was a millwright who fought in the Mexican War. Later, during the period before the Civil War, immigrants from Germany and Scandinavia began to arrive.

The early settlers valued literacy and education, and while it took time for schools to be organized, most farming households were also primary schools. The principal text of a household school was the Bible, from which everyone learned to read aloud. There was also John Bunyan's *Pilgrim's*

Progress, sometimes a volume of Shakespeare, and Christian hymnbooks. Gradually frontier education broadened, with *McGuffey's Eclectic Reader* appearing by the late 1830s.

The Bible molded the ethics and outlook of the community, and many of the settlers could see themselves as Israelites on New World soil. The Old Testament, with its tribal conception of God, could make them judgmental and stern, but life then was also a stern teacher. The life and teachings of the Redeemer, Jesus Christ, softened judgment and brought light and hope to the frontier, including among the missionaries who sought to teach the Indians. Many Indians converted to Christianity in this era, sometimes combining the teaching of Christian values with traditional Indian practices.

Villages and towns began to spring up. There was already a merchant class that had been trading with the Indians and farmers for decades, but now shops and stores appeared, along with churches. People on the farms speak of walking fifteen miles to church, often enjoying a community dinner, then staying overnight with local residents. Men and women would usually sit on different sides of the church's center aisle.

As time went on, there were also newspapers, banks, and lawyers riding circuit and arguing cases at newly-built courthouses. The frontier was armed to the teeth, with every man and boy—and sometimes girls—a hunter, and with posses and militias ready to help the few lawmen protect the peace and ameliorate any outbreak of violent crime or, on the periphery, Indian attack.

Some of the towns were destined to become metropolises of national importance. In 1806, Detroit was incorporated; in 1814, Cleveland; in 1833, Chicago.

Industry

Industrial development was on the way. James Watt's steam engine was used by John Fitch in 1787 to power a boat on the Delaware River, and steam drove a boat built by James Rumsey on the Potomac River at Shepherdstown, Virginia. Steamboat transport through the Great Lakes to the Erie Canal in New York opened the Old Northwest to national and world commerce parallel to the water route down the Ohio and Mississippi Rivers to New Orleans.

Most local and regional trade was handled by horse and wagon. This required decent roads. Road building was among the most common rural activities and, as it still does today, provided part-time or full-time work for multitudes of people. Roads were still in a primitive state until the 1830s, when macadam, a type of packed gravel, was used on the first federally-funded highway, the National Road, that ran from Baltimore to Vandalia, Illinois.

While the 1820s and 1830s saw a boom in canal-building, the future belonged to the railroads. The C&O Canal along the Potomac River in Maryland competed with the B&O Railroad to see which would win the race to the coal fields of Cumberland, Maryland. The railroad won, with the canal later being bought out and used by the B&O for coal hauling until it was finally destroyed in one of the recurrent Potomac basin floods in the 1920s.

But everything depended on iron, as European technology had done since Roman times. Now America followed suit, for it was the mining of iron ore and its smelting into pig iron and later steel that made the modern world possible. Without iron, no ship could be built, and even modern household utensils would be impossible to manufacture.

The iron industry made its appearance in colonial days, with iron furnaces being built all along the East Coast, particularly inland where the forests of the Appalachian Mountains provided a seemingly endless supply of timber. From the forests, charcoal could be made for smelting the iron ore that was dug out of the hillsides. It was these iron furnaces, of which Maryland had several of the most productive, that provided General George Washington's army with cannons, ammunition, and muskets.

An early industrial center was the federal armory at Harpers Ferry, Virginia, today in West Virginia, where the Shenandoah River joins the Potomac, which jointly provided water power for the slitting and rolling mills that produced the metal parts for military firearms.

The trouble with using wood to produce charcoal was that much of the forests of the Appalachians were stripped of trees, used both for the iron furnaces and for building construction and firewood. A consequence was that the land became susceptible to massive flooding and erosion without the trees' root systems to hold the rainfall. In many places the trees did not grow back until the conservation movements of the 20th century.

Fortunately for the iron industry, coal had begun to be mined for production of iron by the 1830s. A massive industry of coal mining for iron and steel manufacture came into existence along the entire Appalachian mountain range. The industry ran from New York and Pennsylvania, including west to Pittsburg, south to Birmingham, Alabama. When in 1859, Colonel Edwin L. Drake drilled the first oil well in Titusville, Pennsylvania, the modern world was truly born.

But all of this activity had to be organized. Industrial processes quickly became too complex to be managed out of a farm or household. All government could do, whether at the municipal, state, or national levels, was stand back and watch.

Modern business corporations were the next big step. But all of it had to be financed. As I wrote in Chapter Two, some of this financing arose from the Baring Bank via the British opium trade in China. The imported gold and silver from that trade capitalized banks, as did the mining of gold in California, the Rockies and Black Hills, and later, Alaska.

Through fractional reserve banking, the reach of the banks grew, seemingly endlessly, and the modern world took off. Until at least the Civil War, a majority of the financing of American development came from British and European banks and investors.

While wages in the U.S. were low compared to the profits of bankers and industrialists, there was still a scarcity of labor which caused American workers' earnings to exceed those available to the working class in Europe. So to meet the industrial expansion of the mid-to-late 19th century, massive immigration was attracted from central, northern, and eastern Europe. So the vast income discrepancy between rich and poor, so prevalent in the U.S. today, had early roots.

The South

The American South covers a vast region that was the home of legalized slavery until the time of the Civil War. I am including in this discussion the eleven states that seceded from the Union in 1861 to form the Confederate States of America—Virginia, Tennessee, North Carolina, South Carolina, Georgia, Florida, Alabama, Mississippi, Arkansas, Louisiana, and Texas. The slave states of Delaware, Maryland, Kentucky, and Missouri that stayed in the Union had a different trajectory, with ties and allegiances to both North and South.

The South relied on plantation agriculture for its livelihood. In Virginia and North Carolina, it was tobacco, which required large numbers of slaves. Tobacco depleted the soil, so that over time, Virginia engaged in breeding slaves and selling them further south. In the mountainous western counties of both states, along with South Carolina, Georgia, and across the mountains into Tennessee, farming was much more diversified, with slavery often confined to a few farmhands and "house slaves."

Counting the enslaved Africans (who in some states made up the majority of the population), well over half the population of the South was engaged in agriculture. In South Carolina the cash crops included rice and indigo. In the Deep South, from Georgia to the Mississippi River and beyond, cotton became the staple crop. In all regions, grains, livestock, and timber were prolific.

The growth and development of the South did create a modest urban manufacturing and financial infrastructure, but nowhere near the extent that developed in the North. The earliest cities were Charleston, SC, which the British founded in 1670, and New Orleans, founded by the French in 1718. Richmond, VA, was incorporated in 1742 but Atlanta was not until 1847. Banking in the South was sparse.

As industry developed in the North, the federal government began to raise tariffs on imported manufactured goods from Britain and Europe. The tariffs protected the prices of U.S.-manufactured products made in the factories and workshops of New England and the mid-Atlantic states. But Southern plantation owners were infuriated by the impact of tariffs on their own cost-of-living. These controversies over tariffs resulted in the first stirrings of Southern secession.

As the cotton plantations grew, their owners increasingly fell into debt both to the merchants who supplied them with manufactured goods and to northern banks out of New York, who were closely allied with bankers in London, especially the Barings and Rothschilds.

One of the unsolved mysteries of pre-Civil War history is the extent to which the New York/London bankers influenced the Southern oligarchs in their decision to secede from the Union in 1860–1861.[2]

Indian Removal

The growth of plantation agriculture in the South faced one major impediment: the presence of Indians who lived on the land and claimed large areas as their farming and hunting preserves. The champion of Indian removal in early 19th century America was Andrew Jackson, a southern planter and military hero, who was elected president of the U.S. in 1828, succeeding John Quincy Adams.

Jackson already had a history of involvement in federal Indian matters, as both a military officer and a commissioner in charge of negotiating Indian land acquisition. When conflict broke out in 1817–1818 between Georgia and the combined community of Indians and blacks who had retreated into Spanish Florida, Jackson and his army invaded Florida to suppress the uprising.

These Indians, who were members of multiple tribes, chiefly Creek, had been designated as "Seminoles." Jackson's army destroyed Indian and black villages across northern Florida, causing the occupants to retreat deeper into the swamps and forests.

After this display of force by the U.S. army, Spain sold the still largely unoccupied Florida region to the U.S. in 1821 for $5 million. The government's failure to bring the Indians to terms led to subsequent conflicts from 1835 to 1858. Eventually, over 4,000 Indians and blacks were deported to Indian Territory in the West, while a small band of Seminoles remained deep in the Everglades until their reemergence in the 20th century.

Although it met with opposition, Jackson's Indian Removal Act passed Congress in May 1830. The act gave Jackson authorization to send commissioners to negotiate removal treaties. The largest Indian tribe to be pushed out of the South, as well as the largest by population in the U.S. today, was the Cherokee Nation. It now numbers about 390,000. There are currently three Cherokee reservations: two in Oklahoma and one in North Carolina. The latter was formed from a remnant of several hundred individuals who avoided removal and received federal recognition in 1868.

Statehood in the South

The first of the new states in the South to achieve statehood was Tennessee in 1796, followed by Louisiana in 1812. As part of the Louisiana Purchase, the region around New Orleans was already part of a flourishing commercial center, so entry into the Union happened quickly.

While slavery was legal, New Orleans already had a substantial community of freed blacks who formed an integral part of the city's life. Subsequent to Louisiana, admission of the states of Mississippi (1817), Alabama (1819), Arkansas (1836), and Florida (1845) followed.

This left Texas as the final state to enter the Union among those that would later secede. The situation of Texas was the catalyst by which the U.S. would embark on its next major phase of expansion that was so monumental it seemed to many to be God-given. The enabling doctrine was called "Manifest Destiny."

Texas

Texas had been an independent nation before it became part of the U.S. in 1845. This resulted from the collapse of the declining Spanish Empire in the Americas, leading to independence first for Argentina, Venezuela, and Chile in 1810, and culminating in the creation of an independent empire in Mexico in 1821. Imperial rule in Mexico was overthrown from within in 1824. Political chaos resulted, with weak and fluctuating leadership being exercised from the new republican government in Mexico City.

Settlers from the American South had already begun moving into Texas, though at first they were outnumbered by the Indian tribes inhabiting the region and Mexican citizens near the Gulf Coast. The most powerful Indian tribe were the Comanches, who were expert horsemen, fiercely controlled much of central and western Texas, and had blocked the Spanish and later the Mexicans from northward expansion.

But with independence, and with a view to increasing the number of settlers in western Texas as a buffer against the Indians, the Mexican government enacted the General Colonization Law in 1824. This allowed immigrants of any race, nationality, or religion to acquire land in Mexico. Grants were already being made to "*empresarios*," the first one being Stephen F. Austin, whose "Old Three Hundred" settled along the Brazos River. Twenty-three other *empresarios* eventually brought in settlers, all but three originating from the American South.

As Americans continued to settle the vast spaces of grassland and prairie, Mexico became alarmed. In 1830 the Mexican legislature passed a law prohibiting further immigration by U.S. citizens. Mexico was also taking steps to centralize national governmental administration, which meant the elimination of Texan autonomy. War was now inevitable, with hostilities recurring over the next several years as privately-raised military detachments under American "filibusters" arrived.

Finally, the Mexicans sent an army under off-and-on-President and General Lopez de Santa Anna, which massacred American captives after wiping out the American garrison at the Alamo in San Antonio, even as the Texans were declaring independence on March 2, 1836.

The war ended with the Battle of San Jacinto on April 21, 1836, with the Mexicans routed, Santa Anna captured, and the Republic of Texas declared. Texas became a U.S. state nine years later, on December 29, 1845. The Mexican War followed within a year, but more on that in Chapter Four.

During the years of Texan nationhood, there was bitter warfare between the white settlers and the Texan Indian tribes, particularly the Comanches. The Texan most effective and persuasive in working to make peace between Texans and the Comanches was Texas's first president, Sam Houston. Houston's conciliatory attitude toward the Indians was not pursued by his successor, Mirabeau Buonaparte Lamour, who moved the capital from Houston to a small town on the edge of Texas hill country named Austin. Lamour advocated extermination of the Indians in Texas.

While most people assumed Texas would soon be annexed by the U.S., Lamour had other ideas. S.C. Gwynne writes: "His dream was to push the borders of his young republic all the way to the golden shores of the Pacific Ocean."[3] Lamour sent a military expedition to capture Santa Fe in New Mexico, but his party was forced to surrender by a combined force of Mexicans and Pueblo Indians. They were marched 2,000 miles to prison in

Mexico City, from which the U.S. government later obtained release of the survivors.

Sam Houston was returned to the Texas presidency after Lamour's single term, but the white population was determined to rid Texas of all Indians. Those living in eastern Texas, including Cherokees who had been removed from the southern states, were attacked and sometimes massacred. The 1867 Treaty of Medicine Lodge Creek, the last treaty made with the Comanches, established a reservation for the Comanches, Kiowas, and Kiowa Apaches in southwestern Oklahoma (then "Indian Territory") between the Washita and the Red rivers. The Comanches are there today, with a tribal government at Lawton, Oklahoma, and an enrolled tribal population of around 10,000.

Texas was by far the largest state in area in the U.S. and remained so until the admission of Alaska in 1959. After the Mexican war, Texas would comprise 268,597 square miles, extending from the coastal plan along the Gulf of Mexico to deep within the Old Southwest where it bordered New Mexico.

Texas grew and prospered, with its principal source of revenue and exports becoming the gigantic herds of cattle that roamed the open range. Cotton culture would come to Texas later. Later still came oil. For now, Texas was a slave state.

My paternal grandfather was Frederick Steele Fitts Cook, who was born in Texas. His birth name was Frederick Steele Fitts. At a young age his father died, and he was adopted by a man named Cook, about whom little is known, except that Cook was a gambler. The Fitts family were all southerners. Besides Fitts, the family names were Steele, Jones, and Everett. They came from North Carolina, Georgia, Tennessee, and Alabama. They ended up in Tioga, Texas, a crossroads north of Dallas, sixty miles south of the Oklahoma border.

Frederick's mother, my great-grandmother, Ida Florence Steele, was born in Tioga in 1874, before Tioga was even a town. Her husband, William Demarcus Fitts, died there in 1904. In the meantime, Ida had moved to California. According to the Tioga website, "In 1882, the first building constructed was a grocery store, and the first schoolhouse was a one-room frame building."

After the Civil War my ancestors from the North, the Hills and Blisses, joined with those from the South, the Fittses and Steeles, to produce the 20th and 21st century generations.

The Eve of Manifest Destiny

By 1845, the two halves of the nation—North and South—were beginning to stare at each other across a border of more than 1,500 miles, extending roughly along the Mason-Dixon Line between Maryland and Pennsylvania in the east, down the length of the Ohio River, jumping the Mississippi, then along the northern borders of Arkansas and Texas, with "Indian Territory" first in Kansas and then in Oklahoma taking shape through the ordeal of the Indian tribes.

As industry grew, steam power appeared on the waterways, while railroads crisscrossed the east and pushed west and south. Iron production was booming. Agriculture was flourishing, with produce being exported as grain to domestic markets and cotton crossing the Atlantic to the mills of England. Grain farming in the Midwest was beginning to explode through the implementation of new horse-drawn farm machinery for plowing, harrowing, and harvesting.

The importation of slaves had been abolished in 1808, but the economies of the southern states still depended on enslaved Africans for much of their working capital. Slaves were assets, plain and simple. Except that these assets required constant whipping to keep them in line. It was slaveholder Thomas Jefferson who said that a man "must be a prodigy who can retain his manners and his morals undepraved" by the South's "peculiar institution." Jefferson also gave investment advice: If a family had some cash, "every farthing of it [should be] laid out in land and negroes, which besides a present support bring a silent profit of from 5. to 10. per cent in this country by the increase in their value."[4]

The National Road and many other roads and turnpikes were busy with foot and wagon traffic. The federal government now had a Military Academy at West Point, New York, to go with the Naval Academy in Annapolis, Maryland, and paid its bills with customs duties from tariffs the South opposed.

But the federal government didn't do much else at a time when state governments were at work chartering the infant corporations that had been called into being to manage the many types of new enterprise. And as always, there was a national shortage of ready cash, leading to numerous small banks going into and out of business, along with scrip-issuing lumberyards, stores, and workshops.

Because horse-drawn conveyances were in such demand, a vast industry had grown up to manufacture wagons and carriages and to raise, feed, and trade in horses. By the end of the 19th century, many people were afraid the country would run out of the horses needed to transport products.

But big changes were coming, and changes often come sooner and faster than people expect. It was a time when the old Federalist Party had long been consigned to the history books. But the manufacture of railroads, large commercial cargo ships, and the newly invented farm machinery was a complicated and expensive undertaking. The fledgling corporations that built them had to be capitalized, necessitating ever larger quantities of investment capital. The big money market of the era was in New York, where European banks and financiers were major players.

With the economic changes, the nation began to divide itself into three broad political factions. The voice of New England and its growing manufacturing base was represented by Senator Daniel Webster of Massachusetts, first a Federalist, then a Republican, but later one of the founders of the pro-business Whig Party. Another Whig, Senator Henry Clay of Kentucky, represented the farmers and expansionists of the West. Clay favored the "American System" of protective tariffs, internal improvements, and the wished-for founding of a new national bank. The spokesman for the South was John C. Calhoun, a pro-slavery Democrat who led the fight in the Senate against high tariffs and first sounded the states' rights alarms.

Crop failures, drops in commodity prices, reckless speculation, and manipulation of the stock market all came together at various times to derail the growing American economy. The effects were often brutal, with people losing jobs, farmers being forced off their land, and railroads, banks, and other businesses going bankrupt.

The underlying cause of financial panics has long been fractional reserve banking and the capital investments derived from the practice. This is because borrowing money created from thin air and lent by usury at compound interest creates a net societal deficit authorized by law where repayment becomes the highest priority for the debtors. The debt is like a vacuum cleaner that sucks in all available resources, leaving people with little or nothing to live on if their economic productivity is the least bit disrupted—even for days or weeks. The banks in turn have only their own judgment to act as a brake on the cumulative debt, but poor judgment is emboldened by their legal ability to seize collateral.

A society that depends on bank loans is never far from bankruptcy. Entire societies and nations can be sucked into the black hole of debt. "Panic" is an apt word for the resulting frame of mind.

Financial Panics

There were three major financial panics[5]—known today as recessions or depressions—before the Civil War.

The first, the Panic of 1819, was triggered by a collapse in cotton prices due to a contraction in credit. The contraction resulted when the Second Bank of the United States tightened lending after its Baltimore branch collapsed under the weight of non-productive speculative loans. Critics of banking had long foreseen the dangers of such speculation.

The Bank tried to save itself by importing $7 million in specie from European banks—at a price, of course. State-chartered banks were forced to call in their own loans in order to remain solvent, and foreclosures of farms and bank failures resulted. The effects of the Panic of 1819 lasted for two years and were felt most in the West and South.

Unhappiness over the economic hardships lasted for years and produced resentment against the nation's financiers that boosted Andrew Jackson into the White House in 1828. The Second Bank in particular took a beating in the press, leading to Jackson's campaign against it a couple of years later.

By 1837, the Second Bank was defunct, with the currency in circulation contracting under President Andrew Jackson's anti-Bank policies. The Panic of 1837 was triggered by the failure of the wheat crop, another collapse in cotton prices—resulting from economic problems in Britain—the bursting of a land speculation bubble, and the lack of readily interchangeable currencies among the variety of unregulated state and local banks.

This time, the effects lasted six years and had a devastating impact. Numerous brokerage firms in New York failed, and a New York City bank president committed suicide. Many state-chartered banks also went under, and the growing labor union movement was effectively thwarted as wages plummeted.

The state bank failures resulted in a withdrawal of European banks from U.S. capital markets, causing much business to grind to a halt. The panic also caused the collapse of real estate and food prices, which was devastating to farmers and plantation owners. The stories told of the Panic of 1837 were repeated a century later during the Great Depression.

Within a scant twenty years more came the Panic of 1857. Prosperity had begun to return with the pre-Civil War industrial expansion in the North and the now-booming cotton business with England and France that benefited Southern plantation owners, along with their creditors at the New York banks. But again, bank speculation, as always a result of their ability to create money beyond their reserve base due to the practice of fractional reserve lending, was out of control.

The Panic of 1857 started with the failure of the Ohio Life Insurance and Trust Company, which was doing most of its business as a New York City bank. Reckless speculation in railroads took the company down, causing stock prices to plummet and a literal panic on Wall Street, as crowds of frantic investors clogged the streets. More than 900 mercantile firms in New York ceased operation, and by the end of the year, the American economy was in a shambles. Recovery didn't begin for two more years, on the eve of the Civil War.

It was starting to become clear that economic crashes inevitably resulted from the vagaries of bank lending and the total dependence of business and industry on the financial sector.

Promoting and Managing Growth

Trouble for many was opportunity for some. Returning to the 1790s, we saw that Alexander Hamilton, the nation's first Secretary of the Treasury, had advocated a rudimentary form of centralized economic development through creation of the First Bank of the United States. A major purpose of the Bank, as Hamilton saw it, was to capitalize what we would call today a national infrastructure, consisting mainly of roads and canals, to benefit economic development for the nation.

But when Jefferson and his party took power, leading to the "Era of Good Feeling" continuing through the Madison and Monroe presidencies, the concept of national economic development virtually disappeared. The exception was the funding of the National Road that was built between 1811 and 1837. Even the C&O Canal was built by a private joint stock company, and the railroads were built entirely with private funding until the 1850s.

While early railroad surveys and construction were financed by private investors, in 1850 the federal government provided a land grant to the Illinois Central Railroad that set the pattern for future railway development that would eventually stretch all the way across the country to the Pacific Ocean. An exception to the lack of public infrastructure funding was the building of the Erie Canal, which opened in 1825 and was funded by the New York state government through bonds and loans.

The federal government contributed nothing to this project. But it was the Erie Canal that opened eastern and foreign markets to the Midwest's agricultural bounty and led to New York's designation as the "Empire State." The canal was the first navigable waterway linking the Atlantic to the Great Lakes, resulting in a huge reduction of the costs of transporting farm products across the Appalachian Mountains. With its splendid natural harbor, New York City also became the leading U.S. hub for foreign trade and immigration.

The concept of national economic development was revived through "The American System" speech made in 1824 by Whig Party leader Henry Clay of Kentucky, who was then Speaker of the House. The Whigs were the heirs of the pro-business Federalists. Clay realized that if the U.S. were to remain a producer only of raw materials and agricultural products for European markets, it would always be economically and politically subservient to foreign powers. Clay saw that the remedy was high tariffs that would protect the nascent American industry from being undercut by cheap European products manufactured by workers who earned poverty-level wages.

Clay was mainly targeting Great Britain, a nation that had moved from mercantilism to its own program of economic nationalism based on that high-sounding term, "free trade." Of course, to the British, free trade meant no restrictions on their own ability to sell manufactured goods anywhere in the world without impediment of unilateral benefit, insofar as they were able to undersell all competitors.

While the British system came to be called "Classical Economics," which later fed into the philosophy of "libertarianism," it was really a system akin to "might makes right." Some might even call it "economic imperialism" or "survival of the fittest." Still, British products were considered by many to be the best in the world, causing the British to feel justifiably proud. The American practice of keeping them out by high tariffs was viewed by Britain as a deeply hostile action.

Henry Clay also saw that in order to develop its manufacturing capabilities, the U.S. would need Hamilton's aforementioned "internal improvements"—again, canals and roads—as well as a national bank. But the Second Bank of the United States chartered in 1816 was compromised in the public mind, because it seemed more beholden to its wealthy investors than to the nation and its needs.

The real issue was that stockholders in the Second Bank, who often bought their shares on credit, saw the price of their stock steadily rise, while the Bank's customers had to repay their loans with interest, on penalty of loss of collateral, property, or even livelihood. So, while Henry Clay had made a good speech to Congress, it was full of ideas whose time had not yet come. But they were not forgotten, becoming the basis for federal legislation in the post-Civil War period. Even then, there was tension between the use of federal funding for infrastructure vs. financing by private banks and investors.

The foremost 19th century figure in favor of government support of infrastructure was Henry C. Carey (1793–1879) of Philadelphia, an economist, sociologist, and chief economic adviser to President Abraham Lincoln. Like Henry Clay, Carey was a proponent of a strong tariff system to protect domestic industry. In 1838 Carey published his *Principles of Economics* that remained a standard textbook for decades and outlined the basic ideas of a *dirigiste* economic system, which includes government intervention for the achievement of national self-sufficiency through application of advancing technology.

But Carey's main contribution, originated by Hamilton and echoed by Lincoln, was the idea of the labor theory of value—that products derive their value from the human labor expended in making them. Carey's 1838 version—preceding Marx's publication of *Das Kapital* in German in 1867—did not exclude managerial labor, but it gave little heed to capitalist claims that the money invested in an enterprise was its driving force and key component.

The labor theory of value would give birth to the modern labor movement but would also move society in the direction of socialism or even Marxism. But there were also those who argued that socialism reflected values from the New Testament that gave rise to many small-scale experiments during the 19th century in communal living, such as among the Shakers. It is not difficult to see that these ideas behind the American System were antithetical to those of bankers, stockholders, and financiers who claimed to have the right to exercise priority when it came to directing and profiting from the Industrial Revolution.

Among the most prominent names in big finance during that era is that of the Rothschilds. Figures connected with the European Rothschild family were deeply embedded in the New York banking structure as the 19th century progressed. The Rothschilds had supplanted the Barings in dominating the world of big finance in Great Britain. Much of their strength came from their ability to leverage parallel Rothschild operations in France, Italy, Germany, and Austria. Strangely, President Andrew Jackson, despite his opposition to the Second Bank of the United States, became a personal client of the Rothschilds, according to the semi-official *Rothschild Archives*.

In 1837, a banker named August Schönberg arrived in the U.S. to represent Rothschild interests during the panic that commenced that year. Schönberg changed his name to August Belmont, and soon the Rothschilds became one of the European financial agents of the U.S. government. Over the next several decades, Belmont became a prominent figure in American society, including politics, culture, and even horseracing. Belmont also became one of the principal power brokers of the Democratic Party, and we'll hear more about him when we discuss the Civil War.

The Independent Treasury

The Panic of 1819 had exposed the Second Bank of the United States as being unable to stabilize the public finances of the nation. While the Bank handled various fiscal duties for the U.S. government, held government funds on account, made business loans, and attempted to regulate other banks by collecting their bank notes and presenting them for redemption, it was subject to intense criticism.

An 1829 report coming out of Philadelphia and issued by the Working Men's Party criticized the entire banking industry as having "laid the foundation of artificial inequality of wealth, and, thereby, artificial inequality of power." The attacks dovetailed with those who opposed paper money in principle, not just the fractional reserve system, who argued in favor of a "hard money" system trading only in gold and silver specie.

President Andrew Jackson, heir to Jefferson and the leader of what by now was the Democratic Party, was strongly influenced by the hard money advocates. While the Bank's president, Nicholas Biddle, was boasting that he was more powerful than the president of the U.S., Jackson declared, "I will kill the bank, or the bank will kill me." Biddle accused Jackson of being "a Marat or Robespierre," naming two famous French Revolution terrorists. The Boston *Daily Advertiser* said that Jackson "is found appealing to the worst passions of the uninformed part of the people and endeavoring to stir up the poor against the rich."

Jackson's attack on the Second Bank obviously made his own Rothschild connections hypocritical. Was the Bank a Rothschild competitor, and were they using Jackson to further their own ambitions to control banking in the U.S.? We'll never know. In any case, the "Bank War" was on.

Hard money and anti-bank activists had meanwhile been proposing the creation of an "Independent Treasury" system, whereby the federal government would store its gold and silver in its own vaults, rather than in the Second Bank. During his second term, in 1833, President Jackson moved in that direction by removing federal deposits from the Second Bank and placing them in state-chartered banks derisively called "pet banks." These banks were authorized to issue paper notes redeemable in specie in denominations of $20 or more.

Jackson was succeeded by Vice President Martin Van Buren of New York, who warned against the U.S.'s becoming a "bank-ridden society." Two months into Van Buren's presidency, on May 10, 1837, several overextended state-chartered banks in New York ran out of hard currency reserves and suddenly refused to convert paper money into gold or silver. Other financial institutions throughout the nation quickly followed suit. As we have seen, the financial crisis, the Panic of 1837, was caused essentially by the overextension of lending by the banking system.

Van Buren now sought to insulate the federal government from the ups and downs of the banking world by proposing the establishment of an

"Independent Treasury." Van Buren said that the system would remove politics from the nation's money supply by requiring the government to hold all of its balances in gold and silver with absolutely no chance of fractional reserve banking getting hold of it. An Independent Treasury, he said, would stop overextension of lending and eliminate inflation. It did in fact eliminate inflation, but at great cost to the nation's producers and workers whose own sources of income and credit now began to dry up with the government's hoarding of hard currency. Any reduction in societal purchasing power results in reduced economic activity.

State banking interests, especially in New York, along with the pro-business Whig Party, fought Van Buren's proposals tooth and nail. The Independent Treasury Act of 1840 was finally passed, but it only lasted a year. When the Whigs won a congressional majority along with the White House in 1840, they promptly repealed the law. The new Whig president, William Henry Harrison, lived only thirty-one days after his inauguration. His successor, John Tyler, was of so little help that the Whigs expelled him from the party. Neither was interested in managing federal government finances, so were content to leave matters to the banks, even though the Second Bank had by now passed into history.

But the Democrats came roaring back in the 1844 elections, retaking Congress and the presidency. Former Speaker of the House James K. Polk was now president and was intent on prosecuting the war that would soon be breaking out against Mexico and on dealing with the impasse that threatened another war with Great Britain over the Oregon Country.

Polk was also determined to beat back Whig high-tariff policy and thwart any attempt to run up federal expenditures through any semblance of Henry Clay's American Plan. Polk revived the Independent Treasury and pushed through tariff reduction. He signed the Independent Treasury Act on August 6, 1846.

The 1846 act provided that the public revenues be kept in the downtown Washington, DC. Treasury Building and in sub-treasuries in various cities. I was shown one of these sites as a historical curiosity in the basement of the old Main Treasury Building on Pennsylvania Avenue when I worked for the Treasury in the 1990s. There was a countertop, a little workspace for the clerks, and behind the counter a very large vault.

The U.S. Treasury was to pay out its own funds as direct disbursements and be completely independent of the banking and financial system of the nation. If, say, a military paymaster wanted to pay his troops, he would take an authorized voucher to the nearest U.S. Treasury vault and be handed a bag of precious metal or a folder containing Treasury Notes. If it were a large bag, he might have an armed guard.

Still, the separation of the Treasury from the banking system was never complete, because Treasury operations continued to influence the money market. Specie payments to and from the government affected the amount of hard money in circulation. This was increasingly the case after the expansion of the money supply following the California Gold Rush that spawned a major economic expansion during the 1850s.

This expansion proved that it wasn't an increase of money in circulation that caused inflation. Rather inflation was caused by an increase in *lending and debt*. There is a big difference between lending and debt and real money. Today both the government and the banking system ignore this difference by calling money owed an asset of the creditor.

Of course, there were times when the money that was sitting in Treasury vaults without being spent was viewed as a kind of hoarding that critics claimed was slowing business expansion. There is truth in this belief, though from the government's point of view, they were being good stewards of public funds.

Success of the Independent Treasury in the Panic of 1857

In 1857, as we have seen, another financial panic hit the country. However, while the failure of banks during the Panic of 1837 caused the government great difficulty, bank failures during the Panic of 1857 did not, as the government, having its money in its own hands, was able to pay its debts. In his December 7, 1857, state of the union message, President James Buchanan said:

> Thanks to the Independent Treasury, the government has not suspended [specie] payments, as it was compelled to do by the failure of the banks in 1837. It will continue to discharge its liabilities to the people in gold and silver. Its disbursements in coin pass into circulation and materially assist in restoring a sound currency.[6]

But the philosophy of fiscal prudence did not survive the Civil War. During the war, Congress passed the National Banking Act of 1863, creating a national banking system, though without an overarching central bank to oversee it. Once again, federal government funds began to be deposited in private banks, with most taxes and payments to the government allowed to be made in national bank notes.

At the same time, however, the government allowed itself to spend its own money directly into circulation through the issuance of Greenbacks. These were not redeemable in specie. It was the Greenbacks that were the measure that saved the Union, to the intense chagrin of the banking system. More on the Greenbacks later.

After the Civil War, the Independent Treasury continued in modified form, as each successive administration attempted in different ways to use Treasury funds to expand and contract the money supply according to the nation's credit needs. Nevertheless, the U.S. experienced almost continuous economic panics of varying severity, taking place in 1873, 1884, 1890, 1893, 1896, and 1907.

In retrospect, we can see clearly the phenomenon of "business cycles" under a system dominated by fractional reserve banking. We continue to have those business cycles today, and they are worsening. With the Independent Treasury, the federal government tried successfully to insulate itself from being held hostage to that system and the inevitable harmful consequences that follow. But the problem of sufficient currency to fuel the nation's commerce had not been solved.

Later, after the Panic of 1907, Congress established a National Monetary Commission to investigate the problem of recurring recessions/depressions and propose legislation to address them. This turmoil culminated in the Federal Reserve Act of 1913, and the end of the Independent Treasury system.

The Federal Reserve Act established the current U.S. Federal Reserve System and authorized the printing of Federal Reserve Notes. Government funds were gradually transferred from subtreasuries to the Federal Reserve. The Independent Treasury Act of 1920 mandated the closing of the last subtreasuries, thus bringing the system to an end. More on the Federal Reserve later.

ENDNOTES FOR CHAPTER 3

1 All material on the Bliss family and descendants has been archived by Johnny Lathrop of Tyler, Texas, and is used with his permission.

2 Judah Benjamin, the Confederacy's Secretary of State, was born in the Danish West Indies, then under British control. Benjamin was believed to have strong ties to the British banking oligarchy. After the Civil War, he fled to London, where he spent the rest of his life. Unfortunately, the historical record says almost nothing about private meetings and discussions among powerful people of the era like Judah Benjamin, the Rothschilds, etc., so historians are reduced to speculation.

3 S.C. Gwynne, *Empire of the Summer Moon* (Scribner, 2011), 75.

4 Ambrose, *Undaunted Courage,* 75.

5 Information on financial panics is from open sources.

6 James Buchanan, "State of the Union 1857 – 8 December 1857" Presidents, *American History from Revolution to Reconstruction and Beyond.* http://www.let.rug.nl/usa/presidents/james-buchanan/state-of-the-union-1857.php

CHAPTER 4

"Manifest Destiny"

A Divine Mandate to Rule

The Manifest Destiny narrative is the idea that a divine power had specially favored Americans with the right to rule over the entire continent of North America "from sea to shining sea"—and maybe even beyond. Future iterations of this narrative would encompass claims that the U.S. is *the* "exceptional" or "indispensable" nation.

Were there, then, any particular responsibilities that devolved upon the holders of this purported supremacy? Or are the fruits of Manifest Destiny just some kind of reward or favor for innate or systemic virtues that the white masters were enabled to use at their discretion? Or does it purport to have a legal basis, as Chief Justice John Marshall affirmed, resting on the "right of discovery" and "right of conquest?"

Where did such an august idea come from? Well, like so many other ideas that have driven Americans to action and distraction, "Manifest Destiny" seems to have started on a crasser level: as a media pronouncement. In 1839, journalist John L. O'Sullivan, an influential proponent of Jacksonian Democratic Party politics, a man described as "always full of grand and world-embracing schemes,"[1] wrote a newspaper article that predicted a "divine destiny" for the U.S. Due to its inherent values such as equality, individual rights, and personal autonomy, the U.S. was preordained "to establish on earth the moral dignity and salvation of man." Use of the term "salvation" might lead us to ask, on the basis of what theological standing or sacred texts did O'Sullivan arrive at this conclusion? He didn't say.

Six years later, in 1845, O'Sullivan published an essay entitled "Annexation" in the *Democratic Review*, in which he introduced the actual phrase "Manifest Destiny." There, he urged the U.S. to annex the Republic of Texas, because it was "our manifest destiny to overspread the continent allotted by Providence for the free development of our yearly multiplying millions."

To use a German word that gained notoriety in the years before World War II, he seemed to have been writing about *Lebensraum*, "elbow room." O'Sullivan used the phrase again on December 27, 1845, in his newspaper *The New York Morning News*. Speaking of the ongoing boundary dispute with Great Britain over the Oregon Country, O'Sullivan wrote of Oregon: "And that claim is by the right of our manifest destiny to overspread and to possess the whole of the continent which Providence has given us for the development of the great experiment of liberty and federated self-government entrusted to us."

So, God gave America the entire North American continent, or at least so O'Sullivan said. But where did Great Britain fit in, with its control of Canada? Well, wrote O'Sullivan, the U.S. was for "republican democracy, the great experiment of liberty." Britain obviously was not. Now that Texas had been welcomed into the fold, O'Sullivan believed California would follow. Then, he wrote, Canada would come along of its own volition, which never happened.

President Andrew Jackson used similar terms when he spoke of national expansion as "extending the area of freedom." But freedom for whom? Certainly not for the Indians Jackson removed from the eastern U.S. Certainly not for the slaves who worked his plantation. Freedom from the British?

O'Sullivan was still pushing his narrative in 1845 when Democratic President James K. Polk was elected.[2] Polk was fully intent on going to war with Mexico over lands in the Southwest and of fighting Britain over Oregon if necessary. Polk's Whig opponents, who had been voted out of office after Martin Van Buren's single term, argued in *The American Whig Review* in January 1848:

> ...that the designers and supporters of schemes of conquest, to be carried on by this government, are engaged in treason to our Constitution and Declaration of Rights, giving aid and comfort to the enemies of republicanism, in that they are advocating and preaching the doctrine of the *right of conquest*. (my italics)

The Whigs did not remark on the fact that, in concert with declarations twenty-five years earlier by Chief Justice John Marshall, "the right of conquest," as indicated above, was the principle by which Europeans and Americans had been disenfranchising Native Americans for centuries.

And who was to say that Manifest Destiny was to stop at the shores of the Atlantic and Pacific Oceans? In 1859, Reuben Davis, a member of the House of Representatives from Mississippi, said:

> We may expand so as to include the whole world. Mexico, Central America, South America, Cuba, the West India Islands, and even England and France [we] might annex without inconvenience... allowing them with their local Legislatures to regulate their local affairs in their own way. And this, Sir, is the mission of this Republic and its ultimate destiny.[3]

Some would argue that now in 2023, Reuben Davis's mission is still the U.S. goal.

The Monroe Doctrine

By 1823, the U.S. felt secure enough for the issuance of the Monroe Doctrine against renewed or further European colonization in the whole of the Americas. Originally proposed by Great Britain to be issued as a bilateral statement, thus making Britain a co-guarantor along with the U.S. against all other European nations, the Monroe Doctrine, as drafted by Secretary of State John Quincy Adams for President James Monroe, emerged as a unilateral U.S. declaration. Though Britain tried to pressure Monroe on the point, the U.S. was averse to sharing. Agreement would have been close to admitting that the U.S. was still part of the British Empire. Here is how the U.S. State Department's Office of the Historian today puts the matter:

> The bilateral statement proposed by the British thereby became a unilateral declaration by the United States. As Monroe stated: 'The American continents...are henceforth not to be considered as subjects for future colonization by any European powers.' Monroe outlined two separate spheres of influence: the Americas and Europe. The independent lands of the Western Hemisphere would be solely the United States' domain. In exchange, the United States pledged to avoid involvement in the political affairs of Europe, such as the ongoing Greek struggle for independence from the Ottoman Empire, and not to interfere in the existing European colonies already in the Americas.

The second part of this declaration about avoiding "*involvement in the political affairs of Europe*" has been conveniently, and grievously, forgotten.[4]

Anyone who cites the Monroe Doctrine without mentioning this provision is being less than candid.

In any case, by 1823 the U.S. was declaring itself the master of the Western Hemisphere and warning Europe to back off. But fighting Mexico was another matter. In the previous chapter, we discussed the annexation of Texas. Now, as the Mexican War approached, the nation was bitterly divided not only over Texas, but over taking any further steps to antagonize what was now the "sister republic" of Mexico.

As we have seen, the Whigs were committed to forestalling the extension of slavery into new territories, and were also taking a stance on principle against any new war of aggression. Future President Ulysses S. Grant served in the Army during the Mexican War but wrote in his memoirs of his belief that the Mexican War was unjustified. Abraham Lincoln, elected to Congress in 1846, was strongly opposed to the Mexican War and spoke picturesquely of President James K. Polk's desire for "military glory—that attractive rainbow that rises in showers of blood."

The Election of 1844

The period following the Democratic presidencies of Andrew Jackson and Martin Van Buren, marked in particular by the demise of the Second Bank of the United States, thus witnessed a period of major expansion under U.S. President James K. Polk. After the War of 1812, the boundary with Canada was still unsettled. To the south, Spain's New World empire had begun to collapse by the early 19th century with independence movements throughout the hemisphere, including Mexico. To the northwest, Russia exerted unmolested control over Alaska, though with no colonization. The British, French, Dutch, and even Danish still had interests in the Caribbean and West Indies. So there was still stiff competition for North American control when Polk arrived.

The election of 1844 was among the most pivotal in U.S. history. In 1840, the Whigs had elected "Old Tippecanoe," William Henry Harrison, who died thirty-one days after his inauguration. Vice-President John Tyler was awakened at his home in Williamsburg, Virginia, with the news that he was now president. During his term he broke with the Whigs—they actually expelled him—embraced his Southern slaveholding heritage, and accomplished little except to promote the annexation of Texas. Henry Clay of Kentucky now ran in 1844 in order to get the Whig program of protective tariffs, national banking, and internal improvements back on track.

Former Speaker of the House James K. Polk of Tennessee was an accomplished career politician who had served as governor of that state. Polk had been Andrew Jackson's protégé—his nickname was "Young Hickory," in reference to Jackson's "Old Hickory"—as well as Jackson's chief congressional ally in the Bank War. But Polk was a compromise candidate—the first "dark horse"—when nominated by the Democrats on the ninth ballot to run against Clay.

Voters had a clear choice. Unlike modern elections, when scarcely more than half the electorate votes, over eighty percent of potential voters went to the polls. As stated above, the Whigs were opposed to territorial expansion, including admission of Texas—another slave state. Democrats were in favor. The Whigs wanted high tariffs. The Democrat's didn't. The Whigs favored restoration of a national bank. The Democrats said, "No way."

Instead, the Democrats promised westward expansion and war. With their proposal to settle the Oregon controversy with Britain, they thereby attracted northern voters eager to begin establishing settlements in that region.

In the national popular vote, Polk beat Clay by fewer than 40,000 votes, a margin of 1.4%. James G. Birney of the anti-slavery Liberty Party won 2.3% of the vote, pulling enough votes away from Clay to make the difference.[5]

With Polk's election, war against Mexico now loomed. It was without a doubt a war of aggression. The U.S. economy had never been stronger, nor had government finances. President Andrew Jackson had accomplished what today would seem impossible: he paid off the entire national debt. The U.S. was not in hock to anyone, including the increasingly powerful New York bankers and the ever-conniving British financiers.

Further, the abolishment of the Second Bank of the U.S. and the Panic of 1837–1843 had been overcome. The U.S. economy was growing, railroads were connecting all the states and even many localities east of the Mississippi, the northern and midwestern industrial base was expanding, foreign markets for cotton were booming, and the Northwest Territories were filling with settlers all the way to Michigan, Wisconsin, and Minnesota.

Moreover, with the national debt having been paid off, the U.S. government's budget during the 1830s had run steady annual surpluses, and the government's cash had been securely protected against financial speculation and even the Panic of 1837 through deposits in Jackson's state-run "pet banks" while legislation for the Independent Treasury was pending in Congress.

State banking was also booming. The Panic of 1857 (see Chapter 3) was yet to come. After the Second Bank closed, a huge expansion of the number of state banks took place, with credit becoming cheap and easy, particularly in the West. The state banks often stood on shaky ground, but even if bankruptcies shuttered a bank, its loans remained in circulation, so fed credit into the growing economy.[6]

In the meantime, government action to enhance the value of gold coins, combined with the discovery of gold in the Southern Appalachians, caused the circulation of specie to grow. Enhancements of silver coinage led to a large increase in circulation, with foreign coins now being removed as legal tender.

The power and reach of the state banks also allowed a huge increase in the budgets of state governments for canal and railroad construction, education, mental hospitals and orphanages, poor relief, and business chartering and regulation. State governments had learned the value of taxation on property as a reliable revenue source, along with the sale of land acquired by the federal government from the Indians. The federal government alone had Constitutional power to make treaties with the Indians but often turned the acquired land over to the states.

Under these conditions, James K. Polk brought energy and determination to the White House, even while pledging to serve only a single term.

Oregon

In his December 2, 1845, State of the Union address, President Polk discussed the controversy over the Oregon Country that he soon would settle with Great Britain prior to the Mexican War. The region had been subject to competing U.S.-British claims and had been explored by Lewis and Clark. What became the U.S. portion of the Oregon Country included what today consists of the states of Washington, Oregon, and Idaho, along with western Montana and much of western Wyoming. The rest would go to what would become Canada.

Polk was able to conclude decades of wrangling when on June 15, 1846, the U.S. and Great Britain signed the Treaty of Oregon establishing the 49th parallel as the international boundary in the Pacific Northwest. Britain also wanted to secure a free right of passage for British vessels travelling down the Columbia River from British Columbia to the Pacific. Polk refused to countenance this compromise of U.S. sovereignty, giving Britain instead the lower part of Vancouver Island that drops below the 49th parallel.

Polk also oversaw ten treaties with Indian nations, seven of which resulted in the acquisition of Indian land. The Kansas, Potawatomi, Chippewa, Winnebago, Pawnee, Menominee and Stockbridge tribes all acquiesced to coercion and nominal payments by agreeing to give up part of their ancestral homelands.

Polk purchased his Mississippi plantation from lands put up for sale from the 1830 Indian Removal Act. Of the nineteen enslaved people Polk purchased during his presidential term, thirteen were children. He died at the age of fifty-three after being afflicted with cholera on his post-presidential tour through the South. Though they were childless, Polk had the benefit of a strong and sociable wife and companion, Sarah Childress Polk.

The Mexican War

But first Polk fought the Mexican War. In 1846 Congress declared war against our southern neighbor. The U.S. argued that Mexico struck first, even though Polk had already moved U.S. forces across the Rio Brazos River toward the Rio Grande into land Mexico claimed. The tactic was similar to how the U.S. dealt with the Indians: invade their land and incite them to attack.[7]

As we have seen, Texas had already been settled by white Americans, declared itself independent, then was annexed by Congress and given statehood in 1845. Through the Mexican War, the U.S. acquired what is today Texas, New Mexico, Arizona, California, and much of Utah, Nevada, and Colorado.

The immediate cause of the war was straightforward. The Americans had begun to move into the region of northern Mexico. They liked what they saw and took it, and Mexico, which had initially invited them in, was too weak to defend it. Much of the land was sparsely occupied, except by age-old Native American cultures and some Hispanics, especially along the Texas border, in central New Mexico, and on the California coast.

The Mexican War was a big step toward what both Hamilton and Jefferson foresaw and desired—an American empire. But trouble loomed as the states of the North and South continued to develop along different lines. There can be little question that the main difference between the two was the existence—and growth—of African slavery. Until the 1850s, the numerical balance between the slave-holding and free states had been kept in balance, which meant an equal number of U.S. senators representing the two sides.

But after Texas, the balance had tilted towards the free states with the admission of Iowa, Wisconsin, California, Minnesota, and Oregon from 1846 to 1859. In the late 1850s, violence broke out in "Bleeding Kansas,"[8] admitted as a free state in January 1861, following the election of Abraham Lincoln. The fires had been fueled with the growing power of the abolitionist movement, particularly in Massachusetts, the success of the Underground Railroad, and the outrage over the Dred Scott decision of 1857, followed two years later by John Brown's raid at Harpers Ferry.

But before we get the Civil War, we'll return to the Native Americans and their struggles to survive.

The Western Montana Tribes

The Native American nations that are spread across North America can be viewed as if one were looking through a repeatedly turning kaleidoscope. None of them ever drew boundaries or printed tribal names on a map. And their locations, mode of life, and even the names of constituent tribes and bands were constantly changing.

Of notable fact is their amazing degree of mobility. If conditions, including climate and the availability of game for hunting, changed, they would pack up and transport themselves elsewhere. It's said of the Plains Indians that a tribe could pack an entire village and be ready to move in fifteen minutes. This would also take place if a more aggressive, numerous, or powerful tribe appeared and drove the former inhabitants out. Often neighboring tribes were at peace with each other; at other times, they fought. Sometimes they exterminated their rivals, but more often, warfare was ritualized, with few casualties and with points being scored merely by touching an enemy, known as a counting coup.

The success of the Indians in upholding a sustainable lifestyle for so many thousands of years can be attributed, at least in part, to their remarkable spiritual closeness to the land, to nature, and to the plants and animals they cultivated or hunted, often viewed not just as nourishment, but as kin.

The domestication of corn by the Indians of Mexico is particularly noteworthy—a process that took thousands of years to develop across the North American continent, with local varieties adapting to local conditions. A similar story took place in Bolivia and South America with the potato. For a detailed look, I recommend Colin Calloway's *One Vast Winter Count: The Native American West Before Lewis and Clark,* along with his other books on Native American history. It is Calloway to whom I owe the metaphor of a kaleidoscope.

But the Indian nation I want to discuss in more detail is the Flathead tribes of Montana. The name is actually a white man's misnomer that the government seized upon to name the Flathead Reservation, the home of the Salish and related Western Montana tribes. I was born in Missoula, not far from the Flathead Reservation, a place where my mother and father both spent time when they were growing up.

I consider the history of the Flathead tribes to be one of the great survival sagas of American history.

Geologic History

At the end of the last Ice Age, called the Wisconsin glaciation, the two great ice sheets that covered much of what is today Canada and the U.S. were retreating. One was the Cordilleran ice sheet, coming down the coast of present-day Alaska, British Columbia, and a bit into Washington state, then spreading across northern Idaho and Montana west of the Continental Divide. To the east was the vast Laurentide ice sheet that covered most of eastern Canada, before dropping down into today's U.S., to what is now the northern Great Plains, the Midwest, and the Great Lakes.

In the region of the southern Cordilleran ice sheet lay the Columbia River Basin, covering 258,000 square miles and including parts of what are today seven northwestern states and British Columbia. Rising in Canada and flowing across the U.S.-Canada border into today's Washington state, the Columbia flows in its 1,200-mile course to the Pacific Ocean through four mountain ranges and drains more water into the sea than any other river in North America. The catalyst for this massive movement of water is the huge amount of precipitation falling on the Pacific Northwest.

Two of the Columbia's longest tributaries are the Snake River and the Clark Fork River, both of which rise in Montana just west of the Continental Divide and wend their way through the Rocky Mountains.

The geologic history has a fascinating subplot: Glacial Lake Missoula and Glacial Lake Columbia.

During the advances and retreats of the Cordilleran ice sheet over thousands of years, the ice would block the steep mountain valleys through which the tributaries of the Columbia River were accustomed to flow. Ice dams of up to 40 miles in width would form across these valleys.

Then, speaking now of the valleys in western Montana of the Clark Fork River, the water from the natural cycles of precipitation, plus melting of the ice sheet at its fringes, would cause the water to back up into the series of valleys that drained into the river's basin. A gigantic lake formed that geologists say held as much water as today's Lake Ontario and Lake Erie combined. Geologists call it Glacial Lake Missoula. To the west, something similar happened; it came to be called Glacial Lake Columbia.

Then, when the ice dam at the head of Glacial Lake Missoula on the Clark Fork River in northern Idaho eroded and burst, a gigantic flood would be unleashed that we living on earth today cannot imagine. It's said that the flow of water out of this lake that was over 2,000 feet deep exceeded the combined flow of all other rivers and streams on the planet. A gigantic wall of water, ice, and debris hundreds of feet tall roared westward, spreading over 16,000 square miles.

Remnants of this flooding can be viewed throughout the Northwest, from Montana westward. Sheer cliffs, gouged-out valleys, "prairie ripples," and "strandlines" are among the many features described in a National Park Service guide on what is called The Ice Age Floods National Geologic Trail. Strandlines are "benches" on the sides of mountains created by the lapping of waves from the Ice Age lakes. These can be seen clearly on mountainsides above Missoula and the Flathead Bison Range.

During the Ice Age, this filling and emptying of Glacial Lake Missoula happened perhaps forty times over a period of 3,000 years. Doubtless such phenomena also took place in Eurasia, as there too the Ice Age must have produced similar ice dams, lake fill, and catastrophic flooding.

And there were certainly human beings living in the vicinity of these periodic upheavals. Some of our ancient human ancestors may have escaped their effects; others, probably not. But it's interesting to pose the question of whether the flooding from these glacial lakes was the origin of the worldwide accounts of a Great Flood, including in surviving Native American legends and those from elsewhere around the world, like the Biblical story of Noah's Flood.

We are living today in a warming period following the end of the last Ice Age. Looking at the Cordilleran ice sheet, geologists believe that it had largely melted by 12,500–14,000 years ago, with only reminders now appearing in the high mountains of the western U.S. and Canada, including the carved mountains and valleys of Glacier National Park.

During this period of more recent glaciation, Native American cultures have lived and thrived across the Western Hemisphere, including in what became Canada and the U.S., though temperatures toward the north had obviously been on the colder side. It was during the end of the Ice Age that the Indians still hunted the "megafauna," including mammoths, mastodons, giant bears, buffalo, birds, and beavers. Horses were still found in America at this time before they died out, as the megafauna seemed to do when the climate changed.

Indian legends extant today contain accounts of some of these giant animals, and mastodon and mammoth bones and teeth have been found at Indian archaeological sites. Indian stories of Thunderbirds may refer to giant birds of this period.

The Salish

As the climate warmed, the Pacific Northwest began to be settled by tribes speaking a common language known as Salishan or Salish. These tribes created an extensive, complex, and culturally rich civilization whose remnants along the Pacific coast with their large village sites, huge totem poles, ocean-going canoes, give-away ceremonies, and long oral history continue to impress. They did not farm or garden but enjoyed a rich hunter-gatherer diet based on shellfish, whale and seal meat, and, above all, salmon. The Columbia River basin was the greatest salmon habitat on earth.

Migrating up the rivers, including the Clark Fork River, the tribes settled in the mountain valleys of eastern Washington, Idaho and Montana. The Montana Indians are today called the Flathead due to a custom of shaping the heads of infants that was practiced in the lowlands to the west in ancient times.

The Flathead Reservation now consists of the Salish, Kootenai, and Pend d'Oreille or Kalispel tribes. The Kootenai, whose homeland was the western side of Flathead Lake, speak a language unrelated to any of the surrounding tribes. Related are the French-named Coeur d'Alenes of Idaho and Washington. In ancient times the tribes included in their aboriginal territory most of Montana and portions of Wyoming, Idaho, Washington, and southern Canada. As they grew in number, they divided into smaller bands. These tribes have lived in the region for *at least* 15,000 years.

Eventually, white man's diseases began to infiltrate the mountains, even preceding the physical arrival of French and English traders. In the 1780s, a smallpox outbreak reached a group of Salish living near present-day Missoula. The camp divided itself into families with smallpox and those without, with the former moving up the Bitterroot Valley and the others going east. By 1782, smallpox had killed an estimated one-half to three-quarters of the Salish and Pend d'Oreille bands.

By now horses appeared again in the New World, brought by the Spanish. They had diffused northwards and were bred in vast numbers by the Indians of the Southwest and Great Plains, reaching all the way into Canada. The first successful horse breeders were the Comanches in Texas. The Blackfeet Indians in the northern Great Plains acquired horses, and the Salish got them from the Shoshone to the south by around 1700.

The Salish often ran afoul of the more warlike Blackfeet, whose reservation today is just east of Glacier National Park. It was with the Blackfeet that Lewis and Clark had their only violent Indian encounter as they were returning home down the Missouri River. The Blackfeet, like the Sioux, had been pushed westward by the coming of the whites to the Midwest.

The Blackfeet gained access to firearms from Hudson's Bay Company traders working out of Canada, leading to an uneven power struggle with the mountain tribes. The combination of disease, firearms, and horses led to profound changes in the northern Rockies. Blackfeet expansion caused the eastern bands of the Salish and Pend d'Oreilles to move their winter camps west of the Continental Divide. The Salish, already decimated by smallpox, were weakened further.

It was the presence of horses that brought the Indians in the Rockies into contact with Lewis and Clark. By the time Lewis and Clark set out in 1803, there were more ponies among some of the tribes than there were Indians. A single warrior might have a dozen or more ponies that he could ride himself or give away as presents to others. A favorite sport of the Plains Indians was to engage in pony-stealing raids against their Indian neighbors.

As Lewis and Clark moved up the Missouri River to its sources deep in the Rockies in west-central Montana just outside what is today Yellowstone National Park, they transported their food and gear in canoes and small river boats. But they knew that when they reached the Continental Divide, they would need horses to cross the mountains before moving downstream in order to arrive at the navigable waters of the Columbia River system, enabling them to proceed to the Pacific.

Sacagawea was a young Shoshone Indian woman, pregnant at the time, who had been purchased, along with one other Indian girl, by a French-Canadian trader named Toussaint Charbonneau. Lewis and Clark hired Charbonneau as a guide and interpreter. Sacagawea, who had been included in the party, told them that her Shoshone relatives, living deep in the mountains, had horses.

The party encountered a couple of Shoshone bands when they reached the source of the Missouri River just east of the Continental Divide in southern Montana. They named the place "Camp Fortunate." The leader of one of these bands was Sacagawea's brother. The Indians did in fact have horses, and Lewis and Clark were able to bargain for a few.

The Shoshones led them across the Divide through the Lemhi Pass into what is now Idaho, then north up the Lemhi to the Bitterroot River until they reached what is now called Travelers' Rest, a little south of the Clark Fork River near today's Missoula. The party's campsite at Travelers' Rest has been located through excavation of their latrine.

On the way down the Bitterroot River, near today's Darby, Lewis and Clark met a band of Salish, part of the Flathead group of tribes. These Indians also had horses, but not to the extent of the Shoshone. They lived in the valleys of the Bitterroot, the Clark Fork, and the Flathead Rivers. They were a peaceable tribe that lived off the bitterroot plant and camas root, a type of lily with a nutritious bulb. They fished, hunted elk, deer, and small game, and traveled annually with the Shoshone east across the Continental Divide to the Montana plains to hunt buffalo.

The Lewis and Clark *Journals* say of the Salish: "They are a timid, inoffensive, and defenseless people."[9] Lewis and Clark did not know that they belonged to the Northwest Pacific family of Salishan Indians, of much different background than the Plains and mountain Indians they had encountered. Sacajawea's Shoshone, though friendly to the Salish, spoke a different language and had moved into the region from the Great Basin ages ago.

The Salish were friendly. Private Joseph Whitehouse called them in his journal, "the likelyest and honestst Savages we have ever yet seen." Sergeant Gass of the Lewis and Clark party wrote: "To the honor of the [Salish] who live on the west side of the Rocky Mountains, we must mention them as exceptions [after citing the loose morals of the other tribes they had met]. They are the only nation on the whole route where anything like chastity is regarded." This did not keep William Clark from fathering a child with a Salish woman.

What the members of the Lewis and Clark expedition did not know was that the Salish had heard of the party of white men about to arrive and had made a conscious decision to allow them to enter their territory and to treat the visitors with generosity, when they easily could have wiped the intruders out.

In *Undaunted Courage,* Stephen E. Ambrose relates of the Salish:

> They were also generous. Although their stock of provisions was as low as that of the expedition, they shared their berries and roots. And they traded for horses at much better prices than the Shoshone demanded, perhaps not aware of how desperate Lewis and Clark were. The captains bought thirteen horses for "a few articles of merchendize," and the Salish were kind enough to exchange seven of the run-down Shoshone ponies for what Clark called "ellegant horses."... The next morning "the Salish galloped out for Three Forks and the buffalo hunt."[10]

Life for the Salish began to change after Lewis and Clark passed through. In 1809 they gained regular access to firearms through the establishment of the fur trade in western Montana by Canadian explorer David Thompson. The most famous of the Hudson's Bay fur traders in the region was Angus McDonald. Some of today's Salish count him their ancestor along with his Native American wife. The peak years of the fur trade lasted until the 1840s and brought Iroquois Indians from the east who intermarried and lived among the Salish people.

As we just saw, the region of present-day Montana where the Salish then lived was west of the Continental Divide. Once Lewis and Clark crossed the Divide in the company of the Shoshone Indians in 1805, they had left behind the U.S.-owned territory acquired through the Louisiana Purchase and passed into the no-man's-land of the Oregon Country.

At various times, Great Britain, France, Spain, Russia, and the U.S. had all laid claim to the Oregon Country. The U.S. based its claim on the Lewis and Clark expedition and on a 1789 voyage by sea captains John Kendrick of Boston and Robert Gray of Rhode Island, who were the first Americans to trade along the Pacific Northwest coast.

Over time, France, Spain, and Russia[11] dropped out of the competition and, as outlined earlier, the U.S. and Britain settled the boundary of the Oregon Country between them in 1846. The U.S. organized the Oregon Territory in 1848, claiming jurisdiction over all the Indian nations west of the Continental Divide. In 1851, the Salish and related tribes lost all claim to their aboriginal territories east of the Rockies due to the Treaty of Fort Laramie—that they had never heard of.

But Americans had been settling the region via what became the Oregon Trail since the early to mid-1830s. The trail left Missouri, followed the Platte River across Nebraska and into Wyoming, proceeded along the Snake River in Idaho, then met the Columbia in today's Washington State before arriving at the Willamette Valley.

Gold-seekers in the 1840s took the Oregon Trail partway before turning south to California, and the Mormons took the trail as far as Utah during their epic journey of 1847. At various times, mainly during the Plains Indian wars in the 1860s-1870s, travelers on the Oregon Trail, including Pony Express riders, were subject to Indian attacks. But there were few reports of travelers on the upper reaches of the Oregon Trail being molested. The Oregon trail also branched off onto the Bozeman Trail into Montana. Once white gold seekers headed north to Montana, the region changed forever.

The Coming of the "Blackrobes"

Just as the first white settlers were arriving in the western regions, so were the Jesuits. The Society of Jesus was founded in 1534 by Ignatius Loyola, a Spanish nobleman. Its chief purposes were to stem the growing tide of Protestantism in Europe and to spread the Christian faith among unconverted peoples around the globe.

The Jesuits worked under their own director, who reported to the Pope in Rome. The Jesuits had been in America as early as the 1560s and established their strongest foothold in New France, with a lesser presence in

New Spain. By the early 1800s, they were moving as missionaries into the American West, including the Oregon Country.

The history of the many attempts by white Christians to convert Native Americans is a vast and extremely controversial subject.

The Salish say that sometime prior to 1700, a Salish prophet, Xaliqs (Shining Shirt), prophesied the coming of the "Blackrobes." Now in the 1820s and 1830s, through the Iroquois, the Salish heard about the Jesuits, whom they too called Blackrobes. The Salish already had an acquaintance with white men from English traders and from memories of Lewis and Clark. But there were other sources of information filtering into the region early in the 19th century.

Much of the following is from a 1950 paper published by the University of Montana in Missoula, now held in their "ScholarWorks" archive on Graduate Student Theses, Dissertations, and Professional Papers. The paper in question is "The History of St. Ignatius Mission, Montana," by Gerald L. Kelly. Another outstanding work is *St. Mary's in the Rocky Mountains: A History of the Cradle of Montana's Culture* by Lucylle H. Evans. She was one of the founders and a director of St. Mary's Mission Historical Foundation, at the site of the first Jesuit mission in the Bitterroot Valley.

The university in Missoula was my father's alma mater, where he graduated in 1950, the year of Gerald L. Kelly's paper. So Kelly and my dad were there at the same time. Much of what you will read here is a paraphrase of Kelly's work, no doubt written with the help of people for whom some of the events he described were part of living memory.

The Jesuits who came to Montana had their headquarters at St. Louis, where a novitiate had been established in 1823. St. Louis was the capital of the northern Louisiana Territory, where William Clark of the Lewis and Clark expedition was serving as Superintendent of Indian Affairs. In 1828, the Jesuits had been appointed by the Pope as the exclusive agents for the Catholic Church among the Indians of the U.S. They recruited most of their workers from Europe, with France, Belgium, and Italy contributing the greatest number.

Many suggestions have been given as to other sources which provided the Indians with their idea of the white man's God. It may have come from traders or transient ministers who had visited the Columbia River country. Baptiste, the name of a "half-breed" from Quebec who came west after the War of 1812 may also have brought Christian teachings. Another story is that of Spokane Garry. He had been selected by Sir George Simpson of the Hudson's Bay Company, along with another Indian named Kootenai Pelly, to be sent to an Episcopal mission school at Red River, Manitoba, Canada. In

1831, Kootenai Pelly died, but Garry returned to his people and read to them from his Bible. Gerald L. Kelly writes that tribes came from all over the area to hear Garry every Sunday when religious services were conducted.

Christian influences were also brought to the Rocky Mountain tribes by bands of Iroquois who had come from the Jesuit mission of Kahnawake on the St. Lawrence River. This group travelled to the Rockies in search of furs while in the employ of the British North West Fur Company. One of the Iroquois bands, under the leadership of Old Ignace, came to the land of the Salish, who were receptive to the Christian teachings. Under the guidance of the Iroquois, they were already learning Catholic rites and prayers. On one occasion, a Salish girl who was dying told those around her of a vision she had of the Virgin Mary. There is a painting of the Salish girl's vision on the wall inside the surviving mission church in Stevensville, Montana.

The Salish, though a peaceful tribe, were not averse to defending themselves, but, as described previously, were subject to harassment by their enemies the Blackfeet and had been decimated by smallpox. So when Old Ignace told them of the "Big Medicine" of the Blackrobes, they began to look for means to secure this medicine. In the spring of 1831, a Salish expedition of four tribal members set out for St. Louis. Some months later, the group reached its destination, where they visited with Superintendent of Indian Affairs General William Clark.

Soon after their arrival, two of the Salish party fell mortally ill and were visited by two Catholic priests. The Indians managed to convey to these priests the Indians' interest in the Christian faith before they died. They were buried in the St. Louis Cathedral cemetery. The remaining two Indians, having failed in receiving any commitments from the priests, returned to the mountains, where their news was received with disappointment.

Word of these Indians and their journey to St. Louis aroused much interest in missionary circles. Nor did the Salish give up. This time the party consisted of Old Ignace and his two sons, who set out in late summer 1835. While at St. Louis, the two sons were baptized, with Old Ignace being assured by Bishop Bosati that a Blackrobe would be sent out as soon as possible. With this, the Indians returned to Salish country.

Eighteen months passed after the return of the second expedition, and still no missionary was forthcoming. A third delegation was now dispatched, again including Old Ignace, but all were killed by the Sioux when traveling down the Missouri.

Undaunted, the Salish sent a fourth party, this time with only two members: an Indian named Peter Gaucher and another named Young Ignace. Leaving in the summer of 1839 and reaching St. Louis, they were reassured with the promise that a priest would be sent out to them the following spring. Peter Gaucher returned to tell the news, while Young Ignace stayed behind to accompany the missionary on the return journey.

Father Pierre Jean De Smet

The missionary was Father Pierre Jean De Smet, a figure renowned in the annals of the Old West.

Fr. De Smet was born in East Flanders, Belgium, in January 1801. He had entered the Society of Jesus novitiate in Whitemarsh, Maryland, which was the first Jesuit novitiate to be located in the U.S. Two years later, he left for another new novitiate at Florissant, near St. Louis, where, says Gerald L. Kelly, "He gained a favorable reputation due to his great physical strength and restless energy." He made good progress in his studies, which included Indian languages, and soon became a teacher at the Jesuit's College of St. Louis.

De Smet was ordained a priest in 1827 and became a U.S. citizen in 1833. Soon afterwards, he left again for Europe and returned with three recruits. From 1838–1839 he established St. Joseph's Mission at what is now Council Bluffs, Iowa, where he was appalled by the disastrous effects of the whiskey trade on the Indians. He also traveled in the region as a mapmaker.

Back in St. Louis, he was directed to undertake the mission to the Salish in what seemed like an unpromising, far-off, isolated location. In April 1840, De Smet and his guide Young Ignace left St. Louis and travelled to Westport in present-day Kansas, where they joined the annual expedition of the American Fur Company to the Green River. At this spot, they were met by a band of Salish warriors who had been sent southward as a welcoming party. Leaving the American Fur Company group, De Smet continued his journey northward and nine days later reached the main camp of the Salish in the Bitterroot Valley.

University of Montana author Gerald L. Kelly writes of what De Smet learned of the Salish/Flathead religion:

> Though more virtuous than most of the Indians in the West, they were still superstitious savages.[12] They imagined the beaver to be a fallen race of Indians, who had been condemned by the Good Spirit to this present fate, but in due time would be restored. Some even maintained they heard the beavers talk with one another and had seen them sitting in council passing judgment on an offender. The Flatheads also had their medicine men, incantations, and charms. They believed in a Good Spirit and a Bad one and in

future states of reward or punishment. Their heaven was a country of perpetual summer where the deceased would meet his wife and children, and where the rivers abounded with fish and the plains teemed with buffalo and horses. Their hell was a place covered with perpetual snows where the departed would be constantly shivering with cold even though he could see a fire from afar. Water, too, was visible, but at too great a distance for the doomed one to wet his parched lips.

Kelly also cites a statement by Father Hoeckan, a later co-worker of De Smet, who had nothing but praise for the Indians' tribal organization:

> The first thing which struck me on my arrival among them was a truly brotherly love and perfect union, which animated the whole tribe and seemed to make them but one family. They manifest great love, obedience and respect for their chiefs, and what is still more admirable, they all, as their chiefs declare, speak and desire but one and the same thing. These chiefs are as much the real fathers of their people as is a good superior father of a religious community.
>
> The chiefs among the Kalispels [one of the Flathead tribes] speak calmly, but never in vain. The instant they intimate their wish to one of their followers, he sets to work to accomplish it. If anyone is involved in difficulties—if he is in want or sickness—or does he wish to undertake a journey, whether long or short, he consults his chief, and shapes his conduct in accordance with the advice he receives. Even with regard to marriage, the Indians consult their chiefs, who can sanction or postpone or disapprove of it, according as they deemed it conducive, or otherwise, to the happiness of the parties.
>
> A man who had a hereditary ailment would not obtain a marriage permit because, says the chief, the village would otherwise soon be filled up with people of that kind, and they would never listen to reason. The chief, in the quality of a father, endeavors to provide for the support of his people. It is he, consequently, who regulates hunting, fishing, and the gathering of roots and fruit. All the game and fish are brought to his lodge, and divided into as many shares as there are families, the distribution is made with rigid impartiality. The old, the infirm, the widow will receive their share equally with the hunter.

Concerning Father De Smet, Kelly writes:

> Such were the people to whom Father De Smet was sent to convert. For two months, he worked feverishly instructing his newly acquired parishioners. He then returned to St. Louis [over 1,600 miles away], as his original purpose was to make a cursory inspection of the tribe and its area and report back as to the advisability of establishing a permanent mission there. When he left the Indians on August 27, he promised to return the following spring with other Blackrobes. He reached St. Louis on the eve of the New Year and immediately began preparations for his return trip.
> In May 1841, he, together with Father Q. Mengarine, Father N. Point, and three lay brothers left St. Louis and made their way northward to the Bitterroot Valley, where their work began. They arrived at their destination on the twenty-fourth of September and established the first mission of the mountains, St. Mary's. They erected buildings immediately and in a few weeks a log chapel was built, capable of accommodating most of the tribe.

The Salish name for the St. Mary's mission location was "Wide Cottonwoods." The St. Mary's structures were the first buildings constructed in Montana by non-Indians. Brother Claessens was a carpenter who led the building of the church. As construction began, he wrote of the Salish workers:

> The women hewed down the timber, assisted by their husbands, with the greatest alacrity and expedition, and in a few weeks we had constructed a log church, capable of holding 900 persons. To ornament the interior, the women placed mats of a species of long grass, which were hung on the roof and sides of the church, and spread over the floor. It was then adorned with festoons formed of branches of cedar and pine.

The plan for the mission included houses with lawns, though the open design made the village susceptible to attack. Meanwhile, a log palisade protected the church. The first Holy Communion took place at Easter 1842. At this time, Chief Victor was the chief of the Bitterroot Salish.

Father Anthony Ravalli joined the mission in 1845. He had been trained as a medical doctor in his native Italy. He proceeded to vaccinate the Indians against smallpox and ran the dispensary. To the extent that the original Salish search for the Blackrobes may have had as part of its motive protection against smallpox, with the arrival of Father Ravalli, the search seems to have succeeded.

The Salish now had the beginnings of a farming community, though many of the tribal members resisted the Jesuits' attempt to convert them to Christian rites. Some, however, took to the prayer practice of reciting the "Lord's Prayer," "Hail Mary," and "Glory Be" several times a day.

According to a 2016 article in *Montana, the Magazine of Western History* by Ellen Baumier, in 1846, "...the fields yielded 7,000 bushels of wheat and a considerable quantum of garden crops."

But support of the mission among the Indians fluctuated, especially among young men who wanted to return to their former lifestyle as hunters, not sedentary farmers. This conflict of lifestyles never went away. The tribe was also influenced to their detriment by the influx of white miners into the Bitterroot Valley who brought alcohol and gambling.

Raids by the Blackfeet closed the St Mary's mission in 1850, when the church was burned to the ground. In 1866, St. Mary's started afresh, now relocated a mile south of the first site. Both Salish and white settlers attended services in the new church which was enlarged in 1879. St. Mary's work as an Indian mission ended in October 1891, when the Bitterroot Salish were forced to move north to the present reservation in the Flathead River watershed, where the Flathead tribes reside today. In 1921, St. Mary's parish was organized and continues as a Catholic parish.

The St. Mary's mission church of 1866 still stands, and the mission complex is open for tours from April through October. The buildings include the church with an attached residence, Fr. Ravalli's house and infirmary, a dovecote, a cabin with Salish artifacts, and a visitor's center that contains a museum, a research library, an art gallery, and a gift shop.

When St. Mary's was terminated as an Indian mission, it had already been supplemented by the St. Ignatius mission, established in 1854 by the Jesuits led by Fr. De Smet and Fr. Adrian Hoecken. St. Ignatius was sited about seventy miles to the north on what is today the Flathead Reservation. St. Ignatius parish continues, with the 1890s church having joined the St. Mary's church on the National Registry of Historic Places.

A year after St. Ignatius was established, the Flathead tribes were forced to sign the Hellgate Treaty, by which they gave up much of their ancestral land to U.S. government representatives who had come across the mountains from the newly-established Washington Territory. Part of the motivation for the government to mandate the Hellgate Treaty was the fact that two years earlier, in 1853, former Army engineer Isaac Stevens, now governor of the Washington Territory, supervised the surveying of a route for the Northern

Pacific Railroad. Unbeknownst to the Indians, the route was to run through Flathead territory.

It was also about this time that Missoula, Montana, about forty miles south of St. Ignatius, was settled at the confluence of five valleys on the Clark Fork River. The word Missoula is based on an Indian expression for "Shining Water." The Army's Fort Missoula was established in 1877 during the strife with the Nez Perce tribe. We shall return to the Flathead tribes in future chapters as representative of Native American life during the course of our history.

ENDNOTES FOR CHAPTER 4

1 Frederick Merk and Lois Bannister Merk, *Manifest Destiny and Mission in American History: A Reinterpretation*, (Vintage Books, 1973), 215–216.

2 As a reward for his patriotism, O'Sullivan was made U.S. Minister to Portugal from 1854–1858 by President Franklin Pierce. When the Civil War came, he shifted his loyalties to the Confederacy, which had its own concept of "Manifest Destiny"; namely, the spread of slavery.

3 *Congressional Globe*, February 2, 1859.

4 This has certainly been forgotten by people like Neocon author Robert Kagan, a founder of Project for a New American Century, who has argued that interference in the political affairs of *the entire globe* was part of U.S. ideology and intention from the nation's founding. It most emphatically was not.

5 Throughout U.S. history, a third party has never won a presidential election; but they have sometimes acted as spoilers.

6 At this time there was no Federal Reserve to crash the economy periodically with draconian interest rate hikes.

7 Arguably, the U.S. followed the same tactic in helping Ukraine incite Russia to invade in February 2022.

8 Between 1855 and 1859 pro- and anti-slavery forces fought a bloody civil war in the Kansas Territory. Slavery was abolished by the Territorial Legislature in 1860.

9 Meriwether Lewis and William Clark, *The Journals of Lewis and Clark* (200th Anniversary Edition (Signet Classics, 2002). https://www.amazon.com/Journals-Lewis-Clark-Signet-Classics/dp/0451528344

10 Ambrose, *Undaunted Courage*, 289–290.

11 Spain's claim stopped at California's northern border. After France sold Louisiana to the U.S., they stayed out of American affairs until helping sponsor Maximilian's attempted takeover of Mexico in 1864. Russia retained Alaska until selling it to the U.S. in 1867.

12 Obviously De Smet reflects typical white prejudices, though the similarities between Indian mythologies and the beliefs of Christianity are obvious to us today. I personally believe that the body of beliefs and practices formed by the ancient shamanic culture that includes the Native Americans can be viewed as a world religion on the same level as those more commonly recognized as such like Buddhism, Hinduisn, Christianity, Islam, and others.

CHAPTER 5

Civil War

Slavery

Lincoln said in his Second Inaugural Address:

One eighth of the whole population were colored slaves not distributed generally over the union but localized in the southern part of it. These slaves constituted a peculiar and powerful interest. All knew that this interest was somehow the cause of the war. To strengthen, perpetuate, and extend this interest was the object for which the insurgents would rend the Union even by war, while the government claimed no right to do more than to restrict the territorial enlargement of it.

Few who fought for the North would have agreed that they fought to free the enslaved Africans, per se. Rather the focus was on the fact that their enslavement was key to functioning of the Southern economy, and that economic system directly clashed with the interests of the industrial North. Most soldiers would have said, and many did, that they fought to preserve the Union, a concept with almost mystical power.

The violence inherent in the slavery system had coarsened the psyche of the South, and while it might be argued that this coarseness could not be allowed to prevail, a similar brand of violence would show up over the coming decades when the U.S. Army turned its attention to the extermination of Indians in the West.

Looking at the Civil War in a larger context, we can see that the U.S. was born in an era of competition for world control that constantly spilled over into violence. When the Civil War broke out, the Treaty of Oregon had settled the northwestern border between the U.S. and Britain's Canada, and the Treaty of Guadalupe Hidalgo had concluded the Mexican War. Now the U.S. itself was about to shatter along its north-south divide even before the new territories had been settled and organized.

Even a black man's liberty on northern soil was called into question by the Dred Scott case of 1857. Scott and his African American family claimed their freedom on the basis of their owners taking them to the free states of Minnesota and Wisconsin, where they lived for several years. The Supreme Court under Chief Justice Roger Tawney of Maryland ruled that as blacks, the Scotts had *no legal standing*, since the Constitution did not grant people of their race citizenship. Therefore, they could not sue in federal court.

Of course, the Constitution says nothing about racial criteria for citizenship. It's a matter of opinion, or however citizenship is defined by the states and the courts. The ruling was viewed as a test case and infuriated not only northern abolitionists but also ordinary people with a sense of justice. The nation was one large step closer to war.

Adding fuel to the fire was the raid by John Brown and his men on the U.S. Armory at Harpers Ferry, Virginia, on October 16-18, 1859. Brown's intent was to ignite a rebellion among slaves throughout the South, which never happened. A slave's main form of resistance was to flee—to the North via the Underground Railroad or south into Florida. Brown was found guilty of treason against the state of Virginia and hanged on December 2, 1859. Six other members of his party were hanged later. Attending Brown's execution was future Lincoln assassin John Wilkes Booth.

Lincoln was the most famous of the many remarkable figures the Civil War brought to prominence. Others I might mention were Frederick Douglass, born in slavery of black and Native American ancestry, who became a noted abolitionist and journalist; Harriet Tubman, who led multiple rescue missions on the Underground Railroad; and Walt Whitman, the poet of the ordinary man.

Bankers and the War

Statesmen and bankers from Europe watched developments closely. Part of the Confederate strategy was to seek the support of Great Britain and France, particularly Britain. The Confederacy knew that a decisive victory, especially in the eastern theater in the vicinity of the two capitals of Washington and Richmond, might enable Britain to step in with military support, or at least use its navy to relieve the North's blockade of Southern seaborne commerce.

Russia supported the North by stationing its own navy in the harbors of New York and San Francisco, standing in the way of British naval involvement. Russia had been defeated in Crimea by a coalition of Britain, France, Piedmont-Sardinia, and the Ottomans just a decade earlier but now warned Britain that a naval attack on the U.S. would be a *casus belli*.

Most of the British bankers favored the South. In an earlier chapter, we had met Britain's Baring family. By the 1850s, however, it was August Belmont, the U.S. representative of the Rothschild banking empire, who now came to the fore.

Nathan Mayer Rothschild (1777–1836) was the head of the Rothschild family in Great Britain. Born in Frankfurt, Germany, he was sent to Britain in 1798 by his father, Mayer Amschel Rothschild, and made a fortune on the London Stock Exchange. During the early 19th century, Nathan Rothschild's London bank provided the subsidies that the British government paid to its allies to fight during the Napoleonic Wars. Rothschild also financed the Duke of Wellington's army in Portugal and Spain.

The Rothschilds' power, like that of every other banker in Europe and the U.S., was based on fractional reserve lending. Again, this was the power the bankers utilized to make loans in excess of the amount of money they held in their reserves, loans which were then multiplied by usury at compound interest. The money the banks lent would circulate as a nation's money supply, translating their power of money creation into political power.

The reserves held by the banks consisted mainly of precious metals and reflected the amount of money their investors and depositors had placed with the bank for safekeeping. Another form of reserves was government-issued bonds. At various times in history, silver was also viewed as valid backing for money, along with gold. This was expressed in U.S. law through the legislated ratio valuing gold to silver at fifteen to one.

As "Bimetallism" increased the money supply, it was viewed as favorable to workers, consumers, and debtors. But gradually during the 19th century, silver lost its value, and gold emerged as the only acceptable form of metallic backing. By the Civil War, the gold standard reigned supreme throughout Europe with the Rothschilds its most powerful beneficiaries.

At times of crisis, including the Napoleonic wars, governments would suspend the right of holders of bank notes to redeem their money in gold. Only France did not do so. Now, with an event as momentous as the American Civil War, the removal of gold redemption as a brake on lending would obviously benefit the Rothschilds in looking for investment opportunities by allowing their banks to create more money out of thin air. By this time the head of the Rothschilds' London bank was Lionel de Rothschild (1808–1879).

As mentioned earlier, August Belmont, who had come to the U.S. from Germany in 1837 under the name of August Schönberg to manage the Rothschilds' American banking interests, had become prominent in U.S. politics. After starting his own bank in New York City by trading on the Rothschild name, Belmont became an American citizen, joined the Democratic Party, converted to Christianity, and married Carolyn Slidell Perry. She was the niece of Oliver Hazard Perry, the War of 1812 naval hero, and the daughter of Commodore Matthew Perry, who opened Japan to trade by his voyage of 1853. Belmont's wife was also the niece of John Slidell, who was a native of New York, but who moved to Louisiana as a young man and became a congressman and U.S. senator. He was one of two Confederate diplomats taken by the United States Navy from the British ship RMS *Trent* in 1861 and later released in a celebrated diplomatic row.

August Belmont's wealth and political connections had propelled him into Democratic Party politics. His first assignment was to serve as campaign manager in New York for Pennsylvanian James Buchanan's run for the presidency in 1852. Buchanan lost out to Franklin Pierce of New Hampshire, so Belmont shifted to making large campaign contributions to support Pierce's successful presidential campaign. James Buchanan persisted by succeeding Pierce as president through the election of 1856. Although Belmont now lobbied for the job of ambassador to Spain, Buchanan denied him the position.

Four years later, as a delegate to the bitterly-divided 1860 Democratic National Convention in Charleston, South Carolina, Belmont supported the presidential nomination of U.S. Senator Stephen A. Douglas of Illinois and made what was called "a powerful speech urging party unity." But Douglas failed to win a majority of delegates because he was not sufficiently pro-slavery. During the 1860 campaign, August Belmont served as Douglas's campaign treasurer. But a split in the Democratic Party assured the victory of Republican nominee Abraham Lincoln. Everyone knew that secession of the Southern states would soon follow.

During the Civil War that followed, August Belmont, now the Democratic Party's national chairman, walked a fine line between support for and opposition to the Union's war effort. Surprisingly, he is credited with helping persuade Britain and France not to support the Confederacy, though no one can guess at his motives for doing so, or his contacts behind the scenes enabling that. For instance, seemingly contrary to his effort to deny British and French support, in an 1863 visit to London, Belmont told Lionel Rothschild that "soon the North would be conquered." He also decried Lincoln's "fatal policy of confiscation and forcible emancipation."[1]

In 1864, with the war now raging, Belmont oversaw the nomination of General George McClellan as Democratic Party presidential candidate. Still on active duty, but cooling his heels at home in Philadelphia, McClellan was Lincoln's opponent in the presidential election. McClellan had been dismissed by Lincoln as commander of the Union's Army of the Potomac in 1862 after he failed to pursue Lee into Virginia following the stalemate at

Antietam. McClellan had spent the next two years sulking as he tried to clear his name for his failure to win a decisive battle, while carping at Lincoln for being too intransigent toward the South. McClellan ran on his famous name without much of an alternative program. He refused to alienate the Army by calling for the Union to recognize Southern independence. Instead, he wanted a negotiated peace, which would nonetheless have made Confederate recognition by Britain inevitable. But Lincoln won the election decisively, with the soldiers in the field overwhelmingly supportive of the president.

Belmont's relative by marriage, John Slidell, besides being a Confederate diplomat, had worked in the New Orleans law office of Judah Benjamin, former U.S. senator and Confederate Attorney-General and Secretary of State. Benjamin, born a British citizen, fled to Britain after the war, where he became a successful barrister in London.

August Belmont's own British connections continued after the war as well. The Rothschilds had tried to persuade some of their younger family members to relocate to New York to take over the family business, but they viewed New York as too boring and provincial. So mainstream history views the Rothschilds as being less influential in American finance than they actually were. Still, their name will recur in the pages of our history, particularly as the allies of J.P. Morgan and the Money Trust that eventually created the Federal Reserve.

It might be argued that August Belmont knew that a split in the Democratic Party in the election of 1860 would lead to war, and he furthered the likelihood of that split and of the war via his support of Senator Stephen Douglas. Belmont himself continued after the war as an influential New York banking magnate. He and his family worked closely with J.P. Morgan in the financial manipulations that produced the numerous banking panics of the post-Civil War era.

So the question of whether the bankers in general or the Rothschilds in particular fomented the American Civil War remains inconclusive.

Financing the Civil War

Two days after the surrender of Fort Sumter in the Charleston, SC, harbor on April 13, 1861, President Abraham Lincoln called for 75,000 volunteers to serve for three months to suppress the rebellion. The states of South Carolina, Mississippi, Alabama, Florida, Georgia, Louisiana, and Texas had already seceded. Following Lincoln's call, Virginia, Arkansas, North Carolina, and Tennessee joined them.

On May 8, 1861, Richmond, Virginia, was named the Confederate capital. From the Battle of First Manassas on July 16, 1861 to the surrender of General Robert E. Lee to Union General Ulysses Grant at Appomattox Courthouse on April 15, 1865, a series of decisive battles took place in the relatively confined locales of central Virginia, western Maryland, and southern Pennsylvania. The carnage was beyond belief.

When Lincoln issued his call for troops, no one had any clear idea of how the war would be paid for. Both sides faced a struggle to find resources to finance their war efforts. When the war ended with Union victory, the nation's economy had been transformed. The South had been crushed and would not begin to recover for decades. The North had become an industrial colossus.

But the problem in wartime lay not in producing guns, munitions, or uniforms, or coming up with volunteers and conscripts to fight. It lay in getting the money to pay for it all, thereby exposing the fatal flaw in modern "political economy," which is the paralyzing confusion over where the money will come from to meet public needs..

In 1861, the federal government's reliance on "specie" went out the window. Available gold and silver to pay for the war did not exist. Under the Constitution, the government could tax or borrow. But who could be taxed and who would provide loans? Or could the government simply print money as bills of credit? No one knew.

Politicians make decisions for which the entire population, sooner or later, has to cough up the money. We remember that the American Revolution was fought over "taxation without representation." The weakness with U.S. democracy was always that people did not *want* to pay taxes. The Whiskey Rebellion of the 1790s and its suppression proved the case.

The collection of customs duties on imports, the main source of federal revenue, could be isolated to customs houses in port cities. Now, however, foreign trade was static. Excise taxes could be extracted from merchants at the point-of-sale of taxed articles like tobacco or salt. Property taxes were left to the states and localities and required a complex system of assessments and collections but were not available to the federal government. Income taxes had long been considered impossible to collect. So it was easier for the government to borrow and rely on growth of trade, along with sale of Indian lands, to pay for war, though many new taxes were in fact imposed despite widespread evasion.

Statistics contained in the following account rely heavily on the classic *Financial History of the United States* by Paul Studenski and Herman Edward Krooss, published in 1952 and reprinted in 2003.

Early in the war, Secretary of the Treasury Salmon P. Chase of Ohio began to use non-interest-bearing demand notes; i.e., paper money redeemable in gold, to pay the salaries of government employees. He then began to ask for loans from banks via three-year bonds at 7.3% for $150,000. The trouble was that the entire U.S. banking system, at least in the northern states, had only $63 million in specie in their vaults.

The banks balked at coughing up all their "hard money," causing Chase to threaten to inflate the currency, thereby allowing the specie to cover the debt. The banks tried but failed to sell the government's bonds to a skeptical public, so Chase issued another $33 million of demand notes, realizing that the government itself could not cover requirements for redemption in specie. Default on these notes would be catastrophic.

Chase was now looking at a fiscal year 1862 federal budget estimate of $532 million; the actual expenditure was $469.6 million. Projected revenues, mainly customs duties, were $55 million, with the actual being $52 million. The overall budget deficit for 1862 was $417.6 million. Union expenditures on the war skyrocketed from $23.0 million in 1861 to $389.2 million in 1862, to over a billion dollars in 1865. A billion dollars—no one had ever seen such a thing.

The Union was facing economic collapse, bankruptcy, and dissolution because the troops, suppliers, and contractors could not be paid. None of these were in a position to work and produce—or kill and be killed—for free. Not in a democracy.

Chase now recommended a restoration of lapsed excise taxes on retail products, both commodities and manufactured goods, large increases in tariffs and customs duties, even as international trade was slowing, and a new national banking system through which government bonds could be marketed and currency values stabilized. These proposals can be recognized as consonant with past proposals by Henry Clay and others, including Henry C. Carey, now an advisor to President Lincoln, for implementation of the "American System." Proposals for enacting the American System would be brought out of mothballs.

But the bankers, all from New York with strong ties to London, didn't buy in to Chase's proposals. On December 30, 1861, the nation's banks *suspended all specie payments* redeeming circulating paper currency or other financial instruments such as bonds. A day later, the federal government opened a gold market in New York to help ease the crisis where at least gold itself could be bought and sold.

But the government was at a standstill. Despite the 1861 skirmish in Manassas and some fighting in the border states, the Union had yet to place a major army in the field. The Confederacy was busy organizing a government in Richmond, while working on its own program of taxes, debt, and military conscription. But in Washington, DC, Treasury funds would be exhausted in 30 days.

The government's credit was so bad that the New York banks, representing both domestic and foreign investors, including Belmont and the Rothschilds, were telling Chase that new government bonds could only be sold at discount rates of 25–50 percent, possibly a *50 percent rate of interest*. The terms were ruinous. The nation's economy would be mortgaged in perpetuity.

Congressman Elbridge G. Spaulding was a banker from Buffalo, NY, descended from Edward Spaulding, an English Puritan, who settled in Massachusetts soon after the arrival of the Mayflower. This was about the time my own ancestor Thomas Bliss had disembarked. Spaulding now proposed the issuance of $150 million in non-interest-bearing Treasury notes *not redeemable in specie* but legal tender for everything but customs duties. Secretary Chase supported the measure, which was passed by Congress on February 25, 1862, as the Legal Tender Act. The money was not issued as a loan, which bank-issued currency would have been.

What were called "Greenbacks" were a true fiat currency, and the Union was saved.[2] Issuance of the unbacked paper by direct government spending was supported by businesses, workers, and soldiers, and immediately began to circulate. Congress authorized a second issue of $150 million on July 11, 1862, and a third of $150 million in January and March of 1863. Substantial amounts of the money came back to the government in taxes, and Chase now had his hands on the phenomenal amounts of cash needed to prosecute the war. Lincoln said, "We gave the people of this republic the greatest blessing they ever had, their own paper money to pay their own debts."[3]

The main argument against fiat currency is that it may cause inflation, and with the Greenbacks this did happen to some degree. In the New York gold market, a Greenback dollar was worth only about 35 cents in gold. But this was at a time when gold was being hoarded. People held onto their gold whenever there was bad news on the war front and spent it when the news was good. And the inflation was due not only to the Greenbacks. The same thing happened with the money the government was borrowing from the banks, which it continued to do.

Eventually, the banks accepted the Greenbacks and even began to buy them as reserves against which they could lend. The government now issued another $50 million in Greenback notes under $5 in value, with $15 million of these small-denomination bills remaining in circulation until 1950!

The government was criticized for not levying enough taxes. So in 1862 the government raised and imposed new taxes, including on luxury items, bank capital, and bank deposits, with a 1-1/2 to 3 percent sales tax on railroad fares, ferry boats, steamship tickets, toll bridges, advertisements, paper checks, sale of stocks and bonds, medicines, cosmetics, and playing cards. An inheritance tax was also instituted.

The government now established its first Office of the Commissioner of Internal Revenue. A tax of twenty-cents per gallon was imposed on liquor, which was almost 100 percent of the cost of manufacture. Federal licensing of liquor sales was also imposed. The income tax was now raised to three percent on $600–$10,000 and five percent on amounts over $10,000. In 1863 the liquor tax was raised to $2 per gallon and the income tax was increased, with deductions allowed for house rents, mortgage interest, home repairs, and capital losses from the sale of land.

From all this we can see that the Civil War introduced much of today's federal government's taxation structure. It also incentivized bootlegging of liquor to the point that illegal booze largely displaced taxed liquor in the market. But consumers took the brunt of the system, leading to consumption of all goods dropping up to fifty percent in 1860–1865 due to inflation and taxation. Sale of Indian lands had dropped to almost zero, and by 1865 the government was still looking at a deficit of almost a billion dollars with the total national debt now at $4.9 billion.

So, toward the end of the war, even more money was needed. Once again, the government refused to mortgage the nation to the banking system. This time a small-denomination bond campaign was launched under the direction of financier Jay Cooke, a Whig lawyer and former congressman from Ohio, who began sending representatives around the country to sell bonds to individuals for as little as $50.

Cooke and his employees sold nearly a billion dollar in bonds that allowed the continued payment of soldiers and suppliers. It was the start of what later became United States Savings Bonds.

Later, Cooke financed the construction of the Northern Pacific Railroad, the line that would run through the Flathead Reservation in Montana. He was forced into bankruptcy in 1873 due to over-lending for the development of Duluth, Minnesota, as a railroad hub. Later he recouped part of his fortune through an investment in the Horn Silver Mine in Utah. But Cooke is remembered due to his work for the government in financing the Civil War.

The National Banking Act of 1863

Congress sought to prevent the banking system from becoming an independent center of political power. With that end in view, in February 1863 Congress passed a National Banking Act to create a uniform currency of national bank notes and to require major banks to invest in government bonds.

But the system paved the way for eventual creation of the Federal Reserve System, which largely ceded Congress's Constitutional power of money-creation to the world of big finance.

By 1864 there were 508 national banks in the system. The fractional reserve ratio of deposits to loans varied from fifteen to twenty-five percent, which still allowed a substantial amount of money to be created out of thin air and lent at compound interest. The new law also taxed notes created by state banks at two to ten percent, which would eventually drive many of them out of business. By October 1866 there were 1,644 national banks.

Reflecting later on his support of the creation of a national banking system, Salmon P. Chase, who would become chief justice of the United States in 1864, had this to say:

> My agency, in promoting the passage of the National Bank Act, was the greatest mistake in my life. It has built up a monopoly which affects every interest in the country. It should be repealed, but before that can be accomplished, the people should be arrayed on one side, and the banks on the other, in a contest such as we have never seen before in this country.

Horace Greeley (1811–1872), founder and editor of the *New-York Tribune*, and co-founder of the Republican Party, said:

> While boasting of our noble deeds, we are careful to control the ugly fact that by an iniquitous money system, we have nationalized a system of oppression which, though more refined, is not less cruel than the old system of chattel slavery.

As we shall see, the banking system that now began to emerge would eventually lead the U.S. to what we see increasingly today as on the verge of bankruptcy.

Lincoln's Assassination

Nevertheless, the actions taken by Lincoln and his government were far from allowing the bankers and financiers to take over completely, often to their anger and chagrin. For this reason, a statement reputedly published in the *London Times* in 1865 has often been cited. Speaking of the Greenbacks, the *London Times* reportedly wrote:

> If that mischievous financial policy, which had its origin in the North American Republic, should become indurated down to a fixture, then that government will furnish its own money without cost. It will pay off its debts and be without a debt. It will become prosperous beyond precedent in the history of the civilized governments of the world. The brains and wealth of all countries will go to North America. That government must be destroyed or it will destroy every monarchy on the globe.[4]

The year 1865, when the *London Times* evidently called for the U.S. government to be destroyed, was the year President Abraham Lincoln was assassinated. His murderer, John Wilkes Booth, had recently returned from a trip to Montreal, the Canadian headquarters of the British intelligence service and where Confederate spy agencies were based.

When soldiers cornered and killed John Wilkes Booth in a Virginia barn two weeks after Lincoln's death, he was carrying a bill of exchange from Montreal's Ontario Bank dated October 27, 1864. A bank book from the same institution, stamped with the same date, was also discovered among his belongings. Booth's account at the Ontario Bank, an institution acquired by the Bank of Montreal in 1906, stayed open with a balance of $455 for an undetermined length of time following his death.[5]

Who, then, was the paymaster for Lincoln's death? It was a question that was never answered *or even investigated*. But Lincoln's death was a great loss. In the words of German Chancellor Otto von Bismarck (1815–1898):

> The death of Lincoln was a disaster for Christendom. There was no man in the United States great enough to wear his boots and the bankers went anew to grab the riches. I fear that foreign bankers with their craftiness and tortuous tricks will entirely control the exuberant riches of America and use it to systematically corrupt modern civilization.[6]

Maryland Blue Ridge Crossroads

Since even a synopsis of the military aspects of the American Civil War is beyond the scope of this book, we'll focus on one small but pivotal area, one that I know fairly well, since my wife and I live there: the Maryland Blue Ridge, which is part of the Blue Ridge Mountain chain that extends from Pennsylvania to Georgia.

The three main formations of the Maryland Blue Ridge, going from east to west, are Catoctin Mountain, South Mountain, and Elk Ridge. Between

Catoctin and South Mountain is the Middletown Valley, sometimes called the most fertile valley in America.

Between South Mountain and Elk Ridge is Pleasant Valley, which broadens around the town of Boonsboro to blend into the expanse of the Cumberland Valley that is part of the Great Valley geographic province of the Eastern Appalachians. This region saw some of the fiercest fighting of the Civil War, to wit, the Antietam Campaign of 1862, the Gettysburg Campaign of 1863, and the Monocacy Campaign of 1864.

The Maryland Blue Ridge has been a crossroads of history, part of the mid-continental transportation corridor following the Potomac River from the Chesapeake Bay and Atlantic Seaboard through the mountains to the American Midwest. The route was traveled for thousands of years by Native Americans, who made their homes along the shores of the Potomac and its subsidiary streams. The Potomac corridor then became a major route for entrance into the frontier by white settlers.

Later, the Potomac corridor was a route for the C&O Canal and the B&O Railroad, the first railroad penetrating the American interior. Later came the first U.S. interstate highway, the National Pike, running from Baltimore, Maryland to Vidalia, Illinois.

The Blue Ridge had been home to Native Americans for at least 13,000 years, as indicated by spear points found along creeks and the C&O Canal.

After 1700, the valleys and hillsides of the Maryland Blue Ridge were settled by Englishmen who received grants of land from the King of England's Maryland proprietor, Lord Baltimore, and by German and Swiss settlers. The Germans were recruited by Maryland agents in Rotterdam and transported through the port of Baltimore. They received cheap farming tracts of about 300 acres, the land often bought on credit. Fredericktown, today just Frederick, was being settled by 1745 through the agency of Daniel Dulany, an indentured servant from Ireland who had become an Annapolis lawyer.

The frontier that ran from Pennsylvania down through Maryland and Virginia was the scene of battles among competing Indian tribes and between whites and Indians until the conclusion of the French and Indian War in 1763. The war began with the trek of General Edward Braddock across Catoctin and South Mountain in 1755. The place where the first British military force ever to enter the Appalachian Mountains is marked today by Braddock Motors, a used car lot, at what is called Braddock Heights.

Accompanied as a volunteer aide by 23-year-old Virginian George Washington, who had been named lieutenant colonial of the Virginia militia, British General Braddock rode to his death in a battle at the French Fort Duquesne in the Pennsylvania wilderness on July 9, 1755. It took several years for the British to recover from the debacle before they drove the French out of the Ohio Valley, then defeated the French army at Montreal, gaining permanent control of what became Canada.

After the Treaty of Paris ended the Seven Years War, the trickle of settlers through the Blue Ridge became a flood. The Germans in the region tended to settle the better land in the valleys, while the looked-down-upon Swiss built their homesteads on the cheaper and less desirable property in the uplands.

Maryland was the only British colony to offer religious freedom to Catholics. Encouraged by Maryland's 1649 Toleration Act on religious freedom, Catholics settled in northern Frederick County, where Elizabeth Ann Seton (1774–1821) set up her girls' school. A young widow, Seton moved in 1809 to Emmitsburg, Maryland, where she founded the Sisters of Charity of St. Joseph's, the first community for religious women established in the U.S. She also founded St. Joseph's Academy and Free School. Elizabeth Ann Seton was declared a saint of the Catholic Church in 1975. Today there is a basilica at her National Shrine in Emmitsburg. Mount St. Mary's University is nearby.

The first national armory at Harpers Ferry, where the rifles were made for the Lewis and Clark expedition, sits on the southern shore of the Potomac, where the Shenandoah River joins it. A little upriver at what is now Shepherdstown, West Virginia, James Rumsey, a mechanical engineer and inventor, built and operated one of the world's first steamboats.

South Mountain was a major route for the Underground Railroad. Some of the many stories and legends of South Mountain were captured in the book *South Mountain Magic* by Madeline Dahlgren, widow of Civil War Admiral John Dahlgren. Mrs. Dahlgren's manor house at Turners Gap is now the South Mountain Inn that sits on the old National Road adjacent to the crossing of the Appalachian Trail.

The Maryland Blue Ridge is one of the most fertile and picturesque areas of America. From deep within Pleasant Valley, Little Antietam Creek heads north and joins the main Antietam Creek at Keedysville, which then curves south to empty into the Potomac just beyond Sharpsburg. On the way, the creek runs through the Antietam Battlefield and under Burnside Bridge. The area surrounding the battlefield is mainly farmland. Many of the farms are owned and operated by Mennonites, originally 18th century immigrants from the German Palatinate.

The Confederate Incursions into Maryland— 1862, 1863, and 1864

The Civil War reached the Maryland Blue Ridge in September 1862. In the western U.S., the Union armies had made steady progress in subduing Confederate forces in Kentucky and Tennessee. With the capture of New Orleans in 1862 and the reduction of the fortress at Vicksburg in 1863, the Union gained control of the entire length of the Mississippi River. But in the east, matters were more tenuous; the South tried to deliver a couple of knock-out blows by its incursions into Maryland.

Antietam

The Maryland Campaign of 1862 was the first of the two major attacks, ending in the Battle of Antietam on September 17, 1862. Earlier that year, the Army of the Potomac under General George McClellan, called "the Young Napoleon," had tried to seize the Confederate capital at Richmond by transporting his force of over 100,000 troops down the Chesapeake Bay on barges and landing them at Fortress Monroe at the mouth of the James River.

McClellan's attack was repulsed by Confederate forces led by General Robert E. Lee. With McClellan's army then recalled to Washington, Lee moved north, where General James Longstreet and General Thomas "Stonewall" Jackson converged to rout a federal force under General Irvin McDowell at the Second Battle of Manassas.

Lee then made the bold move of crossing the Potomac River near Leesburg, Virginia, with about 35,000 troops and setting up camp outside the city of Frederick, Maryland. He invited Marylanders to join the South and overthrow the tyranny of the Union government, though almost everyone that had been a Southern supporter had already left the state for the South. Lee's primary motive, in consultation with Confederate President Jefferson Davis in Richmond, was to draw the Union army out of Washington, defeat it in a pitched battle, then invite Britain and France to recognize the Confederate cause.

Lee's force did not inspire much confidence among the local residents. A Frederick physician later reported the terrible smell of the unwashed Confederate soldiers. Lee had lost many of his troops to desertion, and many were barefoot. Lee's aim was to head for Pennsylvania, then turn and threaten either Baltimore or Washington. With McClellan trying to organize the Army of the Potomac to conduct a pursuit, Lee decided to divide his Army of Northern Virginia into several components. Stonewall Jackson moved his force west across the Blue Ridge to Boonsboro, then re-crossed the Potomac

River at Martinsburg in what is now West Virginia, from which he would sweep down to invest the 11,000-man federal garrison at Harper's Ferry. A second Confederate force would attack Harpers Ferry from the south, a third under General James Longstreet would move to Hagerstown to prepare the planned incursion into Pennsylvania, and a fourth, a rear guard, would hold the South Mountain passes between Middletown and Boonsboro in case the federal army moved faster than expected.

Unfortunately for Lee, a copy of his marching orders wrapped around three cigars was found by an enlisted man in a farm field where the Confederate army had bivouacked outside Frederick. This was the famous "Lost Order" that was immediately handed over to General McClellan who had now arrived after moving his 75,000-strong Union army up from Washington. With Lee already having made it to Boonsboro west of South Mountain and Jackson bombarding the Harper's Ferry garrison, McClellan caught up with the Rebels at the South Mountain passes being defended by Confederate General D.H. Hill. There the Union's Army of the Potomac won its first major engagement of the war to the delight of the Northern press.

Two future U.S. presidents were present at the Battle of South Mountain. Thirty-nine-year-old Rutherford B. Hayes was lieutenant colonel of an Ohio volunteer regiment and was wounded in the arm. William McKinley was a nineteen-year-old commissary sergeant who delivered sandwiches and coffee to the soldiers on the front line.

After the Battle of South Mountain, Lee retreated west toward the Potomac, setting up a defensive line on a ridge just above the town of Sharpsburg, Maryland, on the other side of Antietam Creek. The Battle of Antietam started with Union attacks before dawn on the morning of September 17, 1862. The first attack, at daybreak, into "the Cornfield" was conducted by General Joseph Hooker.

At the bridge to the south that today bears his name, troops were commanded by General Ambrose Burnside. Unfortunately for the Union army, it took Burnside so long to get his men across Antietam Creek that he lost the initiative. His force was met by Confederate troops under A.P. Hill marching up from Harper's Ferry, who saved Lee's battered army from annihilation.

By nightfall, over 27,000 men from the two armies were killed, wounded, or missing. It was "the bloodiest day in American history." The Union army had failed to dislodge Lee from the ridge above the town, so tactically, the battle was a stalemate. After waiting a day to see if the Union army would renew the attack, which it did not, Lee pulled his force back across the Potomac across the fords below Shepherdstown.

General George McClellan has been pilloried ever since for failing to defeat Lee in a battle where his forces greatly outnumbered the enemy. McClellan has been praised for his strategic and organizational acumen but is viewed as an incompetent tactician. Lincoln fired him from his post as commander of the Army of the Potomac after the November 1862 congressional elections. In his place Lincoln named Burnside, who then led the army to a disastrous defeat at Fredericksburg that December.

Gettysburg

The Battle of Gettysburg marked the next major Confederate incursion into Maryland towards the end of June 1863. General Robert E. Lee's Army of Northern Virginia was in much better condition than it had been at Antietam nine months earlier, with Lee now knowing the lay of the land as he passed through the Maryland Blue Ridge into Pennsylvania.

Lee was coming off two major victories over Union forces in Virginia—Fredericksburg and Chancellorsville—where his most effective field commander, Stonewall Jackson, had died after being accidentally shot by his own men.

As depicted in the book *Killer Angels* by Michael Shaara and the movie *Gettysburg* based on it, Lee's march northward was further handicapped by the fact that his chief of reconnaissance, cavalry General J.E.B. Stuart, had seemingly disappeared while riding around the Union army in the Maryland and Pennsylvania countryside. Thus, Lee had only a vague notion of the location of the Army of the Potomac, now commanded by General George Meade, which was moving north from Washington. In some ways the campaign was a repeat of the movement of the two armies prior to the Battle of Antietam. Now, both armies were larger, Lee with 75,000 men and Meade with around 100,000. It would be the largest battle ever fought in the Western Hemisphere.

On June 30, 1863, outliers from the two forces ran into each other just west of the town of Gettysburg, a road and rail hub in the Pennsylvania foothills. As Lee hurried his units toward the town the next day, their arrival was delayed by a holding action on the part of Union cavalry under Brigadier General John Buford.

A little to the north, Stonewall Jackson's replacement, General Richard Ewell, failed to take the heights on Culp's Hill which overlooked the town. These actions allowed the unhindered arrival of the main body of federal forces, which proceeded to occupy the high ground along Cemetery Ridge extending eastward. At the south end of the ridge were Little Round Top and Big Round Top.

Over the next two days, Lee threw his army at the federals strung out along the ridge but failed to dislodge them. The battle ended on July 3, 1863, with the famous but fruitless charge, led by Confederate General George Pickett, of 12,500 Confederate soldiers across a mile of open ground. That assault on Union lines has been called the "High-Water Mark" of the Confederacy. The handful making contact with Union soldiers were killed or repulsed.

Few people have ever understood why General Robert E. Lee made such a catastrophic mistake, and Lee himself never talked about it. The fact was that General "Jeb" Stuart had not arrived at the battle until the second day, reporting to Lee with his cavalry force that had been exhausted by several days of riding and skirmishing beyond the reach of both armies to the east. Stuart then received orders to employ his entire force of several thousand riders to support General Pickett's charge the next day by attacking the Union lines along Cemetery Ridge from the rear.

Lee might have won the Battle of Gettysburg, and possibly the war itself, had the plan worked. But it didn't. Stuart's attack was met by Union cavalry to the north of the main battle around 11 a.m. the morning of July 3. This was known as the Battle of East Cavalry Field; it was ignored in the *Gettysburg* movie which cast Lee's actions in the worst possible light.

Stuart was repulsed by a series of charges led by George Armstrong Custer, a 23-year-old Union cavalry officer who had already been promoted to brigadier general. Later in the war, in 1864, Custer attacked at the Battle of Yellow Tavern outside Richmond where Confederate General Jeb Stuart was killed.

The 1862 Battle of Antietam and the 1863 Battle of Gettysburg were prominent engagements where the action was confined to men in uniform. Both armies were under strict orders to leave civilians alone. Except for foraging for food, where civilians were often paid for their produce even if it was only with Confederate money, the fighting left civilian towns and farms intact.

But by the end of the Civil War, what today we might call terrorist tactics were starting to be used by raiders like those under Confederate Captain John Mosby or Union commanders like General William Tecumseh Sherman in his burning of Atlanta and his March to the Sea.

Monocacy

There was also a third Confederate incursion into Maryland, culminating in the Battle of the Monocacy on July 9, 1864, fought outside Frederick by Confederate General Jubal Early, who was attempting to harass Washington

DC and draw troops away from what by then was the Union's campaign against Richmond. The Union was able to rush enough soldiers north to fend off Early's attack during which President Lincoln went out to Fort Stevens in the Washington suburbs to espy the Confederates in the distance. Here, he was famously told, "Get down you fool!" Early's attack failed, and he retreated to Virginia.

Ancestors in the War

Several of my ancestors, or those of our family archivist, my cousin Johnny Lathrop, as well as others in our extended family, fought in the Civil War.

William Forster

My great-great-grandfather on my mother's side was William Forster, who arrived in the U.S. in 1849, having embarked from Cobh, County Cork, Ireland, during the Irish Potato Famine. Records are scant, but we do know that William Forster, who had settled in Brooklyn, New York, became a gunnery sergeant with the New York Heavy Artillery. Initially, that unit was assigned to the defenses of Washington, DC, but later was part of Grant's army when Lee surrendered at Appomattox. William Forster drew his veteran's pension until his passing in 1893.

Speaking of the Famine, in his book *Empire,* the foremost modern British establishment historian Niall Ferguson is able to admit:

> Direct rule from Westminster [in London] had without question exacerbated the disastrous famine of the mid-1840s, in which more than a million people had died of dearth and disease."[7]

Ferguson does not mention that up to two million more Irish embarked for North America. Thousands died during the trip, including many newborns. Ferguson adds that, "It was the dogmatic *laissez-faire* policies of Ireland's British rulers that turned harvest failure into outright famine."

In fact, huge amounts of produce continued to flow to England and the Continent from farms that had been seized from the Irish during various episodes of British conquest, where the starving Irish population now worked for British landlords as tenant farmers. When the potato harvest failed, large numbers of tenants were evicted for failure to pay rent. I could see, when visiting Ireland in 2013, that such traumas are hard to forget. Particularly moving were the plaques and monuments in the County Cork town of Skibbereen.

Joe Smethurst

Joe Smethurst was born in Morgan County, Ohio, in 1842, and moved with his family to Seneca, Wisconsin, prior to September 1857. This is one of at least eighteen town or cities in the U.S. named after tribes of the Iroquois Indian nation. Joe was an ancestor of my cousin, Johnny Lathrop, on his father's side.

Joe Smethurst worked as a printer before the war on *The Courier* newspaper in Prairie du Chien, about twenty miles away. Prairie du Chien was an early site of French-Indian fur trading and where Black Hawk surrendered to Colonel Zachary Taylor in 1832, ending the four-month Black Hawk War. Wisconsin became a state in 1848 and furnished over 91,000 soldiers to the Civil War. This was almost one out of every seven Wisconsin residents.

Joe joined the 25th Regiment of Wisconsin Volunteers on August 9, 1862 with a three-year enlistment. He was discharged on June 2, 1865. He joined just before Lee and McClellan faced off at Antietam and served until after Lee's surrender at Appomattox. Joe's service started in the western theater of the war.

Joe's wartime diaries commence on February 26, 1863. Having completed his initial training, he moved south by rail, arriving with his unit in Columbus, Kentucky on March 5, 1863. From his diary, Joe spends his days drilling with his unit and performing guard duty. One day he "rolled ten pins." On another, his brother John, in the same unit, had a fight with Dick Bull, who "got his eyes blacked." One day Joe cleans his gun; on another the unit "all went down to the river and took a wash"; on another a soldier dies from unspecified causes, but it was raining so hard "we could not bury him."

By late March Joe was becoming ill. On March 23, he "had the sick headache all day." He was in a hospital on March 31. "Had a shake" on April 1. The next day he "got some medicine that made me very sick," though he felt better in the afternoon. On April 4, "I was sick with the dioreahrea [sic] all night." He was feeling better by the 7th and got more medicine on the 8th. On April 9, "One of the Regulars shot two other soldiers—one fatally." Joe reports more diarrhea on April 13.

On May 4, Joe and three other soldiers started a 30-day furlough. Stopping in Cairo, Illinois, "the bed bugs were so thick I could not sleep." Joe then makes his way home to Prairie du Chien and Seneca. There he "saw all the girls" and "all my old chums."

By mid-June 1863, Joe was back in Kentucky with his unit, drilling and serving guard duty. One night he gets a bad cold from sleeping on the ground. On July 6, the soldiers learn of Meade's victory over Lee at Gettysburg, and the next day, the fall of Vicksburg to Grant. They were told "that there should be a general rejoicing and an illumination [fireworks] in the evening." The soldiers "all had a good old time." The next day Joe was back on picket duty. The monotony was broken by letters from home and picking blueberries.

Joe's unit was part of a federal army that would eventually cross southern Tennessee and defeat the Confederates at Chattanooga in November 1863. Under General William Tecumseh Sherman, the federals would then enter Georgia, burn Atlanta, and conduct Sherman's March to the Sea. They would then turn north into the Carolinas until the war ended. But back to Joe's journey....

Throughout July, Joe continued to drill and stand guard, with some time left to pick apples and help maintain the breastworks. His brother John got "the fever" and "swetz bad." At one point "there was a big scare" that rebel forces might be attacking, but Joe still has yet to see combat. On July 26, some rebel prisoners were brought in, and Joe was detailed to help escort them to Memphis. His brother John was now feeling better. On July 30, Joe "saw a deserter drummed out of camp."

After seven months of inactivity, Joe's unit finally moved out, travelling by train from Columbus, Kentucky, to Chattanooga, arriving after the big battle there was done, with fighting around Chattanooga having stopped in November 1863. For the next few months, Union General William Sherman and Confederate General Joseph E. Johnston would try to outmaneuver each other until a clash at Kennesaw Mountain near Mariettta, Georgia, on June 27, 1864.

Joe's unit then carried out a flanking march south to the Chattahoochee River. Joe writes: "There was a hard battle fought here yesterday, and the rebels charged our brigade and killed about thirty. The rebels left about 300 killed on the field.... One wounded soldier girl."

On July 22, 1864, Joe was under artillery fire for the first time, a year and five months after he enlisted. He writes: "When we were taking our place in the line, the rebel battery commenced to shell our regiment. As it was the first time we had been under fire I felt scared. So did all the rest. Laid down on the ground for about a half hour and then went to throwing up." Joe records a couple instances of fellow soldiers being shot through the head. On July 23, Joe writes, "One of our batteries is throwing shells into Atlanta and set some part of the town on fire."

Joe continued to be involved in the fighting. On July 29, "I fired about 20 rounds." "Milo S. was struck with a spent ball in the leg. Company A took 17 prisoners." But the fight for Atlanta would soon be over. Joe writes, "I think the Johnnies are going to leave our front soon." A few days later, he writes: "If I have a chance to re-enlist for three years I would never take it." On August 6: "I felt bad this morning." On September 18, he is back in Chattanooga, this time in an army hospital. He is then sent to a hospital in Nashville, where the doctor won't give him a furlough. He writes: "The doctor in charge of this ward is an old fool entirely. I could bust his head."

Joe does get his furlough and travels home to Wisconsin. But by January 28, 1865, he is in Savannah, Georgia, having traveled by train via New Jersey. Sherman's March to the Sea has ended, and his army is preparing to move north toward Richmond. Joe reaches Fayetteville, North Carolina, but there are still Confederates in the vicinity. In a clash on March 19, 1865, *Joe is wounded in the leg and taken prisoner.*

After two years and one month on duty Joe is a wounded captive. But the next day, the federals counter-attack and Joe is freed. The last entry in his diary is dated March 31, 1865: "In 1st Division hospital at Goldsboro, N.C. I am almost as good as new. I am going to get a furlough if I can."

Joe Smethurst returned to Wisconsin and married Rose Abigail Mills on August 6, 1866. They had five children and both lived to a ripe old age.

Looking at such accounts, you realize again that the overwhelming motivation for Civil War soldiers—at least on the Union side—was to preserve the Union. The enemy were called "Rebels" or "Secesh"—sometimes "Johnnies." There were also negative sentiments about the Northern Democrats, whose failure to support Lincoln, some felt, actually brought on the war.[8] A faction of Democrats was regarded with such loathing as to be called "Copperheads."

Joe's accounts also reinforce the fact that during the Civil War, disease was responsible for more deaths than battle wounds, the biggest killer being dysentery. Some doctors also said that some of the young men passed away from "homesickness."

ENDNOTES FOR CHAPTER 5

1 Niall Ferguson, *The House of Rothschild: The World's Banker 1849–1999* (Penguin Books, 1999), 116.

2 Various parties, including so-called Libertarians, deride fiat currency as being improper or downright evil. They are wrong. It all depends on how it is issued, how its value is guaranteed, and how it is used.

3 *Money: A Monthly Magazine,* cited in Xaviant Haze, *The Suppressed History of American Banking*, (Bear and Company, 2016), 142.

4 Apparently this famous statement, if it ever existed, has been removed by the *Times* from its archives.

5 Andy Blatchford, "Lincoln Assassin John Wilkes Booth's Canadian Connection," *The Canadian Press*, October 13, 2014. Republished at https://globalnews.ca/news/1611968/lincoln-assassin-john-wilkes-booths-canadian-connection/

6 Bismarck's government followed the "American system" in making large government investments in the development of infrastructure. It was these investments, particularly in railroads, that made the modern unified German state possible.

7 Niall Ferguson, *Empire: How Britain Made the Modern World* (Penguin Group, 2004*)*, 253.

8 August Belmont, whom we encountered earlier, was part of this group.

CHAPTER 6

"The Crime of 1873"

How Close did Britain Come to Intervening in the American Civil War?

The rulers of Britain, including the international bankers—the Rothschilds and others—likely tried to weaken or destroy the U.S. by threatening to enter the war on the Confederate side. By this time the Rothschilds enjoyed a near-monopoly on British war finance, and had a strong vested interest in supporting any military action by Britain, whether against the U.S. or other nations.[1]

Several sequences of events indicated that Britain was eager for the U.S. to break apart. One had to do with Russia. We have seen that the British had led a coalition that included France, Piedmont-Sardinia, and the Ottomans of Turkey in its defeat of Russia during the Crimean War. The war culminated in the fall of Russia's Black Sea fortress at Sevastopol.

Britain had feared that the expansion of the Russian Empire would eventually overwhelm the declining Ottomans and allow the Russians to penetrate the British sphere of influence in the Middle East, including Egypt and Persia, and possibly to move south through Afghanistan to threaten British-controlled India.

After the Crimean War, Czar Alexander II resolved to reform Russian society and modernize its armed forces. In 1861, he issued the Edict of Emancipation, freeing Russia's serfs. On January 1, 1863, President Abraham Lincoln issued his own Emancipation Proclamation, the first step in freeing U.S. slaves.

By 1864, Russia had recovered from the Crimean War and had built the third largest naval fleet in the world, after Britain and France. The Czar, like Lincoln, was also resisting takeover of his nation's economy by the international bankers. On September 24, 1863, Russia's fleet entered New York harbor, with other ships anchoring in San Francisco. Gideon Wells, Lincoln's secretary of the navy, wrote:

They arrived at the high tide of the Confederacy and the low tide of the North, causing England and France to hesitate long enough to turn the tide for the North.[2]

On March 30, 1867, soon after the Civil War ended, the U.S. purchased Alaska from Russia for a price of $7.2 million. Russia recognized that Alaska was too remote for them to colonize, and they knew that selling Alaska to the U.S. would keep it out of the hands of Britain. British seizure of Alaska, moving in from Canada, would have put the British Empire a few miles from Russia at the Bering Strait.

However, another indication of possible British intervention related to its view of the Lincoln government's economic policies as a threat. As historian Anton Chaitkin writes:

Henry C. Carey, creator of the nationalist economic platform of Lincoln's Republican Party, wrote, just before the 1860 election, that the British Empire waged continual political and economic "warfare...for discouraging the growth of manufactures in other countries...for compelling the people of other lands to confine themselves to agriculture...for producing pauperism."[3]

The threat of French intervention in the Civil War came in 1861 when France, under Emperor Napoleon III, with help from Spain and Britain, invaded Mexico and established Maximilian von Hapsburg-Lothringen on an imperial throne. Maximilian was the younger brother of Austrian Emperor Franz Joseph I and commander of the Austrian navy. Mexico had been reeling since its defeat by the U.S. in the Mexican War, with its insolvent government facing a constant need to borrow money. There were business opportunities in mercury, coal, and iron that made the country attractive to the Rothschilds and other investors. But to collect payment on these loans and investments, Mexico's creditors saw the strong hand of raw European power as necessary.

Maximilian lasted only through the end of the Civil War, when President Andrew Johnson invoked the Monroe Doctrine to aid Mexico's republican opposition and threaten invasion. Napoleon III withdrew French troops. After a brief period of armed combat between imperial and republican forces, Maximilian was defeated, captured, court-marshalled, and shot. The Republic of Mexico returned to power, and the U.S. was satisfied that the European powers had been driven out.

A Manufacturing Powerhouse

If Britain expected that the U.S. would be fatally weakened by the Civil War, it was sorely disappointed. The U.S. emerged as an economic powerhouse, enjoying the massive and growing productive capabilities of its farms, mines, and factories.

Vast areas of the continental U.S. in the Midwest and West were open for development. Until now, the chief forms of energy had come from humans and horses, along with wind and water power. But productivity was on the cusp of explosive growth due to the harnessing of power from petroleum. By the 1880s, commercial distribution of electricity would also commence, mainly from hydroelectric facilities.

All that would be lacking was a fair and coherent system of finance that was supported or at least tolerated by all involved parties, including workers, farmers, families, employers, academics, foreign travelers, bankers, investors, politicians, and government officials. Unfortunately, U.S. manufacturing, with its railroad infrastructure, would be taken over by financiers like J.P. Morgan, in league with the Rothschilds and other European bankers. To date, the U.S. has failed to achieve a balanced and fair system of national finance. We shall examine the outlines of such a system later.

More Immigration

The period from the end of the Civil War to the close of the 19th century saw a tremendous increase in immigration from Europe. The largest number came from northern and western Europe, including eight million from Great Britain, Ireland, Germany, and Scandinavia. Immigrants had also begun arriving from southern Europe, particularly Italy, with a little over three million. Between 1880 and 1924, 2.5 million Ashkenazi Jews arrived from Russia, Romania, and Austria-Hungary. About six million Jews now live in the U.S., with roughly half that number living in Israel. As many as 4.5 million Irish also arrived between 1820 and 1930. Today about thirty million Americans have some Irish heritage, four times as many people as live in Ireland itself.

American authorities welcomed the immigrants as a source of cheap labor. About 820,000 also moved south from Canada. This included the family of my maternal grandfather, Carlton William Peilow, whose family moved from Canada to the Upper Peninsula of Michigan. From there, family members moved west to Montana.

Some 244,000 immigrants also arrived from Asia, mainly Chinese, many working on building the railroads in the West. The main center of Chinese life in the West was San Francisco's Chinatown.

Millions of Americans moved west to take advantage of the Homestead Act of 1862, which offered land in the western territories to settlers who would live on and farm it. The law provided 160 acres of land to any citizen who was the head of a household and over 21 years of age. The only conditions were that the settler had to live on the land for five years, build a dwelling, and make improvements.

A Financial Battleground Emerges

After the Civil War, open social conflict broke out, fueled by the accelerating divergence of interests between lower and higher income groups. The American working class rarely became overtly revolutionary, but violence did occur at various junctures. Both sides sought to gain political power, with the wealthy classes usually, but not always, prevailing.

The rich wanted lower taxes, bank lending to service government debt, and elimination of the Greenbacks, which Lincoln had called the "peoples' currency." The lower income groups, growing in proportion with industrial expansion, wanted progressive taxes, with the rich paying higher proportionately in exchange for their privileges, and liberal Civil War veterans' pensions.

Alexander Hamilton's vision of an American empire based on an industrial society was now coming to the fore, but so was the central role of bankers and investors. President Andrew Johnson, who succeeded Lincoln, remarked:

> An aristocracy based on nearly two-and-one-half billion of national securities [i.e., government debt] has risen in the Northern states to assume that political control which was formerly given to the slave oligarchy.[4]

This aristocracy of government creditors, earning their money from industrial growth and banking and multiplying it through lending, increasingly ruled the nation and formed the foundation of the Gilded Age.

At the same time, the end of the war saw drastic cuts in federal government expenditures, slightly higher taxes in certain categories and large cuts in others, and an ongoing budget surplus that the government wanted to use to recall the Greenbacks—but could not, due to popular protest.

By the late 1870s, government expenditures had settled at four percent of national income, after having risen to twenty-five percent during the war. Interest on the public debt, which had reached enormous proportions, was forty percent of all expenditures. Over time, this figure was cut in half, but the national debt would never be eliminated.

Costs of the Army and Navy dropped precipitously. The military pension system became an early form of social security. Civil service and public works expenditures were small, at least at the federal level. Infrastructure continued to be a primary function of state and local governments, as it had before the war. Post-Civil War aid to railroad construction primarily involved private bank loans—$16,000–$48,000 per mile for the Transcontinental Railroad. The federal government also granted the railroads land that had been taken from the Indians.

During this period, Congress engaged in heavy reductions to excise taxes, though without manufacturers or retailers cutting prices. So merchants enjoyed a windfall. The federal inheritance tax and the national sales tax were eliminated.

By now, Karl Marx had made his appearance. Marx's *Das Kapital* was published in German in 1867, and in English some 20 years later. Socialism was well on its way to becoming universally demonized in the U.S., as remains the case today. From this point on, every measure of social or economic improvement was labeled by the rich as "socialism" or "communism."

The wealthy class was finally able to lobby for complete elimination of the federal income tax, which was rescinded in 1872. Stiff taxes on liquor, tobacco, and various trade licenses remained. Protectionist tariffs favorable to manufacturers also continued. "Buy America" continued to be in vogue. This was a bulwark against cheap manufactured goods from Great Britain.

Attempts now were made to convert the Greenbacks to interest-bearing Treasury bonds. But the measure would have reduced their value as currency, so it was politically unpopular and removed from the Congressional agenda during a mild business downturn in 1868. Specie payments had been suspended during the war, with conservatives now lobbying for "resumption."

The post-Civil War period was a time of Republican presidents, starting with the election of Grant in 1868, and continuing through Hayes, Garfield, and Arthur until the election of Democrat Grover Cleveland in 1888. They followed the familiar Republican program, still promoted today, of cutting federal expenditures, reducing taxes on the rich, and favoring business interests.

The government's main activity, which had been fighting rebels, now was fighting Indians and providing a minimal handout of financial support to those it was able to herd onto reservations. But the Army was a small fraction of what it had been during the Civil War. In 1876, 37.8 percent of the federal budget was for interest to lenders on the national debt; 20.1 percent was for the Army and Navy; and 10.7 percent was for veterans' pensions. By contrast, 2.2 percent was for Indian welfare support.

Meanwhile, government bonds were sold to service the debt through private investment brokers who made a living from government commissions. One of the leaders in this enterprise continued to be Jay Cooke of Civil War bond fame. Another was the up-and-coming banker, J.P Morgan. Even though the government was running an ongoing surplus, the net federal deficit was slow to decline from its wartime high of $2.77 billion in 1866, only falling to $1.83 billion by 1884.

Although the role of the federal government in economic matters was negligible, the existence of the federal debt was essential, as national banks were required to purchase federal bonds as a reserve for making loans. Hence the system was inelastic in that an honest effort by the government to pay its debts also reduced the circulation of the currency. Then, as now, the debt was viewed as a necessary source of financial liquidity.

With the economy growing, consumers and farmers were frantic for the government to increase the money supply. This fueled the movement for "bimetallism" or the support of silver as a basis for money along with gold. It also fueled the founding of the Greenback Party, which was active between 1874 and 1889. The party ran candidates in three presidential elections, in 1876, 1880, and 1884, before it faded away.

Corruption

There have been many times in U.S. history when financial corruption was overwhelming. Opening the doors to favoritism, profiteering, and nepotism were among the charges leveled against the First and Second Banks of the United States. But the unprecedented amounts of money hitting the streets with the prosecution of the Civil War produced many incidents of contractors cutting corners or financial agents of the government being caught with their hand in the till. The corruption accelerated during the post-war period.

Scandals erupted involving members of Congress and President Ulysses Grant's brother and brother-in-law, including attempts to corner the gold market, ending in the "Black Friday" panic of 1869. In 1872 the Credit Mobilier scandal disclosed stock handouts involving railroad construction to members of Congress and Vice-President Henry Wilson, and rumors of railroad company bribes also besmirched the reputation of Republican stalwart James B. Blaine.

Civil War-era tax legislation also opened the door to charges of tax fraud and evasion against political figures that have gone on until today. Later, President Garfield's campaign officials were charged with fraud in contract awards. Corruption was pervasive at state and local government levels, including the infamous Boss Tweed ring in New York.

Business Cycles and the Gold/Silver Crisis

As with the pre-Civil War financial panics, we now return to the "business cycle." Such cycles result in economic expansion and wealth during the upswing, but generate chasms of misery and poverty when the crash comes. These cycles have little to do with the willingness of people to work hard and prosper or with the availability of natural resources as essential components of the manufacturing process. Business cycles instead are "exclusively monetary phenomena," borrowing a phrase from modern financial guru Milton Friedman. They are caused by fractional reserve lending, resulting in a system of money creation rooted in debt that circulates as a medium of exchange within the producing economy. Credit expands until economic activity slows down and loans can no longer be repaid. Borrowers then go broke. Their assets are then purchased for pennies on the dollar by the encircling vultures.

Within the U.S. financial system, credit is offered or withdrawn by bankers to whom everyone, including governments, must resort when money is needed. These bankers produce a lot of credit when economic conditions are good, but when loans can no longer be repaid by borrowers, the credit is taken away. The bank itself may fail. So, boom to bust. One way to withdraw credit, of course, is for the banking system to raise interest rates, as the Federal Reserve does today, making borrowing prohibitively expensive.

This is the system that prevails in all Western nations. Something different has developed in large state-managed economies, like those of contemporary Russia and China. Evidence suggests that such systems, while not immune to business cycles, may find it easier to control them. It's the growing strength of these systems and their relative immunity to control by private banks and investment funds that makes the nations practicing them the enemies of the Western financial oligarchy.

Should we be using gold, silver, both, or neither as an attempt to provide backing for the currency the banks lend into circulation? This was a massive economic issue during the 19th century and into the 20th. U.S. economist Milton Friedman discussed this situation in depth in his book *Monetary Mischief,* published in 1994. His discussion takes us from the time when, "The Civil War temporarily ended the reign of gold,"[5] to what is called "The Crime of 1873," when it was alleged that certain British figures conspired "to bribe certain members of Congress and the Comptroller of the Currency" to demonetize silver altogether.[6]

The Coinage Act of 1873 ended the legal status of bimetallism in the U.S. Paper instruments could be redeemed only for gold after "resumption" took place in 1879. Silver was not mentioned in the resumption legislation. The U.S. was now firmly on the gold standard, as were Britain and most other European nations.

The bankers were thrilled. They would use the gold standard to rule the world's economies for the next half-century. The race was also now on to see which nation could hoard the most gold in its vaults. Until World War I that nation was Great Britain.

Obviously, the exclusion of silver as backing for paper money would lead to a contraction in real-life spending power and would therefore crash a rapidly-expanding economy based on credit. From this point on, the restoration of silver was a major political issue in the U.S., leading to the presidential campaign of William Jennings Bryan in 1896 who declared, "You shall not crucify mankind on a cross of gold."

But the deeper reason for agitation by the wealthy class for the gold standard was not just for the price stability they saw would result from limitation of monetary growth. It was actually to make it harder for debtors, to whom the wealthy lent money, to pay off their loans with an inflated (i.e., depreciated) currency.

In fact, the last half of the 19th century saw price *deflation*, with farmers hurt particularly by falling prices of wholesale agricultural products. At one point, the American Banking Association even advised the banking industry to take advantage of the situation to engage in large-scale foreclosure on family farms in order to reduce American farmers to the status of European peasants.

Farmers were able to fight off foreclosure by consolidation of small family farms into larger units and the utilization of mechanized farm machinery like threshers and binders. The trend toward larger farms and industrialization of farming had begun, with greater reliance on banks for operating expenses and purchase of equipment. Distress for farmers from insufficient credit also mean trouble for the vast network of small towns throughout the nation that depended on the farm economy for sustenance.

Many town businesses reacted to the distress by chartering local banks, expanding commercial credit to farms, and by printing and issuing scrip as a local currency. The widespread use of scrip would reappear during the Great Depression of the 1930s. A factor that ameliorated the distress from monetary contraction was the use of paper checks which enhanced the velocity of money. Kiting of checks also became common.

Not by coincidence and taking place along with the Coinage Act, the Panic of 1873 began when our old friend Jay Cooke, head of what was now the top U.S. banking house, went bankrupt when the collapse of bond prices for the Northern Pacific Railroad, which Cooke was financing, caused him to close his doors. Bank runs now began, the stock market crashed, and a worldwide depression was underway.

As if to show that the world was becoming increasingly interdependent, the financial woes in the U.S. and Britain were reflected in a collapse of real estate prices in Germany and central Europe. European banks panicked, and German investors, who had heavily invested in U.S. railroads, withdrew their holdings.

The U.S. government, still in its Republican Party *laissez faire* mode during Grant's second term, could do nothing, especially after Congress opted for the gold standard. So financial collapse had to run its course. It was a decade before the U.S. economy recovered. The Panic of 1873 was so bad it was called, at the time, the "Great Depression."

The Flathead Tribes and the Hellgate Treaty

We return to Montana.

The Hellgate is a canyon at the east end of Missoula, Montana. The Clark Fork River flows through the canyon after being joined by the Blackfoot River that comes down the slopes of the Rockies from the Continental Divide. (The Blackfoot River, an archetypal trout fishing stream, provided the locale of the motion picture *A River Runs Through It.*) My grandparents settled in this area near Seeley Lake, and my mother grew up there.

Hellgate referred to the bones on the ground of Indians who fought battles there. The Hellgate Treaty, signed in 1855 by the U.S. government and the Flathead tribes, mainly the Salish, was the instrument through which the Indians lost much of their ancestral lands. Their ownership was eventually confined to the present-day Flathead Reservation. Even though the Salish were under pressure from their enemies, the Blackfeet, who raided their settlements from beyond the Continental Divide, their most formidable foes were now the whites who had begun to enter Montana to pan for gold. The whites brought the scourges of alcohol and gambling with them.

Meanwhile, Washington Territory governor Brigadier General Isaac Stephens was under orders from Washington, DC, to settle the Indian tribes in the Pacific Northwest on reservations. The U.S. had obtained control of the region through the Oregon Treaty of 1846. Western Montana was now part of the Washington Territory.

Tribal chiefs and Stephens signed the Hellgate Treaty by which the Indians ceded to the U.S. title to the vast majority of their lands west of the Continental Divide. Any claims the Indians may have had to their traditional hunting grounds east of the Divide had already been taken away, without their participation or knowledge, by the Treaty of Fort Laramie signed between the U.S. and the Sioux in 1851.

Through the Hellgate Treaty, the Salish agreed to a reservation of about 1.25 million acres north of the Clark Fork River, extending halfway up Flathead Lake. Along with this cession came a "Conditional Bitterroot Reservation" south of the Clark Fork, with tribal rights to be established later. The tribes also reserved rights on their lost land, including the right to hunt, fish, gather plants, such as the camas root, and pasture livestock on "open and unclaimed lands"; i.e. lands to which individual whites had not claimed title. This was later to include extensive national forest lands surveyed by the U.S. government. But land being claimed and patented by whites in the Bitterroot Valley outside the conditional reservation would be off-limits to Indian use.

Father Adrian Hoecken, a founder of the nearby St. Mary's mission, attended the meetings between the Salish and Brigadier General Stephens and his entourage that was conducted at what was afterwards called Council Grove. Father Hoecken later said that the Salish did not understand a tenth of what Stephens said to them during the "negotiation" of the Hellgate Treaty. We do know that Stephens referred to the Salish condescendingly as "my children."

According to today's Consolidated Salish and Kootenai Tribal (CS&KT) government:

> Tribal understanding of the boundaries of the Flathead Reservation was considerably different from what was actually written in the treaty, particularly the east, west, and northern boundaries.[7]

Stevens also insisted that a single reservation be created for the joint habitation of the three distinct tribes of Salish, Pend d'Oreilles, and Kootenai, with any difference of settlement rights to be sorted out later.

According to the treaty, the Indians were to be paid for their relinquished land in installments. Only about $593,000 was paid, until over a century later, when the Indian Claims Commission ordered the government to pay approximately $4 million in a 1967 judgment, which failed to include interest or penalties for the government's failure to meet the original commitments.

The Hellgate Treaty was ratified by Congress on March 8, 1859. The treaty had included annuity payments to help the Indians resettle on treaty lands, but delays in payment caused the tribes to believe that the government had broken its promises. The annuities were to be paid in the form of supplies like blankets, flannel, rice, and coffee.

Governor Stevens had made a verbal promise of military protection from the Blackfeet, but this promise was not carried out. The government also promised help with education and health care, which were not provided. The Hellgate Treaty was another egregious example of U.S. government treachery.

The Upper Pend d'Oreille and Lower Kootenai tribes moved from the surrounding areas in northwestern Montana to the Flathead Indian Reservation. But some of the Salish continued to reside in the Bitterroot Valley, believing that the treaty had guaranteed their right to do so. Meanwhile, the Montana gold rush of 1864 brought more white settlers.

The Civil War distracted the government from taking further action until 1871, when President Grant issued an executive order stating that the Salish would be "removed" to the Flathead Reservation. Chief Carlo, the son and successor to the late Chief Victor who had originally signed the Hellgate Treaty, refused to agree to the order. Grant sent congressman and future president James Garfield to Montana to negotiate the removal. The Salish say to this day that Garfield forged the mark of Chief Charlo on the treaty document. According to the CS&KT website, "Non-Indians called Chief Charlo a treaty breaker until this outrageous forgery was proved to be true by Senator G.G. Vest in 1883."

Some of the Indians at the southern end of the new Flathead Reservation converted to the Catholic religion under the guidance of the Jesuits, who had established the St. Ignatius mission. A town of that name grew up around the mission, which established a school, a printing shop, and a farming operation that served the Indians in the area. The Catholic church at St. Ignatius continues to operate to this day.

The Salish who had stayed to the south in the Bitterroot Valley around the St. Mary's mission grew progressively poorer as whites continued to flood the area, with the Indians losing the ability to graze their livestock on open range. The Blackfeet continued their raids, and Salish hunting parties that crossed the Continental Divide in search of buffalo returned with less meat and fewer hides each year.

When the Nez Perce Indians came down the Lolo Pass in 1877 on their flight from the Army that was attempting to force them onto a reservation in Idaho, the Salish refused to join their revolt and instead defended the white settlements around Stevensville.

The Flathead tribes never took up arms against the U.S. at any time during their long history. The tribal chiefs consistently made the judgment that no matter how badly they were treated, they would not allow their people to resort to violence. Nevertheless, they have a long record of standing up for their rights and trying to hold the government to its commitments.

Despite their loyalty, Chief Charlo and his band of Salish continued to be pressured by the government to leave the Bitterroot Valley, which they finally agreed to do in 1889. Heedless of its own broken promises, the government offered Chief Charlo new guarantees of housing and farming support if he agreed to move.

In 1890 and 1891 the Salish sold their property and household belongings, and in October 1891, General Henry B. Carrington and troops from Fort Missoula escorted the families from the Stevensville area north to the Flathead Reservation. Fort Missoula had been established during the Nez Perce conflict.

A group of Pend D'Oreille and other tribal members met the arriving Salish at the Jocko Church to help them feel welcome. This time, aided by sympathetic and responsible Bureau of Indian Affairs agents, the government fulfilled its commitments to Chief Charlo to the letter.

The Flathead tribes now lived together on the reservation, which was later opened to white settlement through the allotment process in 1910. We'll hear about this disastrous development in a later chapter. For a time, the Indians were engaged in building a successful subsistence farming and cattle community on reservation lands. They were able to achieve a standard of living comparable to the rest of rural western America.

Meanwhile, the whites set out to extract billions of dollars in wealth from the former Flathead domain, enough to earn Montana the nickname of "The Treasure State," with the state motto, "Oro y Plata," Spanish for "gold and silver."

Wars Against the Indians

With the Civil War ended, the U.S. Army conducted a massive demobilization of the rank-and-file volunteers and conscripts, but an experienced officer corps remained. The Army now had two missions. The first was to occupy the South and maintain a few small military bases to keep the defeated rebels in line and protect freed blacks.

The second mission was to attain victory against the Indians under the leadership of General William Tecumseh Sherman, who became Commanding General of the U.S. Army after Grant was elected president in 1868. With the Civil War over, the Army could now take up its unfinished business of destroying Indian life and culture west of the Mississippi and securing Indian lands for white settlement.

Many of the Indians fought back. General Philip H. Sheridan succeeded Sherman as Army commander in 1888. He said:

> We took away their country and their means of support, broke up their mode of living, their habits of life, introduced disease and decay among them, and it was for this and against this they made war. Could anyone expect less?[8]

Despite this admission, Sheridan could yet declare, "The only good Indians I saw were dead."

Congress declared in 1871 that "henceforth no Indian nation or tribe... shall be acknowledged or recognized as an independent nation, tribe, or power with whom the U.S. may contract by treaty." But the government continued to herd Indians onto reservations where they were considered legal wards of the government.

The Indians' resistance took place throughout the West, from Texas to the Canadian border, through the Plains to the Rockies and beyond, and down the Pacific Coast from the old Oregon Country, across the Great Basin, and throughout California. The only region, other than the Flathead Reservation, that was reasonably quiet was the Old Southwest in New Mexico and Arizona, with settled communities of Pueblos, Navajos, and Hopis.

The Sioux

The literature on the Sioux Indian resistance is vast, with the conflict not yet settled to this day. The Sioux continue to refuse the government's offer of compensation for theft of the Black Hills of South Dakota with its vast gold deposits.

With the removal of most of the Cherokees from the Southeast to Indian Territory in what became Oklahoma and the conquest of the Comanches by the Texans, the post-Civil War focal point of U.S. government attack against the Indians became the Sioux. The government intended to provoke the Sioux into an all-out war that would clear them from the Plains once and for all.

The Sioux had been pushed westward from Minnesota and the Mississippi basin by tribes around the Great Lakes that were being displaced in turn by the pressure of white settlement in the Old Northwest, Kentucky, and Tennessee. The Sioux were a large language family of half-a-dozen distinct tribes, identifiable as a culture for at least three thousand years, one that had acquired guns and horses during the 18th century and settled in the vast area of the northern and central Great Plains.

There they created a complex culture with a rich ritual dimension that focused on the hunting of buffalo, or, technically, bison. Bison were a remnant of the neolithic megafauna that had filled the landscape of North America since the Ice Age. The bison roamed the Plains in the millions. The Sioux were a warrior culture that practiced a nomadic lifestyle. While they had hunter-gatherer origins, they once had cultivated corn and had acquired a somewhat sedentary village life. But now they lived in villages that were quickly broken down and moved as they followed the buffalo herds, though they made more settled camps for the winter, particularly in the Black Hills.

As the whites began to travel west on the Oregon Trail through Nebraska and Wyoming, and later on the Bozeman Trail that branched off toward Montana, settlers and miners passed through Sioux territory. The Army maintained a military stronghold at Fort Laramie near what is today Cheyenne, the capital of Wyoming. The 1851 Treaty of Fort Laramie acknowledged ownership rights for the Lakota Sioux in a large area centered on the Black Hills, with tribes of the Crows, Mandans, Arikaras, Assiniboines, and Hidatsas spread around the northern and western peripheries.

The wars of the U.S. government against the Sioux have been characterized as "a clash of two expanding empires."[9] But the U.S. had the benefit of a well-armed and mobile professional military force, backed by the resources of a populous and growing industrial culture. The Sioux were handicapped by the fact that they needed to defend their homes, since their women and children were constantly threatened with assault.

The Army increasingly engaged in what modern times would recognize as total war, with the fighting against the Sioux devolving into civilian massacres. Thus, the Sioux wars anticipated future American combat against civilian populations around the world in places like the Philippines, Vietnam, Iraq, Afghanistan, Syria, and others.

The peace achieved through the 1851 Treaty of Fort Laramie was short-lived. Settlers, miners, and fortune seekers would soon flood through the Sioux homeland. On August 17, 1854, a cow belonging to a Mormon traveling on the Oregon Trail strayed and was killed by a Sioux Indian named High Forehead. Two days later, Second Lieutenant John Lawrence Grattan of the U.S. 6th Infantry Regiment marched into a band of 4,800 Indians with twenty-nine men and a French interpreter and demanded that High Forehead be surrendered for punishment. A fight broke out, during which a soldier shot the chief, Conquering Bear, in the back. The Indians retaliated by killing Grattan and his entire detachment.

The Army then called in Colonel William S. Harney, who assembled a force of 600 men. On September 3, 1855, at what is called the Battle of Ash Hollow, Harney killed eighty-six Sioux, half of them women and children, and took more women and children as hostages back to Ft. Laramie. The Army's actions were based on the notion of "collective punishment," later made infamous by the Nazis in World War II.

There were five phases of the Sioux wars: the Dakota War of 1862, the Colorado War from 1863 to July 1865, the Powder River War, Red Cloud's War, and the Great Sioux War. It was during Red Cloud's War that Lakota Sioux warrior Crazy Horse came to the fore as a leader of the Indian forces. On December 21, 1866, Crazy Horse and his force of over 1,000 warriors lured a party of soldiers out of Fort Phil Kearny on the Bozeman Trail commanded by Captain William J. Fetterman and wiped them out. Eighty-one soldiers were killed in what was known as the Fetterman Fight.

Despite the violence, whites continued to move through Indian territories, including settlers heading for Oregon, California, and Utah, with the northern Rockies increasingly a target for gold, silver, and later, copper mining. In 1874–1875, gold was discovered in the Black Hills, the most sacred land of the Sioux. By now, George Armstrong Custer had appeared on the scene.

Custer was a career Army officer who had commanded cavalry since the Civil War, where we earlier saw him in action at the Battle of Gettysburg. He had been named a brigadier-general when twenty-three years old, but with wartime breveting over, was now a lieutenant colonel subordinate to higher-ranked officers. He assumed command of the 7th Cavalry Regiment at Fort Riley, Kansas, in July 1866 and fought initially against the Cheyenne.

Custer was suspended briefly for leaving his post to visit his wife, but returned to frontier duty in 1868. On November 27, 1868, under orders from General Phil Sheridan, Custer led an attack on a Cheyenne village in Indian Territory at the Battle of Washita River, killing 103 warriors, several women and children, and taking fifty-three women and children prisoners. At this battle, Custer charged with his entire force into the village of campfires and teepees, killing anyone they encountered. Making war on Indian civilians and families was now part of the Army's standard operating procedure. The Southern Cheyenne surrendered and were moved onto a reservation.

But fate had more in store for Custer. For his entire career he had been a staunch Democrat. A member early on of General George McClellan's staff, he had supported McClellan's 1864 candidacy for president against Lincoln and had political ambitions himself. He was told he might be a shoo-in as U.S. senator from Michigan, but he had higher ambitions. After all, General Ulysses Grant had made it to the presidency and, with Grant being a Republican, Custer thought that Grant was no great shakes.

In 1873, Custer was transferred north to Dakota Territory to protect the workers on the Northern Pacific Railroad which had reached Fargo, North Dakota, in its cross-country trek. In 1874 Custer led his force into the Black Hills and made the announcement that gold had been discovered there. Custer's fame took another leap.

Even as the Sioux lurked in the vicinity, the Black Hills Gold Rush now began, with Custer's name appearing prominently in the headlines. A major gold discovery was very big news. But soon Custer was back in Washington, DC, taking on President Grant's Republican administration.

"Custer's Last Stand"

Events now moved swiftly to "Custer's Last Stand."

In 1875, the Grant administration offered to buy the Black Hills from the Sioux. When the Sioux refused to sell, they were peremptorily ordered to report to government reservations by the end of January 1876. It was a deliberately cruel and deceptive demand, as mid-winter weather made compliance impossible.

This enabled the Sioux to be labeled "hostiles." The Army was ordered to bring them to the reservations in the following year. Custer was to command one prong of a three-part force, with troops under Colonel John Gibbon and more under General George Crook also taking part. With Custer's 7th Cavalry scheduled to set out on April 6, 1876, he was summoned to Washington to testify at congressional hearings investigating corruption on the part of Secretary of War William W. Belknap. Also implicated were President Grant's brother Orville and traders at several frontier Army posts. The traders were accused of price gouging, with kickbacks going to Orville and Belknap.

Custer made several accusations while writing articles for the *New-York Tribune*. In particular, he accused Orville Grant of extorting money. The Democratic press ate it up, Belknap was impeached and his name forwarded to the Senate for trial, with President Grant retaliating by removing Custer from duty. Custer defiantly took a train to Chicago, intending to rejoin his regiment.

Grant ordered Custer's arrest, but the newspapers howled. Custer was supported by Generals Sherman and Sheridan, causing Grant to relent under fears that if the Sioux campaign failed without Custer, the president would be blamed. Grant backed down but insisted that General Alfred Terry lead the upcoming campaign. So Custer would head the force with Terry nominally in charge, though Custer told Terry's chief engineer, Captain Ludlow, that he would "cut loose" from Terry and operate on his own.

The story of the Battle of the Little Big Horn, where Custer's command, part of the 7th cavalry, was wiped out and Custer himself killed by a force of 1,000–2,000 Plains Indians, consisting of Lakota Sioux, Northern Cheyennes, and Arapaho, has been told and retold as one of the great sagas of the West.

Custer intended to follow the usual U.S. Army pattern of a terrorist strike on a large village of Indian women and children, with an engagement against whatever warriors happened to be present. Custer also planned to take civilians hostage, including the elderly and disabled. Unfortunately for Custer, the Indian force under Sitting Bull and Crazy Horse was the largest contingent of armed warriors ever seen in the Plains wars, a force armed, moreover, with almost as many guns as there were Indians. The battle took place on June 26, 1876. There were no survivors among Custer's 268-man force.

There is, however, a "rest of the story," recounted by Stephen E. Ambrose in his book *Crazy Horse and Custer: The Parallel Lives of Two American Warriors*. Before telling the Custer story, let me mention that his two main adversaries, Sitting Bull and Crazy Horse, were later killed in the hands of the Army following their surrender.

The main mistake Custer made was to divide his force in the face of the enemy. This was something that Napoleon was famous for doing, but it didn't always work. Also, Custer had been driving his men so hard to reach the location of the Indian encampment that they were exhausted. Witnesses among the Indians later said that Custer's men were literally shaking with fatigue when they arrived on the scene. Why then was Custer in such a big hurry? Well, maybe Custer needed to get the job done *so he could campaign for president.*

1876 was a presidential election year, and Custer, a Democrat, was being celebrated in the press for standing up to President Grant in the Belknap/Orville Grant corruption scandal. According to Ambrose, "Custer was one of the most famous men in the country and extremely popular to boot."[10]

While the actual deliberation took place in secrecy, Ambrose speculated that Custer was to be the choice of *New York Herald* publisher James Gordon Bennett for nomination to run for president at the Democratic Party National Convention taking place in St. Louis on June 27, 1876. Custer had promised Bennett that he would give the *Herald* exclusive rights to publish his account of the Sioux campaign. The *Herald* had a reporter, Mark Kellogg, riding with Custer during the campaign, against the explicit orders of General Phil

Sheridan. Ambrose writes, "Perhaps Custer hoped that Kellogg could get a report of the battle with the Sioux to the Democrats and to the country before June 27, the opening day of the convention."[11] There were telegraph wires in the region, and Kellogg had already filed a story with the *Herald* about a scouting mission carried out by Custer's subordinate, Marcus Reno.

Was there any other evidence? There is, for after the battle—in some cases, long after—journalists and researchers sought out and interviewed Indians who had fought against Custer, as well as Indian scouts who helped him. Ambrose relates the following:

> One evening shortly before the column moved out, Custer had visited the camp of the regiment's Crow and Arikara scouts, and that visit brings us back to speculation about what may have been said to Custer while he was in the East [i.e., testifying to Congress and talking with the press and Democratic Party leaders]. First, Custer presented his Rhee scout Bloody Knife with several gifts purchased in Washington and told him and the Arikaras of his visit to the capital. Then he said that this would be his last Indian campaign and that if he won a victory—no matter how small—it would make him the Great White Father in Washington. If the Arikaras helped him to a victory, he promised that when he went to the White House he would take his brother Bloody Knife with him. He also told the scouts that he would look after them and see to it that they got houses to live in, and finally promised that as the Great White Father he would always look after the welfare of his children, the Arikaras.[12]

How desperate was Custer in driving his men to their possible deaths in order to defeat the Sioux before the Democratic Party convention? We'll never know. But we do know from modern-day experience how far ambitious people may go to be elected to high political office and how many corpses litter the road to power.

Conclusion

The post-Civil War period has been America's "lost history," yet patterns were established that continue to affect events today—notably, the rule of money over any attempt to establish rational and fair governance. Industry was exploding, while government stagnated in passivity and favoritism. Mediocrity reigned in public life, while the bankers and "captains of industry" grew obscenely rich. The Greenback Party and other reform movements tried to introduce a modicum of fairness to the financial system, but with little success.

The Indians suffered the most, along with formerly enslaved Africans living in poverty in the South. Commentators have pointed out that, had the government made a good faith effort to provide funding to transition the Indians on reservations to an effective system of subsistence farming, the policy might have worked. In fact, the Indian treaties promised as much. But the promises were betrayed, to the lasting shame of white American elites, so the Indians and blacks alike were condemned to back-country poverty that has never been healed.

Sioux Indian chief Sitting Bull put his finger on the problem that has been the bane of U.S. society throughout its history. He said, "The white man knows how to make everything, but he does not know how to distribute it."[13] There is more to this wise observation than you might think.

ENDNOTES FOR CHAPTER 6

1 Ferguson, *The House of Rothschild*, 79.
2 H. Donald Winkler, *Lincoln and Booth: More Light on the Conspiracy* (Cumberland House Publishing, 2003), 143.
3 Anton Chaitkin, "Why the British Kill American Presidents," *Executive Intelligence Review (*December 12, 2008), 28.
4 Studenski and Kroos, 161.
5 Milton Friedman, *Monetary Mischief: Episodes in Monetary History* (Mariner Books, 1994), 57.
6 Ibid., 61.
7 Confederated Salish and Kootenai Tribes (CSKT), *The REZ we LIVE on.* http://therezweliveon.com/myth-busting/taxes/
8 "Philip Sheridan," *Wikipedia.* https://en.wikipedia.org/wiki/Philip_Sheridan
9 Colin Calloway, "The Inter-Tribal Balance of Power on the Great Plains, 1760–1850," *Journal of American Studies,* Vol. 16, No. 1 (April 1982), 46.
10 Ambrose, *Crazy Horse and Custer*, 407.
11 Ibid., 407.
12 Ibid., 405–406.
13 Ibid., 351.

CHAPTER 7

The Gilded Age

The End of Reconstruction

Lieutenant Colonel George Armstrong Custer lay dead on a hilltop in Montana Territory. With his *New York Herald*-sponsored candidacy not surviving until the start of the Democratic Party national convention in St. Louis in September 1876, the nomination for president went, as expected, to New York Governor Samuel Tilden. His opponent would be Civil War hero and Republican Ohio Governor Rutherford B. Hayes.

The two nominations demonstrated that New York City was the bastion of the Democratic Party, with the Midwest the Republican power center. This identification continues today, with New York also the location of the Wall Street banks that formerly had the Southern aristocracy in the grip of debt, while the Midwest was the political bastion of farming and manufacturing interests.[1] By the late 19th century, the Democrats were also gaining the support of immigrant urban voters from Ireland, Italy, and Eastern Europe.

In the 1876 election, Samuel Tilden won the popular vote with 50.9 percent, but disputed votes in three southern states caused the election to be thrown to an electoral commission which the Republicans controlled, which then declared Rutherford B. Hayes the winner. Tilden's percentage of the popular vote was the highest ever received by a loser in a U.S. presidential election.

Some of Tilden's supporters wanted to stage street demonstrations to overturn the decision of the electoral commission, but Tilden disapproved. Instead, the two parties agreed on the Compromise of 1877, whereby the Democrats accepted Hayes as the winner, while the Republicans agreed to withdraw all federal troops from the South, ending Reconstruction and making the Democrats the initiator of the federal government's abandoning further attempts to secure racial equity between whites and blacks.

Many parts of the southern states now had a majority black electorate, where over 1,500 black officeholders were elected. This was rolled back after the 1876 election and the subsequent enactment of the Jim Crow laws.

Politically, this was the beginning of the "Solid South," with southern whites exclusively electing Democratic Party candidates, and African Americans and any remaining Republican Party supporters now shut out of public life.

With hope of political relief gone, black poverty remained endemic. blacks began to migrate to regional and northern cities, while those staying in the rural South worked as sharecroppers on farms and as servants and laborers in the towns. There was also a small black professional class in segregated areas. In Tulsa, Oklahoma, these created a flourishing "Black Wall Street" district until the white terrorist massacre of May 31–June 1, 1921. Fed by internal migrations, relatively sustainable black urban communities took form in Harlem in New York City and in New Orleans, Chicago, and other cities.

The KKK was formed to keep the blacks suppressed, and the Democrats held sway until Richard Nixon's "Southern Strategy" caused the political parties to reverse roles in the election of 1968. By then, the federal government had begun to take an active role in the civil rights movement under presidents John F. Kennedy and Lyndon Johnson. This didn't keep U.S. government-affiliated agents from assassinating black civil rights leaders Malcolm X in 1965 and Dr. Martin Luther King, Jr., in 1968. Gradually African Americans began to be assimilated into centers of public employment, including the military and the civilian federal government.

Neither presidential candidate in the 1876 election showed awareness of the financial causes of the Panic of 1873 that still held the nation in its grip. Tilden and Hayes were both hard-money advocates, with Tilden supporting the gold standard and blaming the Panic of 1873 on a corrupt do-nothing Grant administration. The Republicans pledged to continue the limited-government policies of the post-war period, including the protective tariff.

Rutherford B. Hayes

Few people know much about Rutherford B. Hayes as president. I am familiar with him as lieutenant colonel of the 23rd Ohio regiment of volunteers that charged up South Mountain on September 14, 1862, to dislodge the Confederates from their dug-in positions three days prior to the Battle of Antietam.

I have walked the battlefield at South Mountain many times where the Appalachian Trail runs along the South Mountain ridge. There at Fox's Gap, Union Major General Jesse Reno and Confederate Brigadier General Samuel Garland, Jr., were both shot dead on the same day a few hundred feet from each other. Reno, while dying, is believed to have told an aide he'd been shot by his own troops.

Early in the battle, Hayes was shot in the arm, the bone shattered, and he was carried from the field. He recovered, served with distinction in the war and was later breveted as a brigadier-general. After the war, he became a lawyer, entered politics, and was elected to three terms as Ohio governor before winning the presidency. Once in the White House, he and his wife Lucy were noted as serving only lemonade at receptions in contrast to the usual drunkenness.

Hayes served only a single term, which may be why he is so underrated as a pivotal historical figure between the Civil War and the industrial revolution. Soon after inauguration, he was faced with the Great Railroad Strike of 1877, which began at the B&O Railroad terminal in Martinsburg, West Virginia, and spread to the New York Central, Erie, and Pennsylvania railroads. The main issue was wage cuts imposed by the railroads to recoup financial losses from the Panic of 1873. Hayes sent federal troops to Martinsburg, Baltimore, and other locations, with riots spreading to Chicago and St. Louis.

The only loss of life resulted from clashes between strikers and state militias; federal troops showed restraint. Public opinion forced the railroads to improve working conditions and cease cutting wages. Hayes tried to act as a peacemaker, later writing that:

> The strikes have been put down by *force;* but now for the *real* remedy. Can't something [be] done by education of strikers, by judicious control of capitalists, by wise general policy to end or diminish the evil? The railroad strikers, as a rule, are good men, sober, intelligent, and industrious.[2]

On relations with the Indians, by the time Hayes became president, a national movement had formed to improve their treatment, even as the U.S. Army continued to prosecute the war against the Sioux. The focus of reform was assimilation of the Indians into white culture. This included education.

Indian schools were set up by religious groups on the reservations and by secular institutions sponsored by the Bureau of Indian Affairs. These included the U.S. Indian Industrial School at Carlisle, Pennsylvania, founded in 1879. Many of the schools, including Carlisle, were boarding institutions that involved removal of Indian children from their families and immersion in white ways of dressing, and reading, writing, and speaking in English. Tribal languages were forbidden.

Also in the works was what would become the "allotment" system under the Dawes Act, signed by President Grover Cleveland in 1887, which aimed at replacing reservations with individual Indian properties of around

160 acres each. The Indians were expected to farm their allotments as a route to assimilation into white society. They were supposed to be provided with farm implements, livestock, and seed, but often were not. If they did not wish to farm, they could allow the government to manage their land and lease it to white ranchers, miners, or lumber companies.

These properties would be held by the government "in trust," since the Indians were *not* in fact trusted to manage them responsibly. Thus, Indians remained wards of the U.S. government. Over time, the tribes lost much of their reservation property through sale of what the government classified as "surplus lands," often acquired by white speculators.

A scandal would develop a century later when the government was forced by court action to admit they had no records of the money which the leases on the Indian allotments were supposedly earning. The Flathead tribes of western Montana were forced into opening their own "surplus" land to white homesteaders in 1910.

Besides the Sioux, Hayes oversaw other conflicts with Indian tribes, including with the Nez Perce in 1877, when the Nez Perce refused to move to a reservation in Idaho. The Indians, one of whose leaders was Chief Joseph, commenced a 1,700-mile trek toward Canada. They held off the Army in a series of battles before surrendering within only forty miles of the Canadian border.[3]

Hayes also called out the army against the Bannock Indians in Idaho and the Utes in Colorado. When the Ponca Indians from Nebraska attempted to return to their former homes from Indian Territory, Hayes set up a commission to offer them a choice of domiciles that awarded them compensation for their land rights. In a message to Congress in February 1881, Hayes said he would "give to these injured people that measure of redress which is required alike by justice and by humanity."

Compared with many U.S. politicians both before and after, Rutherford B. Hayes was a man capable of rational, compassionate action under the circumstances of the time. There is nothing of the fanatic, the cruel, the bombastic, or the cowardly in his attempt to do his duty. In an 1887 diary entry, he tried to sum up what he had learned over his long political career:

> In church it occurred to me that it is time for the public to hear that the giant evil and danger in this country, the danger which transcends all others, is the vast wealth owned or controlled by a few persons. Money is power. In Congress, in state legislatures, in city councils, in the courts, in the political conventions, in the press, in the pulpit, in the circles of the educated and the talented,

its influence is growing greater and greater. Excessive wealth in the hands of the few means extreme poverty, ignorance, vice, and wretchedness as the lot of the many.[4]

Hayes saw clearly the evils of big money and political power in combination. But the situation was destined to become much worse over the coming decades. Today it is a national and world catastrophe, fueled by the inequities inherent in the financial system.

The Fate of Silver

The most important political issue of this period continued to be imposition of the gold standard as backing for U.S. currency, a measure promoted most strongly by the wealthy bankers and financiers, especially J.P. Morgan and the up-and-coming New York banker Jacob Schiff in the U.S., along with the Rothschilds, Barings, and other British and European financial magnates abroad.

By the late 19th century, investment by the Rothschilds in the economies of the U.S. and in Britain's "white colonies" of Canada, Australia, and New Zealand was ballooning. Investments were also underway to support British financial interests in South Africa, particularly in diamond and gold mining. It was the gold standard that held back inflation and made these investments fabulously profitable.

A large portion of the Rothschilds' investments was in minerals—not just gold and silver, but also mercury, copper, and lead. The Rothschilds invested in gold mining in California and Mexico. In 1886, they consolidated their mining ventures in what they called the Exploration Company. By the 1890s, they began to focus on South Africa after making a fortune investing in diamonds with British entrepreneur Cecil Rhodes.

After 1895 the Rothschilds' Exploration Company was the main source of finance for the Anaconda Mining Company in Montana, which continued operations until 1982. Anaconda also ran a substantial logging operation in Montana to produce timber for their mines and railroads. My grandfather Carlton "Bill" Peilow worked for Anaconda Lumber in the area around Seeley Lake, Montana. More on life around Seeley Lake later.

The controversies surrounding monetary policy in the U.S. did not go away. As mentioned previously, the Greenback Party was active between 1874 and 1889 and ran candidates for president in 1876, 1880, and 1884. But it is extremely difficult for a third party to sustain momentum, and the Greenback Party, like third parties before and after, faded away.

Still, public confidence in the Greenbacks remained. The Specie Payment Resumption Act of 1875 required that Treasury redeem outstanding Greenbacks in gold, thus retiring them from circulation and restoring a single, gold-backed bank-issued currency. Though the Treasury stockpiled gold in preparation for the public to present their Greenbacks for exchange, few people did. Only $130,000 of the outstanding $346,000,000 in Greenbacks was actually redeemed, with Greenbacks remaining in circulation into the 20th century. Thus Lincoln's "people's money" remained an economic force.

What did not fade away was agitation for silver coinage after the "Crime of 1873" discontinued production of the U.S. silver dollar. Price deflation continued to the point where farmers and debtors alike began to advocate "free silver." This would mean unlimited coinage of the U.S. silver dollar as legal tender, hearkening back to the Coinage Act of 1837.

In 1878, Congress passed the Bland-Allison Act over President Hayes's veto, which restored silver as legal tender and directed the Secretary of the Treasury to purchase silver bullion at its market price in the amount of two to four million dollars monthly and to coin the bullion into U.S. silver dollars.[5] The measure did increase the money supply, and the abundant quantity of silver being mined depressed the price paid for silver bullion by the U.S. Mints.

The Sherman Silver Purchase Act of 1890 introduced a new feature, whereby Treasury would pay for silver with new legal-tender Treasury notes. While the action further increased the amount of silver in circulation, it was still far from the demands of free silver advocates, including politicians from the western and southern states, who now begun to shift their support from the Republican to the Democratic Party.

When another financial panic struck in 1893, the bankers blamed the Sherman Act, and Democratic President Grover Cleveland convened a special session of Congress to demand its repeal. Congress did in fact repeal the silver-purchase and note-issuance provisions of the Sherman Act, although the legal-tender status of silver coins and Treasury notes remained. While he was the first Democrat elected to the presidency since the Civil War, Cleveland remained a "hard money" man.

So, silver did not disappear from the monetary system, while advocates continued to agitate for unlimited silver. In 1896, William Jennings Bryan of "Cross of Gold" fame won the Democratic Party presidential nomination, with free silver the main plank of his platform. However, Bryan's defeat, along with the increasing quantity of gold coming onto the market with discoveries in South Africa and the Yukon, combined with the new cyanide process of gold refining, put an end to silver agitation as a political threat.

Gold now reigned supreme, to the delight of international finance and Great Britain, whose government the Rothschilds and other financiers controlled, with South African gold destined to constitute half the world's production. The "real" king of England during the late 19th and early 20th centuries can be said to have been Nathaniel "Natty" Rothschild, chief lender to Cecil Rhodes and to successive British monarchs.

The Central Financial Problem

The problem of money has been omnipresent throughout the entire history of the U.S., going back to colonial days. No government has ever solved the need for a reliable circulating medium of exchange that was neither inflationary nor deflationary and fair in its issuance and availability to all social and economic classes.

The search for a sound monetary system never included a call for "income equality." Most agreed that income distribution could be left to market forces if the monetary system provided a level playing field. Unfortunately, a monetary system where circulation is controlled by private banks that engage in fractional reserve lending at usurious rates of compound interest makes a level playing field completely impossible.

The tilt causes bank-generated money to roll off the table into the bankers' laps. Not only does fractional reserve banking create in general the most wealth for lenders at the expense of everyone else; it is also vulnerable to specific instances of favoritism and abuse. All this makes the bankers the most powerful political group in the country and has done so to this day.

The U.S., Britain, and France in particular are completely dominated by the financial class in every aspect of political and economic life, which is one reason why by the 1980s and 1990s, post the collapse of the USSR, the "welfare state," also called the "social safety net" began to be annihilated.[6]

During the late 19th century, with the huge cutback in federal government spending after the end of the Civil War, the government ran a budgetary surplus each year from 1866 to 1893. But as we have seen, in order for the growing national banking system to be allowed to issue currency through lending, as specified by the National Bank Act of 1863, the banks were required to purchase U.S. Treasury bonds as reserves.

Why would the government sell bonds if their revenues, mainly from customs duties, exceeded expenditures? Heaven forbid that the government would spend any substantial sums on relief for the Indian tribes the Army was destroying, for assistance in education, or in building decent living and working conditions for the millions of formerly enslaved blacks living in destitution in the South, or for transportation or other types of infrastructure that was left to the free market or to the states and localities to finance. Or maybe to pay off the $5 billion in outstanding debt carried forward from the war years.

Creative financing, however, was not part of the ideology of either party. From Grant onwards, the intent was to slash expenditures to the minimum. Not until the election of Grover Cleveland in 1884, the first Democrat to occupy the presidency since Andrew Johnson succeeded Lincoln in 1865, did that philosophy begin to change to a small degree.

Some of the debt was in fact paid off, but not all. Instead, the debt was rolled over by new bond issues purchased from the government by the banks, then sold at a profit to their customers. But compared to countries like Britain, France, Spain, and the principalities of Germany and Italy, the budget of the U.S. government remained relatively small. This was very irksome to the bankers who wanted more action and who got rich in Europe from financing the activities of always impecunious nation states.

The leading financiers of government borrowing throughout Great Britain and the Continent were the Rothschild banks and their affiliated investment firms like the Exploration Company. The most reliable way for governments to run up debt was through wars, as had happened to the U.S. during the Civil War. Lincoln's government had escaped the worst of the debt trap by issuing Greenbacks and through the small-dollar bonds sold by Jay Cooke. None of the European powers ever implemented such provisions.

The governments of Europe were never so bold or clever as was Lincoln's. The only nation that tried to avoid the bankers' control was Russia. So it's easy to see how the bankers, by the end of the 19th century, became the world's biggest cheerleaders for war.

The linkage between the banking system and war was out in the open for all to see, but few were able to oppose this evil combination. In fact, the system whereby investors get rich off everyone else's misery continues today, not only through war, but from many other types of systemic abuse, including industrial pollution, speculation using retirement funds, leveraged buyouts, and the proliferation of hedge funds and most strikingly, derivatives. Lending on margin for stock trading fed the speculative addiction then as it did before the Great Depression and still does today.

From the bankers' point of view, it doesn't matter what they lend money for, so long as they are repaid with compound interest. Today, if society can't pay, the banks are bailed out by the government through even more public debt. Hello, moral hazard.

Central Events Towards the End of the 19th Century

The following men were elected president through the remainder of the 19th century:

James Garfield, Republican, 1880, succeeded by Vice President Chester Alan Arthur, Republican, following Garfield's death by assassination in 1881. Grover Cleveland, Democrat, 1884. Benjamin Harrison, Republican, 1888. Again Grover Cleveland, Democrat, 1892. William McKinley, Republican, 1896, 1900, assassinated in 1901.

The U.S. Army continued its wars against Native Americans, pushing the remaining tribes onto reservations in the West, and continued to promote slaughter of the buffalo to further the Indian genocide.

The U.S. government now began to create an overseas empire, including the annexation of Hawaii after a coup against Queen Liliuokalani, war against Spain, the acquisition of the Philippines as a Pacific colony, the permanent acquisition of Guam and Puerto Rico as territorial possessions, and the seizure of Cuba as a long-term protectorate, lasting until the Cuban Revolution of 1960.

The U.S. also engaged in machinations in Central America, leading to sponsorship of Panamanian independence from Columbia, the building of the Panama Canal, and furthering of U.S. interests via commercial relations and treaty-brokering among the nations of Latin America.

Industrial Development and Financial Centralization

The period saw the creation of the modern petroleum industry, starting with the founding of the Standard Oil Company in 1870 by John D. Rockefeller, still considered the richest man in history. Rockefeller gained control by purchasing oil refineries and making bargains with railroads for transport of oil based on economies of scale. The creation of reliable supplies of clean gasoline enabled the automobile industry to take off under magnates like Henry Ford.

The Standard Oil Company was ordered broken up as an illegal monopoly by the Supreme Court in 1911. Nevertheless, the Rockefeller family retained controlling interests in the companies that spun off from the original Standard Oil giant. The Rockefellers stood at the center of the U.S. oligarchy of financial interests that have run the country to this day.

The harnessing of electricity and electrification of the U.S. economy through the work of Thomas Edison, Nikola Tesla, George Westinghouse, and many others was equally important. A key event was the adoption of alternating current as the main commercial application of electricity generation, pioneered by Westinghouse Electric, founded in 1886.

General Electric came into being in 1892 through financial support provided to Edison by J.P. Morgan and the Vanderbilt interests. It was formed through the 1892 merger of Edison General Electric Company and Thomson-Houston Electric. J.P. Morgan also controlled U.S. Steel, International Harvester, numerous railroads, and was the key figure in the Banking Trust that would lay the groundwork for the Federal Reserve System in 1913.

Another financial panic took place with the depression of 1893 and persisted until the U.S. economy was reenergized by new federal deficits. These resulted from the buildup of the Navy and from the McKinley Tariff of 1890, passed while William McKinley was a congressman, raising tariff rates by 50 percent.

The tariff was infuriating to the British but popular among U.S. manufacturers and workers.

The Panic of 1893 was set off by the collapse of two of the country's largest employers, the Philadelphia and Reading Railroad and the National Cordage Company. With their failure, a panic erupted on the stock market, businesses closed, and some states suffered over forty percent unemployment.

Congress reacted to the crisis by repealing the Sherman Silver Purchase Act, causing a steep decline in the value of silver. Runs on bond redemption caused the U.S. government's gold stocks to run dangerously low. J.P. Morgan contacted President Grover Cleveland to bail the government out.

But it was the Rothschild's agent, New York banker August Belmont, who had first suggested to Morgan that he make the offer. Nathaniel Rothschild led the Europeans in purchasing the U.S. government's new bond issue. The government thus borrowed $65 million in gold from the Morgan/Rothschild syndicate. The Panic of 1893 also caught William C. Durant, the founder of General Motors, at a vulnerable moment. He had used bank loans to found the company, but with the banks collapsing he was forced to sell out to the Rothschilds and DuPonts, though he came back later to successfully market the Chevrolet automobile.

Today the government bails out the banks by adding to the national debt. Back then, the banks bailed out the government. The government could sell bonds to raise cash, but with the repeal of income tax and no chance of returning to the issuance of Greenbacks, the government was helpless if the market for its bonds collapsed. To cover the risk that Morgan and Rothschild ran in the face of the government's bad credit, they demanded and received a hefty premium for their services.

By 1894 the days of the federal surplus were over. Deficit financing was back. The cause was the depression resulting from the Panic of 1893 and the sharp decline in customs revenues resulting from reduced economic activity. It would be six years before federal revenues returned to previous levels.

In 1899, the government's deficit reached its highest level since the Civil War. The U.S. had opted to pursue a war against Spain, entailing the building of a new navy. War expenditures doubled from 1897 to 1898, then doubled again from 1898 to 1899. The writing was on the wall. The income tax had to be restored, a measure strongly supported by agrarian interests in the South and West and by urban industrial workers. Needless to say, big money opposed it.

Economist David A. Wells published an article in 1880 entitled, "The Communism of a Discriminating Income Tax." Most of the press opposed a new income tax, including the *New York Times* and *New-York Tribune.* One congressman said, the tax "would take from the wealth of the thrifty and enterprising and give to the shiftless and the sluggard."[7]

People began to take to the streets in protest, like the populist Army of the Commonwealth in Christ led by Jacob Coxey, a businessman from Ohio who, with 6,000 followers, marched on Washington, DC, in 1894. The protest leaders were arrested for walking on the grass at the Capitol building.[8]

The income tax that Congress finally passed was a flat two percent on most income over $4,000 but was declared unconstitutional in 1895 by a Supreme Court vote of five to four. When the Spanish-American War came along, Congress increased federal excise taxes and authorized the coinage of 1.5 million silver dollars. But when the federal deficit skyrocketed, the government had to go hat-in-hand to the rich in order to buy the bonds to pay for the war.

The growth of organized crime followed on the heavy influx of refugees into American cities, especially New York. Various criminal gangs roamed the Wild West, while gambling and bootlegging went on in all American cities. The bootlegging was an outgrowth of early federal excise taxes on liquor, where the tax sometimes exceeded the cost of making the product.

Crime accompanied the award of government contracts at all levels, especially on the part of suppliers during wartime. Accusations of profiteering by Indian agents were frequent. But all this pales beside the way in which vested economic interests were able to lobby Congress for favorable treatment, pressuring it to keep taxes down and gold backing for bond sales in place.

By the time of William McKinley's election for a second term as president in 1900, there were many ominous trends. Low farm prices kept farmers near bankruptcy. The big banks, both domestic and international, ruled supreme. Federal deficits had returned. Industrialization had created a vast army of poorly paid workers, including millions in crowded urban slums. Income disparity worsened.

The U.S. now had a growing Navy and major overseas commitments. The U.S. received heavy British investments and substantial spending on American products for the Boer War. Meanwhile, Britain, France, Germany, and Italy had been engaged in carving up Africa and confiscating its vast natural resources, and Germany had launched a naval arms race against Britain. Strife among nations for control of Middle Eastern oil resources had begun. By the end of the 19th century, war in Europe loomed. Where the U.S. would fit in with a system of competing European nations growing more powerful militarily by the day, no one yet knew.

Assassinations—James Garfield

Like Lincoln, President James Garfield (1831–1881) was no friend of the U.S. financiers with their strong British connections. As the Miller Center at the University of Virginia noted:

> Garfield was able to put his financial expertise, which was acquired through his congressional committee experience, to work by recalling government bonds that were paying six percent interest. The Treasury was able to refinance them at 3.5 percent, which saved $10 million annually—about four percent of the overall budget at that time.[9]

This was $10 million per year that the banks would not receive, a large sum at the time. U.S. and British banks were furious. Garfield and his Secretary of State James G. Blaine were also raising alarms about British machinations in the internal politics of nations in South America, putting the Monroe Doctrine at risk.

Charles Guiteau, Garfield's assassin, later hanged, is usually characterized as a "disappointed office seeker." In fact, he had been stalking Garfield for weeks. While he awaited execution, Guiteau dictated his autobiography to a jail officer. He said he bought the pistol that he used in the assassination from "a gentleman."

The Spanish-American War

The U.S. now began creation of a modern Navy, with sufficient warships to maintain a worldwide U.S. presence, mainly in the Pacific, though not yet formidable enough to compete with Britain or France. During President Grover Cleveland's first term, four steel-hulled warships were built.

The U.S. launched the Spanish-American War in 1898 during the McKinley administration. The war lasted only three months and ended in complete U.S. victory over Spain. "Yellow Journalism" worked overtime to supply anti-Spanish propaganda for public and political support. The faked explosion of the *Maine* in Havana Harbor provided the fuse. Accepted today as a major false-flag provocation, the "Remember the *Maine*" incident effectively incited the war.

The Spanish quickly sued for peace after their naval squadrons were sunk in battles off Cuba and in Manila Bay. The new U.S. Navy sailed triumphant. Future President Theodore Roosevelt, as assistant secretary of the navy, helped promote the nation's warships for use in the battle, though when the war began he resigned to become an officer in the Rough Riders cavalry unit, where he won fame in a charge up Cuba's San Juan Hill that cost 200 U.S. soldiers killed and 1,000 wounded. Praising "that splendid little war," he became instantly famous. He was nominated for the Congressional Medal of Honor, but the award was blocked by officers who accused him of headline-grabbing.

After the Spanish were driven out of the Philippines, a Philippine Republic was declared, led by Emilio Aguinaldo. In response, the U.S. annexed the Philippines after the Treaty of Paris that ended the war. The U.S. war against the Philippine Republic lasted from February 4, 1899 to July 2, 1902. After capturing Aguinaldo, the U.S. declared victory.

A colorful figure in Filipino history, Aguinaldo lived to the age of 94. The war between the U.S. and the Philippine Republic resulted in up to a million Filipino civilian deaths, mostly due to famine and disease. In retaliation for Filipino guerrilla warfare tactics, the U.S. carried out reprisals, forcibly imprisoning many civilians in concentration camps where thousands died.

An account of this war may be found in *How to Hide an Empire* by Daniel Immerwahr. Theodore Roosevelt inherited the Philippine War after President William McKinley was assassinated in 1901. Portraying the fight as one against "savages," Roosevelt wrote, "no pity is shown to non-combatants, where the weak are harried without ruth, and the vanquished maltreated with merciless ferocity." To him it was "the most ultimately righteous of all wars."[10] It was also among the most egregious and brutal assaults on human decency ever perpetrated by the "land of the free," even exceeding Operation Phoenix during the Vietnam War or the 2004 destruction of Fallujah in Iraq.

Flexing its muscles elsewhere in the Pacific, the U.S. sent two warships and almost 4,000 men to help an eight-nation force crush the 1900 Chinese Boxer Rebellion. The Rebellion was a delayed reaction to Western victories over the Chinese in the Opium Wars that caused millions of Chinese to become addicted to Western-supplied opium.

During the Rebellion, the Westerners committed numerous atrocities, executed Chinese officials who had the temerity to oppose European and Japanese trade dominance, and laid the groundwork for eventual revolt by China against the West in its future communist revolution. Still today, China has not forgotten its humiliation at Western hands.

Assassinations—William McKinley

McKinley was elected president in 1896 on a platform of high wages and increased tariffs against British and other imports. The purpose, said McKinley, was:

> to preserve the home market…to our own producers; to revive and increase manufactures; to relieve and encourage agriculture…to aid and develop mining and building; and to render to labor in every field of useful occupation the liberal wages and adequate rewards to which skill and industry are justly entitled.[11]

According to McKinley, the 1890 Tariff Act that he sponsored while serving in the House of Representatives, "…gave work and wages to all such as they had never had before. It did it by establishing great industries in this country…. It had no friends in Europe."[12]

The only issue for McKinley and the Republican Party leading up to his 1900 run for reelection was who should replace Vice President Garret Hobart, who died in 1899. When the name of Theodore Roosevelt, former assistant secretary of the navy and "hero" of San Juan Hill came up, McKinley and his leading advisor, Senator Mark Hanna, were strongly opposed. But they yielded under pressure.

McKinley was shot by "anarchist" Leon Czolgosz on September 14, 1901. Czolgosz was a disciple of Emma Goldman, a lecturer and professional agitator whose political headquarters was the Henry Street Settlement House in New York, built in 1893 by Wall Street banker Jacob Schiff, who later paid for Leon Trotsky's transport to Russia to help lead the Bolshevik Revolution. Goldman was arrested on suspicion of complicity in McKinley's murder but later released. Schiff's partner, Sir Ernst Cassell, was personal banker to the British royal family and to the Fabian Society in London, where Emma Goldman had a second headquarters. Belgium's King Leopold II had said, "In England a sort of menagerie is kept to let loose occasionally on the continent to render its quiet and prosperity impossible."

Theodore Roosevelt Ends U.S. Foreign Policy Independence

Whether or not this "menagerie" was also "let loose" on the U.S. through assassination of its presidents, the resultant accession of Theodore Roosevelt to the presidency marked a 180-degree turn in U.S. foreign policy as it concerned Great Britain. The U.S.-British alliance that culminated in World Wars I and II began to be forged with the presidential administration of Theodore Roosevelt.

In a May 1992 doctoral dissertation at Brown University, William Neal Tilchin, in *Theodore Roosevelt and the British Empire, 1901–1907* summarized the historical consensus on Roosevelt's pro-British presidency:

> Where Great Britain and its empire are concerned, additional areas of agreement are apparent. Historians generally realize that Roosevelt believed absolutely in the doctrine of peace through strength: the "righteous" nations should always be well-armed and should take particular care to build up and preserve a preponderance of naval power in order to be able to deter aggression and to defend their interests. *There is broad acceptance of the notion that T.R. saw the United States and Great Britain as the two most righteous nations.* (emphasis added) Correspondingly, it is usually recognized that Roosevelt considered Britain an essential friend for America, and that he cultivated and solidified the Anglo-American rapprochement.... Howard K. Beale labels the developing bond between the United States and the British Empire under T.R.'s presidency as "the foundation of Roosevelt's foreign policy."... He argues at one point that T.R.'s "sense of a common task of Britain and America in ruling 'colonial peoples' with the ultimate purpose of civilizing them...provided the basis of Roosevelt's policy concerning Britain."... David H. Burton concurs. Before, during, and after his presidency, Roosevelt believed that "the most important single consideration about British imperialism is that it had advanced the welfare of mankind."... Raymond A. Esthus... In thinking about Roosevelt

and the Anglo-German rivalry, argues Esthus, one must recognize from the outset a "fundamental consideration that dominated his attitudes and policies in world politics: his conviction that Britain was a friend and Germany was a potential enemy."[13]

The era of an independent U.S. foreign policy was now over, gone for good. It had begun with President George Washington's warning against permanent alliances and Thomas Jefferson's admonition against "entanglements" with European powers. The Monroe Doctrine had also pledged the U.S. to staying out of internal European politics in exchange for no new colonization of the Americas.

Keeping the U.S. free of British/New York banking interests was affirmed by President Andrew Jackson in his war against the Second Bank of the United States, continued by President Abraham Lincoln in his self-generated funding of the Civil War, and furthered by the protective tariffs and budgetary restraint of the presidential administrations of Lincoln's successors lasting until that of William McKinley. Now, with Theodore Roosevelt in charge, nothing stood in the way of total control by the Anglo-American financial elite, followed, as night follows day, by over a century of world war that has not yet ended. Today's U.S. proxy war against Russia in Ukraine is a link in its chain that still binds humanity.

John C. Hill, the Land Run of 1892, and the Oklahoma Indians

John Clark Hill, my great-grandfather on my father's side, was born in Peoria County, Illinois, on May 19, 1866, and passed away in Oklahoma City, Oklahoma, in 1950 at the age of 84. His father was Clark Hill and his mother, Betsy Bliss, both born in Niagara County, New York. She was a direct descendent of Thomas Bliss, my Puritan blacksmith ancestor, who arrived in Massachusetts from England in 1636.

We possess two interviews with J.C. Hill about his life story. One is a biography by a field worker for the Works Progress Administration. In this section I will be quoting from these interviews. My great-grandfather's life story presents a vivid picture of life on the Western frontier toward the end of the 19th century. I consider him a true representative of an American type for the purposes of this book.

Throughout the West, land that had been "reserved" to the Indian tribes was now being divided up. The Indians were given "allotments" of 160 acres for heads of households and 80 acres for individuals. The government's intention was for these plots of land to become family farms. This was part of an attempt, cheered on by reformers, to help the Indians "assimilate" into American life. But in many cases, allotment was a way to destroy the reservations. Indian land was viewed as valuable for mining, timbering, and grazing.

The problem of allotment was most acute in Oklahoma, which had been established by Congress as Indian Territory, with a number of large reservations being formed. This was where the Indians who had been expelled from the Southeastern U.S. were sent, including the Cherokees, along with Shawnees from the Ohio Valley, Comanches from Texas, and tribes such as the Arapahos from the Plains.

The government now began to sell off the "surplus" reservation land that had not been allotted to individual Indians. This was how the famous "land rushes" or "land runs" in Oklahoma came about. The government set dates for the land to be opened, and anyone who showed up, including foreigners, could dash across a line at the sound of a bugle and plant a flag on their claim. Prior to this, the government had staked out parcels.

There were five land runs, plus a land lottery and land auction. When it was over, President Theodore Roosevelt on November 16, 1907, proclaimed Oklahoma the 46th state. Congress had passed a law to merge the former Indian and Oklahoma territories into a single legal entity. The land runs took place in 1889, 1891, 1892, 1893, and 1895. The land was to be settled under the provisions of the 1862 Homesteading Act, with parcels available to be bought and sold.

John C. Hill rode in the third land run on April 19, 1892, on what had been 4.3 million acres of Cheyenne and Arapaho land in western Oklahoma. The land was largely uninhabited. J.C. was living at the time in Kingfisher, north of Oklahoma City. He heard about the land run taking place to the west, and borrowed a horse. Both Oklahoma City and Kingfisher had been prairie until occupied in the land run of April 22, 1889. I'll allow John C. Hill to tell his own story, drawing from the two first-person accounts. He starts by telling about his family that had lived in Niagara County, New York, before coming to Illinois.

> My grandfather, John Hill, had a nice country home in Niagara County.... He fought in the War of 1812. He received a land grant for his services in the army and selected 160 acres near Princeville, Illinois. This farm has remained in the family since the original patent was issued. My father came to Illinois with my grandfather when seven years old. They came by boat through the canal to Chicago, which was then only a fort and was called Fort Clark.

J.C. lived in Illinois until he was eighteen. He then went west to Sidney, Iowa, which was 350 miles away, to learn the printer's trade.

> I stayed there two years and then got a job with the *Caldwell News* at Caldwell, Kansas. Here I met several Oklahoma cattlemen, as Caldwell was a sort of headquarters for them. Major Dunn, Ras Williams, Colonel Cragin, and John A. Blair all belonged to the Cherokee Strip Livestock Association, and the *Caldwell News* did all the printing for this association. John A. Blair was the secretary of this association which had leased all of the Cherokee Strip from Chief Bushyhead, who was chief of the Cherokees, at 1-1/2 cents per acre. Every year the chief would come into Caldwell driving a team of horses to a buckboard and would carry all the money in gold back in his buckboard.

He continues about life in Caldwell, a cowboy town where herding cattle was the staple business:

> Each cattleman had a brand, which was registered and advertised. Ras Williams and his cowboys drove a big herd of cattle over the old Cantonement Trail to Pond Creek, where they always rested several days and grazed their cattle before driving them into Caldwell. One time there were sixty [train] carloads to be shipped to Chicago.
>
> There were big feeding pens at Eldon, Iowa, where they unloaded the cattle, fed and watered them, and rested a couple days. Then they re-loaded and shipped them on into Chicago. Mr. Williams took me on this trip. Several men were allowed to go along free to look after the cattle. The loading chutes at Eldon were almost too narrow for the long horns, and we had to turn the heads of some of them sideways to get them by. This was my first experience in loading and unloading cattle, but after that I made several trips to Chicago with different cattlemen.

The Cherokee Strip Livestock Association was famous in its day for running the last of the open range in the region for grazing cattle. The Cherokee Strip in southern Kansas was 60 miles wide and contained 6.5 million acres. Adjacent to it was the Cherokee Outlet in Oklahoma. The federal government wanted the Cherokee Outlet for white settlement, so purchased it from the Indians in 1891. It was opened for settlement in the 1892 land run.

J.C. writes:

J.W. McCloud wrote the original treaty for opening the Cherokee Outlet. I printed enough copies of the original treaty for each senator to have one. This treaty was finally adopted, and Mr. McCloud wrote it with few changes.

Only twenty-three years old, J.C. writes:

I came to Kingfisher [Oklahoma] in May 1889 and went into the newspaper business there. I called my newspaper the *Kingfisher Journal.* I had a man of the name of Sims to help me. I edited the *Kingfisher Journal* for a year and then sold it to Captain Admire, who changed the name of the paper to the *Free Press.*

Then he took part in the 1892 land run. About 10,000 people participated, including someone in a hot air balloon and others with buckboards.

When the Cheyenne and Arapaho Opening was staged on April 19, 1892, I went to the livery barn [in Shawnee] and hired an old gray cow pony for which I agreed to pay $25 per day. This cow pony was scared to death at the noise around him, but when the guns were fired and I got him started, he ran like a race horse. I finally got him stopped and stuck a stake in as quickly as I could. I took the bridle off the pony and let him graze while I walked about admiring my claim. In about twenty minutes, Bob Lyle, an old friend of mine, came along. His horse had given out, so I told him to take my horse and to be sure and run a mile before he stopped and stick his stake in, and in about thirty minutes Bob returned for his horse.

Before long, J.C. Hill and Edna Belle Hubbard, his sweetheart from Kingfisher, were married. He says it was the first wedding in town. Her father, Colonel Hubbard, owned the only hotel, the Hubbard Hotel, and Edna's mother and "some of the ladies from the hotel" rode out to witness the race. They rode in a wagon called a hack which was pulled by a single horse. At one point when the horse was pulling the hack uphill, it stalled, rolled backwards into a creek, and tipped over. No one was injured, "only slight bruises," and the party treated the incident as a joke.

J.C. kept the land for fifteen years. He says:

I proved up my claim and sold it in 1907 for $3,500. At the opening of the Enid townsite I secured a business lot in the center of the east side of the square at Enid, and I sold this lot the next day for $200. My next venture was in the grocery business in Shawnee [a town southeast of Oklahoma City], and I stayed in this business for ten years. During that time the Shawnee Indians traded with me, and I learned their language well. I have extended credit to the older ones to the amount of $1000 and never lost a penny. Chief Three Fingers was one of my best customers. I have visited and eaten in his home many times. He was educated and very intelligent. Many of these Shawnee Indians went to Haskell Institute and obtained good educations, not only in books but in music and art.

J.C. continues:

An amusing incident happened one day. A smart aleck of a young traveling man came to Shawnee. It was lunchtime, and we were all in the restaurant. One of our smart, talented, full-blood Cheyenne girls had finished her lunch and had walked over to the piano. This girl's name was Nell. Mr. Smart-Aleck walked over to the piano and said, "Sit down and play." Nell was looking at some music, and she didn't look up but just kept her eyes on the music. The young man kept motioning with his hands up and down the keys, but Nell didn't say a word. Finally he walked off and said, "That dumb Indian couldn't play Chopsticks." In a few minutes, Nell sat down and played Chopsticks, then started playing classical music, one piece after another for about thirty minutes. I never saw a more surprised look on anyone's face than there was on his. He walked over to me and said, "Could you beat that?"

J.C. says of the Shawnees: "The Shawnees were the last tribe of Indians subdued, and I found them honest and sincere." He adds: "I have attended many of the Indian ceremonial dances. I knew them so well that they invited me to many of them."

He also speaks of Indian religious meetings conducted by white preachers:

The protracted meetings were interesting. They were held for the Indians. A white man would preach, and three or four clans of Indians with their interpreters would be in attendance. The Kiowas would be there with their interpreter, and the Cheyennes with their interpreter, and the Arapahos with theirs, and as the minister would preach, he would stop and all of the interpreters would translate to their tribe at the same time.

J.C. told how his father-in-law became mayor of Kingfisher:

During the early times the town folks decided to select a mayor by his looks, so they got a big box and put different men, all strangers, upon the box for the crowd to say "no" or "yes." The crowd turned three or four men down until finally they put my father-in-law, Colonel G.E. Hubbard, a big fine-looking man, on the box, and the crowd hollered, "He'll do!" That was the way Colonel Hubbard was elected mayor.

J.C. told many other stories that available space prevents repeating. He told of hunting trips around Oklahoma, disputes he'd witnessed over land claims, and a story about once riding home from a trip when a panther attacked his horse, inflicting such pain that the horse ran full-speed all the way back to town. On arriving at the stable, he discovered that the horse had deep claw marks on its hind-quarters.

He also told how the Indians would pass on their allotments to their heirs, who would then sell them to the whites, thereby losing their stake in the land. J.C. didn't think much of the government's allotment policies, though he had taken advantage of them in the 1892 land run.

J.C. lived to a ripe old age and witnessed many changes in the world. Later in life he supported his family through various occupations and by trading in oil leases. In his old age he was grateful to FDR for introducing Social Security. He died when I was three years old.

St. Ignatius Mission, the Flathead Tribes and Allotment

We are fortunate that so much documentation of Flathead tribal history is available, including books published by today's tribal cultural committees, the Salish and Kootenai College, and other sources.

Members of the western American Indian tribes intermingled by moving about, intermarrying, leaving the area, then returning home. The makeup of the Indian population was fluid . But often in the history of the Flathead tribes the name McDonald comes up. This is due to the early presence of Angus

McDonald (1816–1889), a Highland Scot descended from the MacDonalds [sic] of Glencoe, Scotland, who emigrated to Canada in 1839 and worked for the Hudson's Bay Company. Angus became the factor, or lead trader, at a trading post in Colville, now in Washington state.

Angus married an Indian woman as was the practice among white traders. Catherine Baptiste was a Nez Perce whose homeland was in Idaho. They were married in both civil and Catholic ceremonies and had twelve children. Duncan (1849–1937), his eldest and most well-known offspring, at the age of eighteen took responsibility for running the Hudson's Bay Company store at Fort Connah, Montana. This is on the Flathead Reservation, just north of the Jesuit mission at St. Ignatius.

Many descendants of the McDonald clan live on the Flathead Reservation today and are tribal members and leaders. There is a display of artifacts belonging to Angus McDonald in the Ninepipes Museum near Charlo. Duncan McDonald's Fort Connah trading post continued to be overseen by the Hudson's Bay Company until 1872. It was the last of the Hudson's Bay trading posts within the boundaries of the U.S.

After the Hudson's Bay Company departed, Duncan ran his own store further south on the reservation. He also built a hotel and restaurant employing Chinese waiters near Ravalli where the Northern Pacific Railroad came through. Today on U.S. 93, a place called the Four Winds Trading Post has a log building which it says was Duncan McDonald's restaurant that was later transported to that location.

After working the store for a number of years, Duncan sold his businesses, which included a ferry on Flathead Lake, and became a successful rancher. His father Angus had formerly maintained large herds of cattle and horses and is said to have been the founder of the livestock industry in the region.

Duncan also wrote articles on Indian life and lore for newspapers in Montana and authored a detailed history of the Nez Perce War of 1877. Duncan traveled to Canada to interview Indian survivors of that tragedy. This was the last instance of violence between the Indians and the U.S. Army in western Montana.

Duncan's home at Fort Connah was near the Jesuit mission at St. Ignatius, where the Ursuline Sisters had come from Montreal to operate a girls' boarding school, and where the priests ran a school for the Indian boys. The mission also served as a community center where meetings and festivals were held, the biggest one being the feast at Christmas. Duncan was Catholic, though he didn't attend many services, but his wife Louise Quil-soo-see, or Red Sleep, was devout, as were many other Indians.

The girls' boarding school run by the Ursuline nuns continued until the 1970s, with St. Ignatius today a Catholic parish serving the native and white populations of the Flathead Reservation.

The Jesuit fathers at the mission ran a large farming operation with help from the students and Indians in the area. Some of their farm products, including hogs and flour from the mill, were sold to Duncan McDonald to market at his store.

The mission also operated a printing press with the help of Indian students. The Jesuit priests studied the Indian languages and devised a system of transliteration which they used for printed translations of Catholic religious texts.

The period from the Civil War to the end of the 19th century saw the Flathead tribes becoming settled on the reservation created by the 1855 Hellgate Treaty, which amounted to over 1.25 million acres of valleys and hills, surrounded by the Mission Mountains to the east, Flathead Lake to the north, the Salish Mountains to the west, and the hills above the Clark Fork River to the south.

Though part of the arid region of the Rocky Mountain West, the reservation offers excellent prospects for livestock grazing on the prairies, gardening aided by irrigation, and extensive fir and pine forests on the mountain slopes. The reservation is a natural basin among the surrounding mountain ranges, with the Flathead River exiting the basin where it flows into the Clark Fork River near the town of Paradise.

The Indians were able to create a somewhat prosperous lifestyle as cattle ranchers as long as the open range was available for grazing. Other Indians worked for hire for the reservation agency or other local employers. Many Indians were cattlemen. Others had large horse herds.

The reservation was under the jurisdiction of an agent at Arlee headquarters, the town closest to Missoula to the south. The agent was in charge of all legal and enforcement matters pertaining to the tribes. The agent also leased tribal lands for timber harvesting and livestock grazing to non-tribal members. The most popular agent was Peter Ronan, with his wife Mary Ronan keeping house and entertaining visitors. Mary was renowned for her long beautiful brown hair.[14] Today's reservation town of Ronan was named for them.

Led by Duncan McDonald, the Indians were able to lobby the government to fire a previous agent they considered dishonest and abusive. Reservation politics were a constant preoccupation during the time the Indians were under the control of the Bureau of Indian Affairs. The Indians did not hesitate to complain about misbehavior by the agent or his employees.

Duncan McDonald himself was considered by some whites as a "trouble-making half-breed."

One problem that was never solved was that of whites bringing alcohol onto the reservation where it was banned by the government. The conflicts over funding between the St. Ignatius mission and the government was another. The mission school was supposed to receive money for partial support of its boarding students, but often the payments were not forthcoming. The city of Turin in Italy had "adopted" the mission and sometimes sent funds.

The first powwow held at Arlee in 1898 was a major event which then became an annual affair. Whites from the region were welcome, and both my parents went to the powwows when young. My dad made his own teepee and said he loved to listen to the old men chant.

The building of the Northern Pacific Railroad was particularly intrusive for life on the reservation. The government had surveyed a route that would pass through the southern part of the Jocko area of the reservation. A book about Duncan McDonald states:

> The federal government did not get around to negotiating with the Flathead Reservation tribes for the sale of the railroad right-of-way through the reservation until August 31, 1882.[15]

This was for a project that began in 1870 and that was supposed to be completed by the laying of the last track and the driving of a "golden spike" in 1883.

Duncan was asked by the government to host a big to-do to celebrate the completion of the Northern Pacific in September 1883. A train pulled into the Ravalli Station with Henry Villard, president of the Northern Pacific Railroad, U.S. Senator George F. Edmunds, Lord Norwood, representing British investors, and other dignitaries on board. An abusive, though unnamed, man from New York, angrily blamed Duncan when a wagon turned over and rolled the senator and the British lord in the mud. Duncan, who had prepared a feast of fried chicken and champagne, said, "It was the funniest sight I ever looked at." Senator Edmunds recalled:

> This is the greatest trip I have ever had. I never saw such scenery as this, and I never in my life saw anything half as funny as our spill in the mud. I haven't dared laugh before for fear of hurting the feelings of somebody. But I have got to laugh now.[16]

But implementation of the Dawes Act of 1877 was looming. The Indians in Oklahoma got their allotments and saw the rest of their land carved up for white homesteading in the late 1880s and early 1890s. Montana was more remote. The Flathead tribes retained their reservation land, open grazing of livestock, and their ability to freely hunt and roam in the mountains through the turn of the 20th century.

But all that would change. Congress passed the Flathead Allotment Act in 1904. Construction of the Flathead Indian Irrigation Project, using the Mission Mountains as the water source, was also authorized by Congress. Thousands of acres on the reservation were reserved for town sites, schools, and a National Bison Range. Individual tribal members were given allotments of either 80 or 160 acres of land per household. Indians said of each other, "He's been allotted."

Then in 1910, the unallotted tribal land was opened to white homesteading. Opening the reservation to whites was disastrous for Indian standards of living. Some left the reservation for West Coast cities. Others stayed to make the best of an increasingly dire economic situation with rural poverty setting in. Probably the worst consequence was the elimination of open grazing for tribal members' livestock.

But there were also heroic efforts by tribal members to work their way out of hardship, efforts which began to bear fruit as the 20th century proceeded. More on this later.

ENDNOTES FOR CHAPTER 7

1 It was these interests that were instrumental in electing Donald Trump as president of the U.S. in 2016.

2 Harry Barnard, *Rutherford Hayes and His America* (American Political Biography Press, 1994), 446–447.

3 One band of Nez Perce did in fact make it to Canada.

4 Charles Richard Williams (ed.), *The Diary and Letters of Rutherford B. Hayes* (1922), 354.

5 The silver dollars would be introduced into circulation through the banking system via U.S. Treasury payments on bond issues.

6 Today, this has devolved into what might be called a "war" on the working class and the poor. Compare today's slashing of food stamps and Medicaid while more than $100 billion is spent on the proxy war against Russia in Ukraine.

7 Studenski and Kroos, 223.

8 Ibid., 221.

9 Justus Doenecke, "James A. Garfield: Domestic Affairs," UVA Miller Center. https://millercenter.org/president/garfield/domestic-affairs>

10 Daniel Immerwahr, *How to Hide an Empire: A History of the Greater United States* (Farrar, Straus and Giroux, 2019), 101.

11 Ibid.

12 *Wikipedia.*

13 William Neal Tilchin, *Theodore Roosevelt and the British Empire, 1901–1907* (Brown University Doctoral Dissertation, May 1992).

14 Ellen Baumler, *Girl From the Gulches: The Story of Mary Ronan* (Montana Historical Society Press), 2003.

15 Robert Bigart and Joseph McDonald, *Duncan McDonald, Flathead Indian Reservation Leader and Cultural Broker, 1849–1937* (Salish Kootenai College Press, Pablo, Montana, 2006), 50.

16 Ibid., 185.

CHAPTER 8

"Rule Britannia"

Why Study the British Empire?

The significant role of Britain in any account of U.S. history is unavoidable, from the founding of the country right up to the present day.

The U.S. was formed from thirteen British colonies running down the Atlantic Coast of North America from what is today Maine—then part of Massachusetts—to Georgia. The U.S. fought two wars to secure its freedom from Britain: the Revolutionary War and the War of 1812.

In 1823, the U.S. promulgated the Monroe Doctrine, declaring the Western Hemisphere to be off-limits to further European colonization, which obviously included Britain. This included a pledge by the U.S. to stay out of European politics. *That pledge has been forgotten.* U.S. meddling in European politics has been a major component of 20th century history, due in large part to British pressure to come to its rescue in its wars against Germany. On the other hand, during the American Civil War, the Confederacy had tried but failed to gain Britain as an ally.

After the Civil War, the U.S. saw heavy financial investment by the British, with the most prominent names being the Barings and the Rothschilds, with August Belmont serving as the U.S. agent for the Rothschilds. Britain also successfully pressured the U.S. to adopt the gold standard, though silver as currency never entirely went away.

The U.S. continued to the end of the 19th century to pursue a foreign policy based on its own precedents and interests. Even then, the U.S. was potentially a rival power to Britain, but never to the extent of France, Spain, Italy, Denmark, the Netherlands, Russia, and particularly Germany.

In 1914, World War I broke out. The Central Powers of Germany, Austria-Hungary, and the Ottoman Empire faced off against the Triple Entente, an alliance of Great Britain, France, and Russia. Italy was lining up with Germany but switched sides. Britain was aided in the war by substantial numbers of troops from its Empire, especially from Australia, New Zealand, Canada, India, and some African colonies.

Britain and its allies came close to losing World War I, though Britain never admits this. They were rescued by the fact that in 1917, after three years of steadily eroding neutrality, the U.S. sent its army overseas. The extent to which the U.S. turned the tide in World War I war is debated to this day. But the fact remains that more than 140 years later, after revolting against Britain, the U.S. stepped in as its savior. It is crucial to understand how and why that happened.

How, then, to approach a topic as large and complex as the British Empire?

Looking at this history, one of my main sources will be an Englishman and the other a German. I have already cited Niall Ferguson, author of *Empire* and other books, including histories of the Rothschilds, as a source in earlier chapters of this book. On the German side, my source is Ludwig Dehio.

Less well known than some other German historians, Dehio was an archivist in the Secret State Archives of the Weimer Republic from 1922 to 1933. When the Nazis came to power in 1933, he left the government to become director of archives of the House of Hohenzollern, which was the German family of the deposed German Kaiser Wilhelm II, then living in exile in the Netherlands. Dehio was not a Nazi.

After World War II, Dehio emerged as editor-in-chief of the *Historische Zeitschrift,* a German historical journal, serving until he retired in 1956. Probably there was no person in post-war Germany with greater knowledge of European history. He published several books, including *Germany and World Politics in the Twentieth Century* and *The Precarious Balance: Four Centuries of the European Power Struggle*. It's the latter book, published in 1962, that is used here as a source, dealing as it does with the relationship from Elizabethan times through World War II between Great Britain and the Continental European nations.

Dehio considers the U.S. an integral part of this history. He concluded in his post World War II research that the wars among the nations of Europe over the past four centuries have destroyed the European system of states. The U.S. has been the beneficiary.

Origins of Britain

The origins of the British nation are lost in the mists of time. After the glaciers retreated at the end of the Ice Age, the British Isles took shape and became subject to successive waves of immigration from across the English Channel and the North Sea. There was a long period of habitation by the Celtic peoples who migrated from central Europe, followed by the Romans who treated Britannia as an agricultural province and military outpost. The word "Britannia" is derived from "Pretannia," used by the Greek historian Diodorus Siculus for the Pretani people, believed to live on the remote islands on the northwest fringe of Europe.

But what was called England came to be inhabited by West Germanic peoples: the Anglo-Saxons, who first arrived in 449 AD. As had the Celts before them, the Anglo-Saxons were subject to raids from the Vikings, another Germanic race. Then the Anglo-Saxons were conquered in 1066 by the French-speaking Normans, but these were also Germanic, having originated in Scandinavia before settling down in Northern France. The Normans spoke French and made the Anglo-Saxons a class of serfs while massacring the Anglo-Saxon nobility.

England has deep Germanic roots, and even today has a royal house whose forebears arrived from Germany in modern times. The real name of the House of Windsor is the House of Saxe-Coburg and Gotha. In conquering Wales, Scotland, and part of Ireland, or at least fighting the Irish to a standstill, in 1801 the British declared the existence of the United Kingdom of Great Britain and Northern Ireland. The UK would rule over much of the world for the next century.

Elizabeth I and Her Spymaster

The story of modern Britain begins with Henry VIII (1491–1547) who threw off the regime of Papal Catholicism. His Protestant daughter, Elizabeth, ascended the English throne, and kept her Catholic cousin, Mary Queen of Scots, locked up and facing execution. When the execution took place in 1587, Phillip II of Spain, ruler of the greatest empire in Europe, sent his Armada to invade England, with a view to killing its queen, and stamping out the Protestant heresy.

The defeat of the Spanish Armada in 1588 saved England. The English then decided once and for all that they could never allow themselves to be subject to invasion by another European power. This determination has been the centerpiece of Britain's foreign policy to this day.

At the risk of offending my British friends, I'll say this: The chip on the British shoulder has remained rather large. Charles Darwin (1809–1882) was the ideal prophet for a nation whose psyche was based on "eat or be eaten," "kill or be killed," and "might makes right." Or to a people whose confinement on an island with a relatively small population led to (a) continuous emigration and (b) the "science" of eugenics.[1]

England defeated the Spanish Armada with a fleet of highly maneuverable warships and privateers under Admiral Charles Howard, 1st Earl of Nottingham, and Sir Francis Drake, whose fleet was lurking in ambush at the mouth of the English Channel. Their deadly work of cutting the Spanish fleet to pieces was completed by violent storms that crushed the Spanish ships on the rocks as they circumnavigated the British Isles on their return home.

The British would not have been ready for the Spanish attack, were it not for an extensive network of spies, including double and triple agents reinforced by intrigue and assassination. The network was run by Elizabeth's principal secretary (an office that later became Secretary of State), Sir Francis Walsingham (1532–1590), known to history as Elizabeth's "spymaster."[2]

Son of a well-connected London lawyer, whose admission to the gentry had been purchased from the crown, Walsingham was a King's College, Cambridge scholar, a Gray's Inn lawyer, well-traveled on the Continent, and versed in ancient and modern languages. He was a Puritan, part of a Protestant *intelligentsia* that included his son-in-law, Sir Philip Sidney, the poet Edmund Spenser, and occultist John Dee, men with a mystical attachment to the destiny of England as a world power.

Dee wrote in his 1577 *The Brytish Monarchy*:

> A petty Navy Royall of three score tall ships or more, but in case fewer…seemeth to be almost a mathematical demonstration, next under the merciful and mighty protection of God, for a feasible policy to bring and preserve this victorious British monarchy in a marvellous security. Whereupon the revenue of the Crown of England and wealth public will wonderfully increase and flourish; and then…sea forces anew to be increased proportionately. And so Fame, Renown, Estimation, and Love, and Fear of this Brytish *Microcosmus* all the whole of the great world over will be speedily and surely be settled."[3]

John Dee is said to have been the first to use the term "British Empire." Fundamental to Dee's prescription for the attainment of "fame, renown, estimation, love, and fear" was, of course, the violence of armed warfare—"three score tall ships or more." Corollaries of armed warfare, all practiced then as now, are terrorism, assassination, aggression, assault, piracy, murder, and extortion. As an occultist and "conjurer" to the Queen, Dee claimed to receive his insights from the whispers of spirits. For, this he had a "spirit mirror" that he used with the aid of mediums.[4] John Dee was "a man possessed."

In combating Catholic plots to assassinate Elizabeth and place Mary, the daughter of James V of Scotland, on the throne, Elizabeth's principal adviser, William Cecil, 1st Baron Burleigh, with his main assistant, Walsingham, placed in motion England's first major covert action around

1559. This involved secret payments to anti-Catholic Scottish rebels, authoring anonymous propaganda pamphlets, leaking disinformation to political figures, bribing or kidnapping foreign diplomats, and running a web of spies throughout Europe, including within the Vatican. Spies would routinely be executed if discovered.

In 1570, Pope Pius V excommunicated Elizabeth and declared her deposed as queen. This made her and the land she ruled fair game. The last straw was Mary's execution based on secret messages to and from supporters that Walsingham had ferreted out. These letters supposedly called for Elizabeth's assassination. Aware that Phillip II and the Spanish laid plans against England, Walsingham sent raiders to harass Spanish shipping, attack ports and coastal cities, and gather information about Spanish naval strength. Months before the Armada sailed, Walsingham possessed documents that listed every Spanish vessel, their size and location, and their carrying capacity of armed men.

To know who makes the decisions in Britain, you have only to look at the Privy Council, which is the body of top officials reporting to the King or Queen. Every prime minister, cabinet minister, military commander, or top official is inducted into the Privy Council through a ceremony by which the new member kneels on a stool and kisses the hand of the monarch.

The new member also takes an oath—not to the nation, nor to the laws and constitution, but to the person of the monarch. The inductee swears "by Almighty God…to be a true and faithful servant" unto the reigning monarch, to "keep secret all matters committed and revealed unto you," and to "assist and defend all Jurisdictions, Pre-eminences, and Authorities," etc.

It's this age-old tribal loyalty, with treason being punishable by a gruesome death, that is the basis for British governance. Nothing comparable exists in the U.S. In fact, it was to get rid of such institutions as the monarchy and Privy Council that the American Revolution was fought. It's also why the U.S. Constitution banished all trappings of nobility.

The European System of States

Ludwig Dehio in *The Precarious Balance: Four Centuries of the European Power Struggle,* writes of the power struggle growing out of the European system of states. He writes, "Those with a sense of history will never cease to ponder the mystery of this abounding vitality of the Western world which was now spreading across the globe."[5] He notes:

> The fate of the system of states was decided by England. Her key role was due to her island position, but also to her maritime

power. Under Elizabeth I, nation and state braced themselves with marvelous resolve.[6]

Dehio explains that it was through an alliance with the French and the Dutch Calvinists that Elizabeth "piloted England for the first time to the leadership of a kind of Continental coalition against a dominant Continental power."[7] This power was the Spain of Phillip II, a Habsburg and son of Holy Roman Emperor Charles V, who at one time had been the most powerful monarch of Europe before his empire was split into Spain, the Netherlands, and the Holy Roman Empire remnant.

Dehio writes that it was the decline of France from its medieval dominance of Europe that now cast England in the role of Spain's opponent. From this time on, *England became the adversary of every subsequent Continental superpower*, a role that lasted through and beyond World War II, if we view Great Britain as a partner to the U.S. in the Cold War against the Soviet Union and again today, acting as head cheerleader in goading the U.S. in its proxy war against Russia in Ukraine.

But what about the sovereignty of the nations that Britain sought to manipulate or dominate? The Treaty of Westphalia that ended the Thirty Years War, which pitted Catholic states against Protestant in Central Europe and the Holy Roman Empire, pledged to uphold the sanctity of sovereign nations against future attack or coercion.

But Britain was not a signatory to the Treaty of Westphalia. To Britain, the principle of inviolable sovereignty for its actual or potential enemies was alien.

Holland's Rise

Holland was the second emerging naval power of the era. Before England began to accumulate an empire, Holland had begun to build its own. In order to attack Spain, England supported the revolt of the Spanish Netherlands—today's Belgium—and sent troops to the Continent to fight. But then France began to revive, seeing the reestablishment of a strong monarchy as the path to a return to prominence.

Meanwhile, England and Holland were at each other's throats. The Stuart kings of England were not interested in naval affairs, but when Oliver Cromwell gained power, he built up the English navy, defeated both Holland and Spain at sea, and "acquired a title to world power."[8] Four Anglo-Dutch wars were fought in the 17th century, all at sea, along with the Anglo-Spanish War of 1654–1660 by which England acquired the valuable Caribbean properties of Jamaica and the Cayman Islands.

Unfortunately, the attempt by republican forces to take over the English nation during the era of the Civil Wars, Cromwell, and the Commonwealth ended in failure. The "Restoration" of King Charles II in 1660 began the modern era of rule over Britain by the fabulously wealthy despots known today as "the royals." This collection of unbelievably rich and privileged people are mistakenly viewed today as "figureheads." As we shall see, their elevation to moral dominance has caused the power and talent of the British to serve, not humanity, but a cult of primitive tribal dominance.

Louis XIV

France under Louis XIV (1638–1715) had its own cult of the Sun King and Europe's first professional standing army. It was built with a modern and efficient central taxation system that no other nation could match. France was now evolving toward its *grand siècle*, But France was never able to match the fighting strength of Britain's fleet, so it remained a Continental land power. No matter; Louis still went to war against both Holland and Britain.

Holland was never able to return to its glory days, but when William of Orange, the Dutch Prince, was invited with his English wife Mary, daughter of Charles I, to assume the British throne through the "Glorious Revolution" of 1688, a merger of English and Dutch sea power now took place. Holland lost its status as a world power as England continued for the next two-and-a-half centuries building its own empire. Dehio writes that by the time of William of Orange, "Britain's character as a maritime nation had settled into its final shape."[9]

The merger between Britain and Holland also brought the system of bank-centered deficit financing to Britain that Holland had perfected. It was a system that Holland had acquired from an earlier influx of Italian bankers, many Jewish, from the declining Venetian empire. Under this system, banks floated government bonds that were purchased by wealthy domestic and foreign investors, with the money being used by the government for its naval fleet and war efforts. Of course, such lending was inflationary and required constant economic growth to pay off the loans.

After William of Orange came to England, the bankers founded the Bank of England in 1694. The bonds used for the financing of war as well as the world's greatest fleet made the British Empire possible. Eventually, the U.S. would imitate both in its own rise to power.

The union of English and Dutch interests was also facilitated by the merger in India of the British and Dutch East India Companies. During the ensuing century, the British East India Company became a state in and of itself, with its own army and administration, driving out the French, and making India the crown jewel of British overseas power.[10]

Deeply alarmed, Louis XIV realized that England was a menace and sought to destroy it. In 1692, the French sent an invasion force via the French fleet; it was decisively defeated by England and Holland at Cape La Hogue off the coast of Normandy. The French had intended to restore James II to the English throne. The plan failed disastrously.

England fought the War of the Spanish Succession (1701–1714) to prevent Louis XIV's grandson from inheriting the throne of Spain and uniting France and Spain. This war is considered the first "world war," with major campaigns in Spain, Germany, Italy, and at sea. Again the British sent an army to the Continent under John Churchill, 1st Duke of Marlborough, to lead a coalition against the French. For the first time, England and Prussia were allied in a war. France was exhausted by the war, and Holland never recovered.

But England was able to preserve a balance of power in Europe that had been threatened by the French revival. England's navy proved strong enough to protect its overseas trade, control its home waters, and launch attacks on its enemies. England also turned its back on acquiring territory on the Continent, instead focusing on maritime outposts like Gibraltar and on colonial power in America and India. By 1700 the American colonies had a population of a quarter million, while the French in North America had only a tenth of that.

Russia and Prussia

Europe now became witness to the rise of Russia under Peter the Great (1672–1725). Combining the patience of the East with Western technical methods, Peter, says Dehio, "sent the power of Russia soaring upwards."[11]

Peter learned the art of war by fighting the Swedes, whose Baltic empire now began to decline. He created the modern Russian army, aided the Orthodox church in exchange for its support, and built a bureaucracy through judicious taxation. Peter's adulation of Western civilization was rapid but not shared; it "concealed the smothered discontent of the outraged soul of a people."[12] From Peter the Great onwards, Russia was subject to sudden revolts that culminated in the Bolshevik Revolution of 1917.

Britain and Russia now became the two flanking powers of Europe. Russia's growth was based on wars against Sweden, Poland, and Turkey, with Peter crushing Sweden at the Battle of Poltava in 1709. He also established relations with China and moved toward Central Asia and Persia. Britain had its own eyes on Persia and the Middle East as possible connecting points between India and the Mediterranean. *Britain now became alerted to Russia as a future adversary.* The English diplomatic service commenced a secret espionage operation in Russia that persists until today.

France had come back from decline after the death of Louis XIV to compete with Britain for overseas territories. Its motives were trade, wealth, and above all, control of foreign markets for domestic manufactured goods. This was especially so between the end of the War of the Austrian Succession and the Seven Years War during the period 1748–1756. Britain had gone to the aid of Austria against the two Bourbon powers—France and Spain. Meanwhile, Russia too was opposing any bid by other nations for Continental supremacy.

Dehio writes: "A tendency on the part of any nation to dominate the old Continent would always cause the flanking powers to shelve their antagonism and join forces."[13] So Britain and Russia would join their strength but only when it was convenient. There would be constant maneuvering and intrigue, spy vs. counter-spy, to enable Britain's mastery in weaving favorable coalitions.

Progressive victories of Britain and Russia over France, Spain, Sweden, and Poland gave space for the rise of the small Prussian state in northern Germany under Frederick the Great. Dehio calls Prussia "…a trough between two waves…. A country on the culturally backward fringes of the Protestant world now shot into prominence, a power based on a military civilization."[14] Prussia would mold the German character during the formation of the German Empire under Wilhelm I and Bismarck.

Britain succeeded in its Continental wars by subsidizing its allies rather than fighting itself. As Eric Berne wrote in *Games People Play*: "Let's you and him fight." Britain had money to spare in financing its allies' armies both from trade and from the seemingly bottomless pit of government bond sales; i.e., public debt. Such proxy wars clearly foreshadowed today's campaign in Ukraine against Russia led by the U.S. but involving other NATO nations, most notably Britain.[15]

Britain vs. France and the Seven Years War

In the Seven Years War (1756–1763), Britain and France would fight for the future of a continent—North America—and a subcontinent—India. Britain's victory on both fronts was due to its mastery of the sea. Britain could project power thousands of miles away in a manner that was impossible for France. But soon Britain's American colonies broke away.

It appeared that the center of gravity of world power might someday shift to the United States. During the American Revolution, all the Continental powers joined against Britain to help make America free. The usually disunited states of Europe sought to humble Britain and create a new Transatlantic counterweight.

Meanwhile, Russia advanced to a new level of prestige. It started when Catherine the Great (1729–1796) seized the throne from her husband, Peter III, who would shortly be assassinated. The partitions of Poland followed, and a Russian fleet destroyed the Turkish navy in the Aegean Sea. Britain, now alarmed, began to side with Sweden and Prussia against Russia, but Parliament refused to authorize a new Continental war. With the carving up of Poland and the weakness of the Ottoman Turks, Prussia now worked toward maintaining the balance of power in Eastern Europe.

French Revolution and Napoleon

Meanwhile, Britain was becoming the first modern industrial state. The Industrial Revolution began with harnessing of power from coal. Over forty years, there was a ten-fold increase in coal-fired iron production. Iron was the foundation for domestic manufacturing, building of railroads, construction of ships, and forging of armaments.

It was Britain's creation of the factory system as applied to cotton and the utilization of limited liability corporations to protect investments that enabled it to become the leading world power by the end of the 19th century. From 1760 to 1820, Britain's population doubled, a development aided by improved methods of medicine. It was industry that gave these people a livelihood. Those who turned to petty crime for a living were transported to the penal colony of Australia.

Meanwhile, in France, centuries of authoritarian rule and abuses by the elites produced a middle-class explosion. With the French Revolution, a political party, the Jacobins, came into being that could keep the masses in ferment. To divert the anger of the mob, they launched into foreign aggression.

France attacked its eastern neighbors to secure new borders by pushing its control to the Rhine on the German border. Napoleon became First Counsel of France in 1799 and Emperor of the French in 1804. With Napoleon, terror and war boosted each other. Dehio says: "Napoleon raised the power of the state of Louis XIV to the level of the new age."[16]

Britain knew that war and possible invasion again loomed. Nothing could match Napoleon's force of almost one million men, but France was still trying to build up its navy. Britain began to pump vast funds into Europe, providing stipends to any nation that had a chance to stand up to the French hurricane. With its own navy, Britain organized a trade war against France.

One of Napoleon's objectives in opposing Britain was to move toward India, which he had already attempted in a failed invasion of Egypt. Next,

Napoleon tried to forge an alliance with Russia after Russia and Britain had fallen out over various shifting alliances and British moves to attack Russia's capital, St. Petersburg, via the Baltic Sea.

After Czar Paul I, the son of Catherine the Great, had broken with Britain, he was courted by Napoleon to plan a joint expedition against British India. Paul had dispatched an army of 22,000 Cossacks to attack India when he was assassinated by members of the Russian nobility.

It was alleged that the assassination was abetted by the British ambassador to St. Petersburg, a military officer named Charles Whitworth, who had been arranging cash subsidies to pay Russia for the use of troops against France. Czar Paul had abruptly dismissed Whitworth, who returned to England and became a Privy Counselor. But British intrigue in St. Petersburg continued.

On December 2, 1804, Napoleon crowned himself Emperor of the French. His coronation took place in the cathedral of Notre-Dame in Paris. Wearing robes of satin and diamonds, he strode up the aisle wearing high-heeled shoes and carrying the scepter of Charlemagne. Pope Pius VII travelled from Rome for the enthronement, during which Napoleon placed a crown of gold laurel leaves on his own head as the Pope watched.

By now Britain was war-weary and made overtures to Napoleon suing for peace. In fact, France could have enjoyed peace at any point through Napoleon's conquests, except for his lust for power. Such has been the case with so many empires, and Napoleon continued to covet India.

Seeing that Napoleon was determined to restore the French fleet, Britain restarted the war. Napoleon now decided to invade Britain, but his plans proved impracticable. So, France renewed the war on the Continent, causing Britain to form the Third Coalition with Austria, Russia, Sweden, and Naples-Sicily. Britain also began to impress American seamen into its navy, causing U.S. President Thomas Jefferson to cut off trade with both Britain and France.

The British fight to the death against Napoleon was now underway. Previously, Britain could have lived with Napoleon, due to its maritime trading power and overseas possessions, including India, but no more. Britain and France met offshore at Cape Trafalger in Spain for the greatest sea battle of the age. Lord Nelson, the British admiral, died, but the battle was won. Next, Britain under the Duke of Wellington invaded Spain, and Napoleon again courted Russia to join him. But Russia was reeling by the cutting off of trade with Britain under Napoleon's Continental System.

Most of the battles against Napoleon were fought by Britain's continental allies, not by British troops. Russia now refused to submit to the

Continental System. So, in order to defeat Britain, Napoleon had to conquer Russia before Russia sent its own forces west. Napoleon's 1812 attack on Russia was a preventive measure. Russia deliberately drew Napoleon deeper into Russia, and the French army failed to master the great distances of the steppes.

The Battle of Borodino was inconclusive, with historians still asking why Napoleon did not commit his Imperial Guard. The only possible answer was that he could not afford to lose it so far from home. The French occupied Moscow, but after it burned, they launched a retreat in the dead of winter. French losses were catastrophic. The British now launched another coalition, with Napoleon's final defeat taking place at the hands of the British and Prussians at Waterloo in Belgium in 1815. But it was Russia that had dealt the decisive blow as it would later do against Hitler.

Britannia Rules the Waves

After 1815, Britain enjoyed decades of unbroken peace, power, and prestige. The Congress of Vienna and Holy Alliance had firmly returned power in Europe to the traditional monarchies and aristocracies that had been overthrown by Napoleon. In 1838, the crowning of Queen Victoria inaugurated Britain's modern age. It had monopolies on sea power, sea-borne trade, and manufacturing exports. Britain's profits from India grew with the export to Europe of tea and gems.

Its conquests now extended to China through the humiliation of that country through the Opium Wars from 1839 to 1860. According to Niall Ferguson, writing in *Empire*, opium accounted for a staggering forty percent of British exports from India to China. The profits from opium were remitted to London by the East India Company to pay the interest on its enormous debt to wealthy bondholders.[17] Britannia ruled the waves but was ruled in turn by big money.

The fabled City of London, legally a private corporation at the heart of the capital, had become the wealthiest square mile in history and may still be. Within the City's confines, the Baring and Rothschild interests held sway.

But once again Russia loomed as a rival in central Asia. Britain countered by allowing France to survive as a major Continental state, with a view to France's interests in North Africa and the Middle East helping counter the Russian overland advance.

The "Great Game," with Afghanistan as the hinge, was underway, followed by Russia's humiliation by Britain, France, Piedmont-Sardinia, and the Ottoman Turks in the Crimean War, when Russia's Black Sea port of Sevastopol fell after a prolonged siege in 1856. Meanwhile, the British navy protected the independence movements against Spain that were taking place in South America.

By the mid-19th century, Britain stood supreme. The defeat of Napoleon was not only a victory over France but against the international revolutionary impulse. While the revolutionary year of 1848 saw agitation against established governments on the Continent, Britain seemed untouched. Incipient revolts among farmers, workers, and the Irish were easily quashed.

In the U.S., the victory of the North in the American Civil War may have disappointed British conservatives, but no one was surprised at the growing strength of the industrial colossus across the waves. In fact, a coalescence of U.S. and British interests had already been evident with the Monroe Doctrine and peace on the Canadian border.

The Rise of Germany

Germany was the newest industrial power, where Chancellor Otto von Bismarck proved adept at playing Britain, France, and Russia off against each other. Prussia, "the wave between two troughs," rose up to dominate most of what was once the Holy Roman Empire. Bismarck was able to obtain Russia's acquiescence in Prussia's war against Austria and in its defeat of France in 1870. Wilhelm I was crowned Emperor of Germany at Versailles in France, and the German Empire was born.

The rise of Germany coincided with a new race for colonies in Africa. This was enabled by industrialization and saw the opening of the diamond and gold mines by Britain in South Africa and the invention of the Maxim gun. This early machine gun allowed European forces to obliterate any native opposition to their imperialistic advances. By the time of the Berlin Conference of 1884, almost the entire African continent had been gobbled up by Britain, France, Germany, Italy, Belgium, Portugal, and Spain. The only independent African state was Ethiopia.

Germany held three large colonies: German East Africa (Tanzania), German Southwest Africa (Namibia), and Kamerun (Cameroon). Germany also began to cultivate its relations with the Ottoman Turks, supporting the modernization of Turkey with loans and technical expertise and laying plans for construction of a Berlin to Bagdad railroad, which conceivably could outflank the British in their sea route to India through the newly built Suez Canal. The German presence could also threaten British interests in Middle Eastern oil.

Bismarck was able to "establish the new Germany as a Continental power state *par excellence*."[18] Through his Kulturkampf, Bismarck abolished the Socialist Party, while German industry became bent on the acquisition of power, which it accomplished by building a military machine on the old Prussian base. Bismarck was a realist in his drive to make Germany a modern economic state through authoritarian methods. Bismarck was forced from office in 1890, leaving Kaiser Wilhelm II in the hands of younger and vainer men and less inclined to compromise in Germany's search for a role in world power.

But money was a problem. In Germany there was a long tradition of regional governments borrowing through their banking systems to pay for wars and infrastructure. Selling bonds to the rich was a way to extract their money, but at a price—the loans had to be repaid with interest. The alternative was to expand exports and extract wealth from the new colonies. But doing these things meant building ships, also costly. And then, new merchant ships would have to be defended by new warships. And building warships would place Germany in direct competition with Britain. So the groundwork was laid for a new general European war by the last decade of the 19th century. Meanwhile, industry was developing at breakneck speed, with revolutions starting to take place in communications, transportation, and military firepower.

These changes transcended national boundaries and mitigated divisions among nations, causing many to think that maybe a really big war was not such a great idea. But habits die hard. The nations of Europe had been fighting each other "forever." Hatreds lay close to the surface, and it was not difficult for rabid national presses and the growing mass media to stir them up.

Something else was new: the U.S. was now a player on the international chessboard. With acquisition of the Philippines, it had extended its reach to Asia and had built a semi-respectable fleet. Its young men had shown in their Civil War that they could kill, a fact that impressed Churchill.

Great Britain was alarmed at the lightning-fast pace of Germany's advance. Its massive economic growth had allowed it to create an army that dwarfed Britain's. Germany as a naval power was catching up. So Britain now reached a rapprochement with France and an alliance with Russia, after leaning toward the Japanese against Russia in the Russo-Japanese War of 1904–1905. In 1904, Britain, France, and Russia joined in the Entente Cordiale. Dehio writes, Great Britain "rebuilt on a world-wide scale the grand coalition of earlier centuries against the dominant power on the old Continent."[19] This power was now Germany.

With regard to France and Russia, neither posed a threat to Britain any longer. But the shifting alliances had to be explained to the public. After all, Britain was a "democracy." So the British press began cranking up its propaganda machine to demonize the new ogres, the monstrous Germans, the "Huns." Of course, it was embarrassing insofar as the roots of the British royal house were German. But by 1917, the House of Saxe-Coburg and Gotha had been renamed the House of Windsor, the name deriving from a castle in the London suburbs dating from the time of William the Conqueror.

Only the U.S. was unaccounted for in Britain's planning. But with the accession of Theodore Roosevelt to the White House after President McKinley was assassinated in 1901, Britain had a friend it could count on. American support, however, did not come for free. Dehio writes: "The price Britain paid for American backing was her attitude of indifference toward construction of the Panama Canal."[20] Eventually the price would be much higher: subservience of the pound to the dollar. But not yet.

South Africa

There are times and places in history that are so pivotal that the events which transpire reverberate far beyond their origin. So it was with the British colony of South Africa at the end of the 19th century.

When Europeans began their voyages of discovery, the Portuguese arrived in South Africa first, but the initial settlements were Dutch, starting in 1652. Their purpose was to raise food for Dutch ships traveling to India. Gradually, more Europeans arrived and began to farm the rich soil. They also imported over 70,000 slaves from East Africa and elsewhere.

During the war against Napoleon, the British took over to keep the French from using South Africa as a base to threaten their holdings in India, which was always Britain's Achilles heel. Britain brought in its own troops of settlers, causing the Dutch to migrate inland, undertaking what was called the Great Trek into the Transvaal. The Dutch themselves were now called Boers, which simply means "farmers." The British called the area it now controlled the Cape Colony, after the Cape of Good Hope, and paid Holland six million pounds in compensation.

Inland, the Zulu tribe was carrying out a massive ethnic cleansing of the African plains, killing from one to two million other black Africans, thereby weakening resistance to the Europeans. Nearby, the Boers were establishing the South African Republic, lasting from 1852 to 1902, which gained British recognition when they fought against the indigenous Basotho people, resulting in a small area then set aside as Basutoland. A second Boer-dominated republic was the Orange Free State. Total white population of the Boer regions by 1900 was around 400,000.

In 1866, a fifteen-year-old Dutch farm boy named Erasmus Jacobs found a small shiny pebble on the banks of the Orange River on a farm being leased from local natives. The pebble was sold and resold and turned out to be a 21.25-carat diamond. Digging continued and the word spread. By 1873, 900 claims had been registered on land that became the Kimberley diamond mine. The town of Kimberley, named after a British Lord, was established just outside the Orange Free State and in 1877 was annexed by the British and incorporated into the Cape Colony. South Africa was elevated to worldwide importance.

Cecil Rhodes

Enter Cecil Rhodes. Born in Herfordshire in 1853, Rhodes was the son of a Church of England clergyman and the grandson of a Middlesex brick manufacturer. A sickly child, he was sent by his father at the age of seventeen to the healthier climate in South Africa, where he intended to join his older brother Herbert on his cotton farm in Natal. Separately administered from the Cape Colony and situated along the southeastern coast, Natal had been annexed by the British in 1843.

When Cecil arrived, twenty-six-year-old Herbert had already departed Natal for the diamond fields of Kimberley. After working the cotton farm for a season, Cecil traveled overland for almost 500 miles and joined Herbert. At the mine, the two carried out the heavy on-site work for what was now their joint operation. Another partner, Charles Rudd, carried on with the now-successful mine while the Rhodes brothers traveled to explore the Transvaal, home of the Boers.

When Cecil suffered a mild heart attack, he returned to England to rest and enroll at Oriel College, Oxford. After only a term, he went back to Kimberley, where Rudd was running the diamond mining business that would become the De Beers Company. But to reach this level of production, Cecil Rhodes needed cash, and a lot of it. For this he turned to the greatest financier of the age, Nathaniel Rothschild, whom we have already seen at work bailing out U.S. President Glover Cleveland and investing in Anaconda Copper in Montana.

According to Niall Ferguson, "the Rothschilds were strongly attracted to gold and silver mining."[21] But it was in diamonds that Cecil Rhodes first needed help. In 1885 Rhodes approached Nathaniel Rothschild for financial support in consolidating the multiple companies that he had acquired and those that were still operating independently in Kimberley. Over time, Rothschild acquired more shares in De Beers than those held by Rhodes. This gave him substantial leverage in dealing with Rhodes, not only in diamonds, but later in gold. Soon, De Beers was in control of ninety-eight percent of world diamond output, with profits by 1886 running at forty percent annually.[22]

Then in 1886, gold was found in the Witwatersrand in the Boer-controlled Transvaal. It was the largest gold deposit ever discovered. Rhodes expected Nathaniel Rothschild to support him in getting his hands on the gold. Here, British imperial ambitions would now come into play. Again, Nathaniel Rothschild, who was deeply imbedded in imperial politics, was key. Already, the Rothschild banks in London, Paris, and Frankfurt were in charge of marketing four major bond issues for the British-protectorate government in Egypt. But it was in South Africa that the future of the British Empire was being forged.

In London, Nathaniel Rothschild was heavily wired into the British government via such figures as Benjamin Disraeli, Randolph Churchill, and others. Of Randolph Churchill, the *Rothschild Archive* boasts:

> The father of Winston Churchill was an intimate of the Rothschild family. He formed a close association with Nathaniel, 1st Lord Rothschild, on whose behalf he reported on the development of the mining industry in South Africa. Churchill was a frequent guest at Rothschild houses. The Rothschilds made extensive loans to Churchill.

In 1902, Nathaniel Rothschild, a close friend of King Edward VII, was made a Privy Councilor. Kneeling on the King's footstool and kissing the King's hand, he had truly arrived.

But for the final act to be played out in the South Africa story, decisive action was required. In 1890, Rhodes, who had been a member of the Cape Parliament since 1880, was named prime minister of the Cape Colony under Baron Henry Loch, the Crown-Appointed Governor and High Commissioner. It was time, Rhodes believed, for Great Britain simply to seize the Boer Republics, the Transvaal, and the Orange Free State, and control them by force.

The Boers had already defeated the British earlier, through a short conflict in 1880–1881, known today as the First Boer War. The Boers were heavily armed and superb marksmen. The British redcoats were easy targets. The British called a truce after suffering their first decisive defeat in the field since the American Revolution, here too by white men who could shoot, and the Boer republics retained their autonomy. Now, almost two decades later, the Boers would be an even tougher nut to crack. Their numbers had grown, and, seeing what was coming, they had laid in artillery and powerful German-made Mauser rifles.

But here we'll stop our account and resume it in the next chapter, as the Second Boer War can be considered as much a prelude to World War I as a conflict of importance in its own right. In the meantime, we'll take a look at Cecil Rhodes in a broader context of what his fortune and proclaimed mission meant for the future of the British Empire, and the future of the U.S. as well.

Rhodes's "Confession of Faith"

During his hiatus back at Oxford in 1877, Cecil Rhodes envisaged his life's work as not merely to become fabulously wealthy, but to launch the British Empire on a mission for world dominance. That year, he wrote his "Confession of Faith."

Following is the amended statement. The spelling and grammar errors were in the original. Due to its historical importance, I am including much of the document from where it appears on the University of Oregon website:

> It often strikes a man to inquire what is the chief good in life; to one the thought comes that it is a happy marriage, to another great wealth, and as each seizes on his idea, for that he more or less works for the rest of his existence. To myself thinking over the same question the wish came to render myself useful to my country. I then asked myself how could I and after reviewing the various methods I have felt that at the present day we are actually limiting our children and perhaps bringing into the world half the human beings we might owing to the lack of country for them to inhabit that *if we had retained America* (italics added) there would at this moment be millions more of English living.
>
> I contend that we are the finest race in the world and that the more of the world we inhabit the better it is for the human race. Just fancy those parts that are at present inhabited by the most despicable specimens of human beings what an alteration there would be if they were brought under Anglo-Saxon influence, look again at the extra employment a new country added to our dominions gives. I contend that every acre added to our territory means in the future birth to some more of the English race who otherwise would not be brought into existence.
>
> Added to this the absorption of the greater portion of the world under our rule simply means the end of all wars, at this moment had we not lost America I believe we could have stopped the Russian-Turkish war by merely refusing money and supplies.

Having these ideas what scheme could we think of to forward this object....[23]

We can see right away that a central theme of Rhodes's beliefs was how much better off Great Britain would be if it had only "retained" America. Rhodes now writes that he wants to form a secret society to achieve his aims.

> The idea gleaming and dancing before one's eyes like a will-of-the-wisp at last frames itself into a plan. Why should we not form a secret society with but one object the furtherance of the British Empire and the bringing of the whole uncivilised world under British rule for the recovery of the United States for the making the Anglo-Saxon race but one Empire. What a dream, but yet it is probable, it is possible. I once heard it argued by a fellow in my own college, I am sorry to own it by an Englishman, that it was good thing for us that we have lost the United States. There are some subjects on which there can be no arguments, and to an Englishman this is one of them, but even from an American's point of view just picture what they have lost, look at their government, are not the frauds that yearly come before the public view a disgrace to any country and especially theirs which is the finest in the world.
> Would they have occurred had they remained under English rule great as they have become how infinitely greater they would have been with the softening and elevating influences of English rule, think of those countless 000's of Englishmen that during the last 100 years would have crossed the Atlantic and settled and populated the United States. Would they have not made without any prejudice a finer country of it than the low class Irish and German emigrants? All this we have lost and that country loses owing to whom? Owing to two or three ignorant pig-headed statesmen of the last century, at their door lies the blame.
> Do you ever feel mad? Do you ever feel murderous? I think I do with those men. I bring facts to prove my assertion. Does an English father when his sons wish to emigrate ever think of suggesting emigration to a country under another flag, never—it would seem a disgrace to suggest such a thing I think that we all think that poverty is better under our own flag than wealth under a foreign one.

Again, it's a secret society that Rhodes envisions as the means of achieving his aims. Obviously, to form such a secret society will require a lot of money and very high-level support within the British government and society. Through Nathaniel Rothschild, Rhodes will acquire both. He adds that: "I contend that there are at the present moment numbers of the ablest men in the world who would devote their whole lives to it." He adds that not only should the secret society seek out capable young men who are inspired to serve its ends, but it should also control the press and use it to further its propaganda.

Rhodes's Will

Rhodes couldn't make his objectives more clear. But he was always the victim of borderline health. And when he wrote this document he was only twenty-four years old. He also was likely aware that his flourishing diamond business in South Africa promised wealth and influence if it could be harnessed.

So he decided to put his wishes into a will. At the time, his estate was only 10,000 pounds. But it would grow phenomenally. One result was the establishment of the Rhodes scholarships. But was his estate ever used for its primary purpose, the creation of the all-important secret society? Rhodes revised his will numerous times. The final executor was to be Lord Nathaniel Rothschild, Privy Councilor, and the most important, powerful, and influential financier of the age. The following is an excerpt from the draft of September 9, 1877:

> To and for the establishment, promotion and development of a Secret Society, the true aim and object whereof shall be for the extension of British rule throughout the world, the perfecting of a system of emigration from the United Kingdom, and of colonisation by British subjects of all lands where the means of livelihood are attainable by energy, labour and enterprise, and especially the occupation by British settlers of the entire Continent of Africa, the Holy Land, the Valley of the Euphrates, the Islands of Cyprus and Candia, the whole of South America, the Islands of the Pacific not heretofore possessed by Great Britain, the whole of the Malay Archipelago, the seaboard of China and Japan, *the ultimate recovery of the United States of America as an integral part of the British Empire* (italics added), the inauguration of a system of Colonial representation in the Imperial Parliament which may tend to weld together the disjointed members of the Empire

and, finally, the foundation of so great a Power as to render wars impossible and promote the best interests of humanity.

The stage was now set for a century of world wars.

ENDNOTES FOR CHAPTER 8

1 Today, of course, Britain is faced with the problem of mass immigration from its former colonies and other Middle Eastern and Asian nations.

2 Stephen Budiansky, *Her Majesty's Spymaster: Elizabeth I, Sir Francis Walsingham, and the Birth of Modern Espionage* (Penguin Books, 2006).

3 Guido Giacomo Preparata, *Conjuring Hitler: How Britain and America Made the Third Reich* (Pluto Press, London & Ann Arbor, 2005), 1.

4 John Dee's downfall came after his wife became pregnant in an "angel-sanctioned" wife-swapping escapade. He died in poverty, shunned by the court of King James I. His "spirit mirror" is in the British Museum. "Ten Facts You Might Not Know About John Dee," *Get History,* March 10, 2020. http://www.gethistory.co.uk.

5 Ludwig Dehio (author), Charles Fullman (translator), *The Precarious Balance: Four Centuries of the European Power Struggle* (Alfred A. Knopf, 1962), 21.

6 Ibid., 50.

7 Ibid., 54.

8 Ibid., 71.

9 Ibid., 80.

10 It was wealth from the looting of India that created the modern English landscape of estates and formal gardens.

11 Dehio, 92.

12 Ibid., 96.

13 Ibid., 114.

14 Ibid., 111.

15 According to RT.com, an "unnamed military-diplomatic source" claims that the UK is training 100 Ukrainian saboteurs/assassins to spread terror among African nations suspected of harboring sympathies toward Russia. "MI6 plans to send Ukrainian mercenaries to Africa—media," *RT.com,* August 16, 2023.

16 Ibid., 141.

17 Ferguson, *Empire,* 166.

18 Dehio, 220.

19 Ibid., 235.

20 Ibid., 239.

21 Ferguson, *The House of Rothschild,* 294.

22 Ibid., 358.

23 Cecil Rhodes, "Confession of Faith" (1877). https://pages.uoregon.edu/kimball/Rhodes-Confession.htm

CHAPTER 9

The Money Trust

Overview

By the end of the 19th century, many people in America and Europe had no doubt that humanity was facing a bright future. Much of this had to do with technology. In his publicity for the 1900 World's Fair in Paris, French commercial artist Jean-Marc Côté predicted transatlantic air travel in balloon gondolas, flying automobiles with flapping wings, videoconferencing, robotic barbers and salon workers, and use of the newly-discovered radium to heat houses, among other future innovations.

While automation has come a long way since 1900, though maybe not as far as Côté predicted, our response today may be, "So what?" Has technology contributed to making people happier? Has material and medical advancement made up for over a century of the most devastating wars in history and the continually rising environmental destruction, etc—to say nothing of overwhelming anxiety?

These are some of the questions we'll now be exploring. By the first decade of the 1900s, the world had been neatly divided up among the European powers and their colonial empires. White Europeans seemed to be reigning supreme, though off in the jungles, while ransacking dark peoples' resources, they often collided with each other.

By now, the U.S. had gotten into the colonizing act. Soon the entire continental U.S. would be divided into forty-eight states, with Indians, blacks, and most immigrants placed at the bottom. A brand new central bank, the Federal Reserve, which bankers had been agitating about for decades and which should have ended those pesky financial panics, is soon to be created.

Life is good, right? The U.S. was exploding economically, particularly its steel industry. The behemoth was U.S. Steel, controlled by banker J.P. Morgan. The U.S. government tried to break up U.S. Steel at the same time it was going after Standard Oil, but failed. From 1899 to 1908, led by the big corporations, U.S. economic growth averaged a staggering 14.2 percent per year.

The U.S. was doing so well that it was becoming a creditor, rather than a debtor nation. One of its borrowers was Great Britain. The purpose of their borrowing? To prosecute the Second Boer War.

The Second Boer War

The governments of Europe were now preparing for the biggest war in history—World War I—in which forty million people would die over a four-year period. Every nation accepted war as a routine factor in how they did business. But an industrial-sized war would be somewhat novel. Not only would the U.S. become involved in the war, its participation on the side of an exhausted British-led alliance against a just-as-exhausted Germany and its allies would prove decisive. Then, twenty years later, everyone would decide to do it again. Understanding that the Second Boer War was one of the preludes to World War I helps us to grasp the part played by Cecil Rhodes in his ambition to "recover" the U.S. for the British Empire.

But now a decisive new element appeared on the scene: mechanized warfare. The first advanced manifestation of the application of modern industry to killing people was the invention of the machine gun. The premier example was the Maxim gun, invented in 1884 by an American, Hiram Stevens Maxim. How was its production financed? Niall Ferguson writes:

> When the Maxim Gun Company was established in November 1884, Lord Rothschild was on its board. In 1888 his bank financed the 1.9 million-pound merger of the Maxim Company with the Nordenfelt Guns and Ammunition Company.[1]

Not only were the Rothschilds invested in diamonds and gold, they also financed the weaponry to allow the British to seize and profit from the potential wealth. We are also seeing, if not the birth, at least the infancy of the military industrial complex—and its connections with big finance.

With the Maxim gun, the Europeans had what they needed to subdue Africa, and the British were quick to put it to use. We have already seen how Cecil Rhodes, with Nathaniel Rothschild's money, gained control of the South African diamond mining industry. In 1886, after the world's largest deposit of gold was unearthed at Witwatersrand in the Transvaal, Rhodes and Rothschild moved quickly to dominate that too.

But first they had to deal with the Matabele Kingdom, whose land lay beyond the Limpopo River, where more gold fields would be found. Rhodes had gained mining rights for his Rothschild-financed United Concessions Company, but hostilities broke out when the Matabele king, Lobengula, balked at what seemed to be a complete British takeover. At the Battle of Shangani River in 1893, Lobengula's force of 1,500 warriors was wiped out. It was the Maxim gun that did the dirty work. An eyewitness wrote:

> The Matabele never got nearer than 100 yards led by the Nubuzu regiment, the king's body guard who came on yelling like fiends and rushing on to certain death, for the Maxims far exceeded all expectations and mowed them down literally like grass. I never saw anything like these Maxim guns, nor dreamed that such things could be.[2]

Meanwhile, the Boers continued their own resistance to British power by refusing to grant voting rights to British prospectors, the *Uitlanders*, in the newly-discovered gold fields. Four years later, the Second Boer War broke out as British citizens continued to migrate to the Boer republics. By now, a new face had appeared on the scene in South Africa: Alfred Milner (1854–1925), who in 1897 was appointed Governor of the Cape Colony and High Commissioner for South Africa.

Milner had Anglo-German roots and had been a brilliant student at Oxford who qualified to practice law, but instead became a journalist, working as executive assistant to W. T. Stead. An advocate of "imperial idealism" as editor of the *Pall Mall Gazette*, Stead was the founder of British tabloid journalism—and also a close friend of Cecil Rhodes. Stead later died on the *Titanic* in 1912. His protégé, Milner, chose a career of government administration, starting with the post of assistant secretary of finance in Egypt. Back in England, he became chairman of the Board of Inland Revenue, Britain's tax collection agency, and during his tenure he inaugurated Britain's inheritance tax. He was also inducted into the prestigious Most Honourable Order of the Bath.

Now regarded as one of the most clear-headed and competent civil servants in Britain, Milner began his term as head of the British colonial government in South Africa by traveling through the disputed regions held by the Boer republics and the most heavily-populated native regions to study the crisis. He even learned Dutch and the Boers' Afrikaans language. He determined that only complete submission by the Boers to British imperial policy would secure and stabilize South Africa. The Boers refused to submit, and so the war began.

The Second Boer War was a bitterly-fought series of bloody battles by two determined foes. The Boers also had Maxim guns, along with German-made Mauser rifles. Following a military standoff, the Boers resorted to guerrilla warfare. The British responded by rounding up Boer civilians and interning them in what historians agree were the world's first concentration camps. In fact, the British invented the term. They set up forty-five tented camps for Boer internees and sixty-four for the black Africans who fought with them. Most of the Boers in the concentration camps were women and children, with most of the 28,000 captured Boer men sent overseas to prisoner-of-war camps, where 26,370 died from starvation, disease, or neglect. Of the blacks, around 20,000 died. The British deliberately created deadly conditions in order to force the Boer fighters to surrender.

The British also engaged in scorched-earth warfare against Boer farms, burning crops and killing livestock. The strategy worked. The last of the Boers surrendered in May 1902, and the war ended with the Treaty of Vereeniging on May 31, 1902. Having won the war, the British now offered generous terms to regain the support of the now-subdued Boer people. In order to win, the British had countenanced losing over 125,000 men killed, missing, or wounded. Boer losses were around 30,000, with over 46,000 civilians dead.

The war was so expensive that for the first time in over a century the British government had to borrow from abroad to fight a war in its own empire. This was a harbinger of things to come. Half of the new war bonds were sold in the U.S., with J.P. Morgan handling the sale to American investors. The war saw a substantial jump in the British national debt.

The British populace largely supported the war, with newspaper propaganda playing a heavy role, emphasizing Boer mistreatment of the maltreated British men who had innocently flocked to the Boer republics to earn a living and raise their families. In addition to troops from the British Isles, thousands of Canadian, Australian, New Zealand, and Asian Indian soldiers traveled to South Africa to fight for the Empire. On returning home, these troops were hailed as heroes.

The Boer War evoked much criticism of British imperialism in Europe. In Belgium, fifteen-year-old Jean-Baptiste Sipido, a tinsmith's apprentice, fired shots at the Prince of Wales, the future George V, who was passing through Brussels. Sipido accused the Prince of causing the slaughter of thousands during the Boer War. A Belgian jury found Sipido not guilty, provoking outrage in London.

After the war, Alfred Milner spent three years as governor of the new Transvaal and Orange River colonies, working on the rebuilding of the regions that had been devastated. In March 1902, having always suffered from poor health, Cecil Rhodes died at the age of forty-eight at the seaside town of Muizenberg.

The cost of rebuilding the former Boer republics was enormous. Milner began by levying a ten percent tax on the annual production of the gold mines, and gave attention to the repatriation of the Boers, organizing land settlement by British colonists, building schools and court systems, hiring police, and constructing railways. But the Boers lived on. Today, there are more people in South Africa who speak the Afrikaans language than English, though a majority of the population speak native African languages.

Milner also recruited a team of young lawyers and administrators, most of them Oxford graduates, who became known as "Milner's Kindergarten." Thus began a cadre of disciples who would follow Milner back to England and form the core of Cecil Rhodes's "secret society" for the elevation of the British Empire to world dominance.

The Russo-Japanese War

Increasingly, relations among Britain, the U.S., and Russia must be understood if we want to grasp today's geopolitical crises. Thus, despite the fact that Britain and Russia were allies during World War I and World War II when Russia was part of the Soviet Union, Britain has *always* viewed Russia as a potential enemy. Russia is simply too big and powerful a Continental European power to be tolerated.

As Russia continued its expansion in the 17th and 18th centuries under Peter the Great and Catherine the Great, it began to push into regions that Britain saw as its own spheres of interest, most obviously India, which Russia threatened with its push into Central Asia. Other potential points of conflict included any movement Russia might make through the Caucasus toward Persia; Russian force being applied through the Black Sea against Turkey at Constantinople, the point of transit from the Black Sea to the Mediterranean; or Russia's influence among Slavic peoples in the Balkans, particularly Serbia; or its push into Eastern Europe against Poland; or any advances in the Baltic Sea region where the Scandinavian nations faced Russia, which was building a modern fleet to sail from St. Petersburg to enter the North Sea and the Atlantic.

All these were sensitive nodes for Great Britain. This was why the British applied so many diplomatic resources to its Russian problem. In the big picture, in order to maintain the desired balance of power in Europe and the world, Britain needed Russia to counter the rising power of Germany, but didn't want Russia to become so powerful that Britain itself would be threatened. So it was always a balancing act.

In Britain's view, Russia was also getting too big for its britches in the Pacific. The Russian Empire had spread across Siberia to the Pacific coast, with the Trans-Siberian Railroad then under construction. Wanting to obtain a warm-water port, Russia had its eyes on Korea, where it desired a protectorate, and on Manchuria, where it had leased harbor facilities at Port Arthur.

Britain wanted no further Russian expansion anywhere, but particularly in the Pacific, where they and other European powers had established themselves in Indochina, Hong Kong, Malaysia, Singapore, Indonesia, and, of course, Australia and New Zealand. Always there was India.

But Japan too was a rising Pacific power, with ambitions on the Asian mainland. In 1902 Britain and Japan signed an agreement, specifying that if any nation allied itself with Russia in a war against Japan, Britain would enter the war on Japan's side. This treaty would keep France and Germany out. Kaiser Wilhelm II, meanwhile, egged on Russia's Asian adventures, urging Czar Nicholas II to do his duty of upholding the white race against the "Yellow Peril."

Bolstered by British support, including collaboration with British intelligence, Japan attacked the Russian navy at Port Arthur on February 8, 1904. Within less than two years, Russia threw over a million men and two-thirds of its warships against Japan, which fielded 1.2 million men of its own. To win the war, Japan borrowed heavily from U.S. and British banks. The world was stunned at how thoroughly the Japanese put the Russians to rout.

Warfare in the Pacific would never be the same. It was the first time an Asiatic nation was able to stand up to a Western power, so the war gave solace to people elsewhere that had been oppressed and humiliated. Given Britain's role in furthering Japan's victory, it was astonishing that within less than a decade from the debacle of the Russo-Japanese War, Russia and Britain would be allies. But, as stated previously, Britain needed Russia in any war against Germany, and Germany had angered many in Britain by rooting for Russia so vociferously against the Japanese.

But time now to return to the U.S., where the "Indian problem" had not been solved.

Trouble in Flathead Country

The following account is based on *The Politics of Allotment on the Flathead Reservation* by Burton M. Smith, the Confederated Salish and Kootenai Tribes website, books published by Salish Kootenai College, and other sources.

Peter Ronan (1839–1893) was the Bureau of Indian Affairs agent for the Salish, Pend d'Oreilles, and Kootenai tribes of the Flathead nation in northwestern Montana from 1877 until his death. After a string of short-term incompetents and corrupt time-servers, Ronan and his wife Mary were viewed as two enlightened people who had the Indians' best interests at heart. It was a period of rapid cultural and economic change for the tribes, as hunting and gathering resources declined and the numbers of the surrounding white population exploded in the region around the growing city of Missoula. But the Flathead Reservation itself was peaceful.

Previously, Ronan and Kootenai Chief Eneas had worked to avoid open conflict with white settlers encroaching on the northern boundary of the reservation. Despite repeated provocations, Eneas was able to keep the peace and struggled to get equal justice for Kootenai victims of marauding white criminals. Ronan also worked to relocate the Bonners Ferry Kootenai and Lower Pend d'Oreille Indians onto the Flathead Reservation and to secure land allotments to Flathead tribal members who chose to remain in Idaho and Washington.

But the U.S. government, after driving the Indians west and herding them onto reservations, did not know what to do with them. The Indians might still keep livestock, but were subject to illness, lack of tools and implements, malnutrition, and other privations.

There were whites, both inside and outside the government, who wanted to improve the Indians' lot. Schools to teach them "civilized" arts like sewing, housekeeping, blacksmithing, carpentry, and printing began to be established. But more than anything else, it was the Dawes Act of 1887, similarly cast as well-intentioned, that moved to provide Indians with settled allotments, while opening "surplus" land to the whites, that was destined to cause the most changes.

In 1893, Peter Ronan passed away unexpectedly from a heart attack. The Indian agents who followed him on the Flathead Reservation were proponents of commencing the allotment process then and there. But to implement the Dawes Act, congressional authorizations applying to specific reservations were required. In 1895, Congress appointed a "Crow, Flathead" commission to negotiate for reservation lands, but tribal leaders refused to give up any.

In 1901, a U.S. government delegation led by Commissioner of Indian Affairs Charles Hoyt met with tribal leaders to discuss an offer to buy part of the northern end of the reservation at Flathead Lake. Chief Charlo said, "I will not sell a foot." Kootenai Chief Isaac said, "My body is full of your people's lies. You told me I was poor and needed money, but I am not poor. What is valuable to a person is land, the earth, water, trees, and all these belong to us.... We haven't any more land than we need, so you had better buy from somebody else."

On the reservation, agriculture had begun to flourish. In 1902, statistics showed there were 25,000 cultivated acres with 120,000 bushels of grain,

25,000 tons of hay, and 20,900 bushels of vegetables produced by tribal members. There were also 25,000 horses, 27,000 cattle, and 600 bison owned by tribal members, though the bison would soon be sold to Canada.[3]

Full-blooded Indians favored horses, while those of mixed ethnicity leaned toward cattle, with a successful ranching culture having formed. Successful ranching, however, required an open range for grazing. Raising fenced-in or feedlot livestock requires a different approach, one that was alien to the Indians and required capital that was not available. There were no banks from which to borrow, and the Indians wouldn't go hat-in-hand to a bank, anyway.

The big change came in 1903 when Montana Congressman Joseph Dixon of Missoula introduced a bill in Congress to impose the 1887 Dawes Act on the Indians of the Flathead Reservation. Dixon was a Quaker from North Carolina who had come to Montana to study law and make his fortune. He entered Republican Party politics, married the daughter of a Missoula businessman, and became rich.

Dixon saw his congressional mission as securing Indian allotment. Dixon and his political and business supporters saw the Indian reservations as large areas of what looked to them like empty land, with vast potential for agricultural, timber, and mineral exploitation. In other words, they saw dollar signs. Nothing could be more important, and nothing would be allowed to stand in the way.

Dixon first succeeded in opening the Crow Reservation in eastern Montana. His proposal to open the Flathead Reservation next was a big hit among Montana businessmen and politicians, especially in the Missoula area. Later, Dixon would become a U.S. senator representing Montana and a leading supporter of Theodore Roosevelt. When Roosevelt ran for president in 1912 against his former protégé William Howard Taft on the Progressive, or "Bull Moose," ticket, Dixon served as his campaign manager. In 1904, after Dixon's heavy lobbying of fellow members of Congress, both the House and Senate passed the Flathead Allotment Act, which Roosevelt promptly signed, setting the course for what would eventually lead to loss by the tribes of more than sixty percent of the reservation land base.

Under the Flathead Allotment Act, Indian heads of households would receive 160 acres, with single adults getting 80 acres. Two rounds of allotment were conducted. The first saw 2,390 tribal members eligible to receive allotments amounting to 245,000 acres out of 1,245,000 acres available, with the total allotments being 228,434 acres. The second round of allotments to tribal members took place in 1920 and consisted of 124,795 acres. The remaining grazing and agricultural land was now considered "surplus" and opened to white homesteaders. Later amendments to the act would seize additional tribal land for town sites, the Bureau of Indian Affairs agency office, churches, reservoirs and power sites, and 6,100 acres for public schools.

The Flathead tribes had no say. It was all decided by Joseph Dixon and the Montana whites working the legislative process. When the Indians disagreed, Dixon would cite the Supreme Court "Lone Wolf" decision of 1903, which found that Congress had the right to pass unilaterally any law it wished in abrogation of any standing Indian treaty. The Bureau of Indian Affairs, for its part, said it had no objections to Flathead allotment. The white supporters of allotment argued that it would benefit the Indians to stop relying on government handouts and stand on their own two feet.

In 1905, Chief Charlo traveled to Washington, DC to try to get an audience with President Theodore Roosevelt to stop the allotment program. Roosevelt was notoriously anti-Indian and would not talk to Charlo. The Indians sent a second mission to Washington in 1906 that also failed.

In 1908, Congress passed the Flathead Irrigation Project, supposedly to aid the Indians in making their transition to farming. But the project would actually benefit non-Indian farmers and ranchers receiving homesteads and harmed many Indian subsistence farms where irrigation ditches had been dug by hand. The project would be paid for by assessments on property, so many Indians had their allotments seized to settle debts when they lacked ready cash. The injustice rankles on the reservation to this day, when water rights remain a sensitive issue.

Finally, in 1910 the Flathead Reservation was officially opened to non-Indian settlement, with land that had been declared surplus now sold to white homesteaders. Chief Charlo, who had opposed any type of allotment, died that year. Most tribal members had chosen land for their own allotments close to the mountains where wild game still roamed, so prime farmland in the center of the Flathead River valley was taken by the whites.

When the land was opened to white settlement in 1910, 81,363 white applications were received for 1,600 parcels of land. Lottery winners were awarded only 600 tracts, leaving 1,000 still available. These too were taken later in what the tribes to this day consider a "land grab." The homesteaders paid the government, which was obligated to share the proceeds with the tribe but far below the fair market value of the land. Decades later the tribe received some compensation for the land taken, but it required court action.

The government's allotment had long-term effects. As of the 2010 census, the total population of the Flathead Reservation was 28,324. Of these, 9,186 identified as Native American, while 19,221 identified as other ethnicities, outnumbering Indians by 2:1. Over the years, the tribal government has succeeded in recovering a substantial amount of reservation land, and the Flathead Reservation has survived.

The "Great Rapprochement"

We have seen how Cecil Rhodes desired to found a "secret society" that would help "recover" America for the British Empire. Of course, the U.S. and Britain had always had relations in matters of business, banking, and finance, as exemplified by the careers of George Peabody (1795–1869) and Junius Morgan (1813–1890), both of whom headed financial institutions with major branches in both nations. The Rothschilds had major investments in the U.S., and when J.P. Morgan took over the family business in New York and London, he and the Rothschilds worked closely together well into the 20th century.

Another financier whose interests were closely tied to British bankers was Jacob Schiff, whose father began as a broker for the Rothschilds. Born in Frankfurt, Germany, Schiff came to the United States after the American Civil War. Later he was invited to join the firm of Kuhn, Loeb & Company and brought his British connections with bankers like Ernest Cassel with him. Schiff became the foremost Jewish banker in the U.S. and retained a special interest in aiding Russian Jewish refugees. He was an early supporter of the Zionist movement that sought to establish a Jewish national state somewhere outside Europe. He also became a director of numerous corporations, including the National City Bank of New York, Equitable Life Assurance Society, Wells Fargo, and the Union Pacific Railroad.

So there had been nothing unusual about Anglo-American collaboration, at least in the world of finance. But in politics, matters were different. As we have seen, the Republican Party worked to keep Britain at arm's length and, from the time of Lincoln, tried to keep U.S. government finances out of the hands of the Rothschilds and other British financial magnates. But this became more difficult as the U.S. moved toward the gold standard.

Nonetheless, what is called the "Great Rapprochement" took place. The phrase refers to the convergence of diplomatic, political, military, and economic activity of the U.S. and Britain from 1895 to 1917, leading up to American entry into World War I. In 1901, British newspaper magnate W. T. Stead and friend of Alfred Milner, published a book entitled, *The Americanization of the World*, a massive two-volume tome that advocated a merger of the English-speaking nations. This would allow Britain "to continue for all time to be an integral part of the greatest of all World Powers, supreme on sea and unassailable on land, permanently delivered from all fear of hostile attack, and capable of wielding irresistible influence in all parts of this planet." U.S. steel magnate Andrew Carnegie is said to have told Stead, "We are heading straight to the Re-United States."

Stead's perspective was identical to that of Cecil Rhodes and Alfred Milner. After Milner returned to Britain from South Africa, he was out of government for a time, defending his reputation against critics of his imperialistic policies in South Africa and securing his personal financial position. He became a bank director and an investor and manager in the Rio Tinto Company, one of the world's largest extractors of copper and zinc.

Milner was also the founder in 1910 of *The Round Table: A Quarterly Review of the Politics of the British Empire*, whose purpose was to promote the cause of the British Empire. The journal is still in existence, renamed in 1966 *The Round Table: Commonwealth Journal of International Affairs*. An organization of the same name also came into being. The Round Table was an elite group of the most notable British imperialists, many of whom had ties to the same level of elitists in the U.S. As executor of Cecil Rhodes' will, it would have been Lord Nathaniel Rothschild who appointed Milner as the Round Table's head.

If there was ever a "secret society" as Cecil Rhodes envisaged, this was it. But little is known of the Round Table, even today. It was the subject of considerable elucidation by American scholar Carroll Quigley in his book *The Anglo-American Establishment* published in 1981. The book cites names, especially in reference to what became the Milner Group, though it is extremely sketchy on how British Round Table members interacted with the Americans they wished to influence.

The real facts of Anglo-American imperial governance remain shrouded in secrecy. The role played by the British "royals," probably the world's richest and most privileged family, is particularly obscured from view.

Creation of the Federal Reserve

The U.S. saw two major power struggles during the late 19th and early 20th centuries. One was between the heads of the big corporations and the most powerful bankers. The other was between the bankers and the federal government.

The bankers took over industry by forming trusts. These were financial holding companies that controlled the key corporations, such as the Steel Trust, the Railroad Trust, the Standard Oil trust, the Tobacco Trust, the Sugar Trust, and, most importantly, the Money Trust. By 1904, forty percent of the capital invested in U.S. manufacturing was controlled by trusts. It was also a period of growth in union membership, which reached over two million members by 1904. Indeed, it was only through joining unions that workers had a chance at gaining any benefit from the rampant industrial growth.

President Theodore Roosevelt built his reputation as a progressive through his "trust-busting" initiatives based on the Sherman Anti-Trust Act of 1890, and aided by rulings from the Interstate Commerce Commission, established in 1887. Roosevelt's activism was considered a marked departure from the period of minimal governmental intervention by the post-Civil War presidents through McKinley..

The trust that President Theodore Roosevelt failed to break up was the Money Trust. According to Nomi Prins, writing in *All the President's Bankers*, there were actually *two* Money Trusts, sometimes but not always competing with each other. One was the Rockefeller family trust, consisting of John D. and his brother William Rockefeller, James Stillman of the National City Bank of New York, E.H. Harriman, director of the Union Pacific Railroad, and the aforementioned Jacob Schiff, head of the Wall Street firm of Kuhn, Loeb & Company.

The other Money Trust, the "inner group," as it was called, was J.P. Morgan's, which included Great Northern Railway CEO James Hill and George Baker, Sr., head of the First National Bank that later became part of Citigroup. James Stillman also belonged to Morgan's trust, along with that of Rockefeller. He was able to straddle the two sets of interests.[4]

The Money Trusts were the progenitors of what economists today call the "FIRE" economy—finance, insurance, and real estate. The rise of the Money Trusts indicated a major shift of emphasis from earning money through industrial development to the making of money for its own sake. This was accomplished by the familiar method of fractional reserve banking based on usurious rates of compound interest.

The most successful application of the Sherman Anti-Trust Act was the breaking up of the Rockefellers' Standard Oil Company in 1911. But this breakup did little to slow the money-making activities of the Rockefeller family, who had already leveraged their profits from petroleum to diversify into a number of other business lines, including railroads, chemicals, mining, insurance, utilities, medicine, and banking. It was the National City Bank that came to be most strongly associated with the Rockefeller fortune. This bank would someday become Citigroup. Later the Rockefellers gained control of the Chase National Bank.

The Rockefellers were to parlay their wealth into political influence by a union cemented by marriage, with Abigail, the daughter of U.S. Senator Nelson Aldrich (1841–1915) of Rhode Island, to the only son of John D. Rockefeller—John D. Rockefeller, Jr. Aldrich was descended from 17th century Massachusetts immigrants and had himself married into wealth. After brief service in the Union army during the Civil War, he became a partner in a wholesale grocery firm before entering politics. A Republican, he won election to the Rhode Island legislature, then the U.S. House of Representatives, then the Senate. By the 1890s, Nelson Aldrich was one of the key senatorial power brokers through his position on the Senate Finance Committee, where he promoted a strong tariff policy to protect U.S. manufacturing interests. But most notably, Nelson Aldrich became the *de facto* founder of the U.S. Federal Reserve.

It must be asked: is it possible to educate the public on who is really pulling the levers from behind the scenes by means of the press? While what Theodore Roosevelt called the "muckrakers" appeared on the scene to inform people of corporate and financial abuses during the progressive era, early on, the financiers promoted their power-seeking by purchasing the big U.S. newspapers and magazines, including, most notably, top-tier publications like *The New York Times*. Control of the press and later of all other forms of mass media was becoming an obsession with the controlling elite that has continued.[5]

By now, financial power had begun to shift away from Great Britain toward the U.S., where J.P. Morgan played a central role. In the 1890s, the Bank of England approached Morgan for a bailout of London's Barings bank, which was crashing due to bad investments in Argentinian bonds.[6] This followed Morgan's aid to Great Britain in the selling of government bonds to American investors in order to fight the Boer War and the assistance provided by Morgan and the Rothschilds to President Grover Cleveland by selling him gold to cover a U.S. Treasury shortfall, an action repeated by Morgan aided by James Stillman, in buying foreign gold to cover a shortfall for the McKinley administration.

American investment capital, aided by the government, was also reaching out to control Latin American companies through what was then called "Dollar Diplomacy," as well as to China for the construction of the Hankow-Canton Railroad, to Sweden and Germany, and for railroad loans to Russia.

In 1903 and 1907 financial panics erupted when corporate stocks crashed, banks called in their loans, and workers were thrown out of their jobs. With the money supply drying up, local bank clearinghouses began issuing their own emergency currency. It was now the Republican Party and its newspaper outlets that were most aggressive in criticizing big finance and in calling out the politicians who spoke for it.

The Panic of 1907 began with a run on banks heavily invested in copper. As had happened in the past, Wall Street's speculative lending was vastly overextended. The U.S. Treasury offered banks low-interest-rate bonds and imported gold from London. The U.S. balance of trade shot up with favorable crop yields, and the panic stopped after a few weeks before too much damage was done. But for both bankers and politicians, "The panic of 1907 was the last straw."[7] A European banker accustomed to working with central banks in Britain and France declared the U.S. "a great financial nuisance."

It was at this moment that Senator Nelson Aldrich stepped in. Aldrich walked a fine line as a Republican politician at a time when the reputation of the nation's banks was in the gutter. He could not create the appearance of favoring the banks with an institutional bailout, yet he was firmly in the camp of the financiers through inclination, temperament, and his family alliance with the Rockefellers. As chairman of the Senate Finance Committee, he was free to act, especially since it was the waning days of the Roosevelt Administration, and Roosevelt's designated successor William Howard Taft, himself a patrician, was firmly on-board.

Aldrich's first accomplishment was the Aldrich-Vreeland bill of May 1908, authorizing the formation of a National Currency Association, allowing national banks to make loans based on emergency currency, setting a minuscule interest rate of one percent to the government for its deposits, and creating a Monetary Commission of nine senators and nine congressmen "to make a comprehensive study of the necessary and desirable changes in the money and banking system."[8]

Aldrich and his banker friends, with Morgan always at the center of the scheming, now went to work behind the scenes. Their objective was to create an institution—a central bank—that would give bankers backup when they ran out of gold or cash in financial emergencies and that would also offer an umbrella under which a unified currency could operate on equal footing with the British pound, French franc, or German deutschmark.

Aldrich and others traveled to Britain and Europe to study national banks in their native environments. The German banker Paul Warburg had emigrated to New York in 1902 to work for Kuhn, Loeb & Company and now took the lead in drafting a secret plan that would allow the central bank to exercise "a method of creating currency in downturns."[9] Aldrich's goal was to make the U.S. and New York the leading financial center in the world. The British, for their own reasons, would go along with it.

Aldrich, Morgan, Stillman, and Rockefeller put together a plan for a group selected by them to meet at an exclusive club at the offshore resort on Jekyll Island in Georgia. The series of meetings took place from November 20–30, 1910. Present were Aldrich, assistant treasury secretary A. Piatt Andrew, bankers Henry Davison and Arthur Shelton, president of the

Rockefellers' National City Bank Frank Vanderlip, and transplanted German banker Paul Warburg, a rising star with New York's Kuhn, Loeb, and with powerful Rothschild connections.

The product of their meeting was the plan for the Federal Reserve System, which they initially called the Federal Reserve Association. The word "bank" was not included. Benjamin Strong, future Federal Reserve governor, and Frank Vanderlip wrote the final report. It circulated within Congress and was reported in the press as the Aldrich Plan. President Taft enthusiastically supported it.

But the real mastermind behind the Aldrich Plan, though not revealed until publication in 1931of a book by Elisha Ely Garrison entitled *Roosevelt, Wilson and the Federal Reserve*, was Baron Alfred Rothschild of London.[10] At the time, all of the big U.S. banks, including Morgan's, maintained close relationships with the House of Rothschild, principally through the Rothschild control of international money markets through its setting of the price of gold. Each day, the world price of gold was set in the London office of N.M. Rothschild and Company.[11]

But a battle lay ahead. Why, critics asked, should the banks have access to readily created currency furnished by the government to step in and save them, in instances most likely caused by the banks' own speculative forays in pursuit of greater profits? Wouldn't that just be a monumental invitation to abuse?

Minnesota Congressman Charles A. Lindbergh, Sr., father of the future aviator, had introduced a resolution to investigate the money trusts and their role in the passage of the Aldrich-Vreeland Act which had led to the Aldrich Plan. Lindbergh's resolution resulted in hearings before Louisiana Democrat Arsène Pujo, chairman of the House Banking and Currency Committee. Even J.P. Morgan himself would be called before Pujo's committee which, unsurprisingly, pro-finance organs like *The New York Times,* derided. But the writing was on the wall. The Aldrich Plan might not make it through Congress.

Meanwhile, one of the strangest episodes in the history of U.S. presidential politics was unfolding. President William Howard Taft was running for re-election as president against Democrat Woodrow Wilson, former president of Princeton University and current governor of New Jersey. Taft had been a faithful successor to the wildly popular Theodore Roosevelt, and there was no reason on the face of the planet that Taft should not have been re-elected, especially since Wilson was relatively unknown at the national level and did not receive his party's nomination until the 46th convention ballot.

But on the Republican side, former President Theodore Roosevelt had broken with Taft for reasons no one has ever been able convincingly to explain. Much has been made in the historical literature about the differences between the two men, claiming that Roosevelt's views were more "progressive" than the more conservative Taft. But in some respects, the opposite was true. Taft wanted more of the trusts broken up than did Roosevelt.

Roosevelt had rejected calls to run again in 1912. After all, he had already served as president for a majority of William McKinley's term plus four full years of his own. There was no national emergency calling out for his return to the presidency. In fact, a party split could only be harmful to government stability and to the Republican Party's hold on power. Amid considerable acrimony, Roosevelt had lost out against Taft in the party convention and had no reason now to refrain from folding his tent and going home.[12]

But at the last moment, Roosevelt decided to run on a third-party ticket, which was called the Progressive Party, or, more colloquially, the "Bull Moose" Party, the "moose" obviously being Roosevelt himself. It would have been impossible for Roosevelt not to have known that splitting the Republican Party between two equally-matched candidates could have only one result: Woodrow Wilson would be elected president.

It was a set-up. The Republicans intended to lose. Why? Obviously, they were being told to do so. Who could have told them? Only their bosses—the bankers.

Woodrow Wilson, who was born in Virginia and grew up in Georgia, was the first Southerner elected president since Zachary Taylor in 1848 and the second president, after John Tyler, to have been a citizen of the Confederacy. He was also the only president to hold a PhD. A capable scholar, but a chameleon in regard to his political views, no one quite knew where he stood on anything. The only certainty was that he was a racist, later going out of his way to assure that positions filled with the federal government would not go to blacks.

During his 1912 presidential campaign, Wilson bitterly attacked Wall Street. He claimed that the Republican Party had given their campaign contributors—the financiers—"special favors and monopolistic advantages."[13] He said, "The Republican Party has put the intelligence of this country into the hands of receivers in Wall Street offices." Wilson claimed he would "break up that little coterie" of financiers and politicians "that has determined what the government of the United States should do." In another speech he said:

> The great monopoly in this country is the monopoly of big credits.
> So long as that exists, our old variety and freedom and individual

energy of development are out of the question. A great industrial nation is controlled by its system of credit. Our system of credit is privately concentrated. The growth of the nation, therefore, and all our activities are in the hands of a few men.

Wilson was elected in 1912 but with only 41.8 percent of the popular vote. It was the lowest proportion of the popular vote since Lincoln won with 39.8 percent in 1860. But 1860 was a time of extreme crisis with votes splintered among four candidates and the nation on the cusp of dissolution. In 1912, the only crisis was that the bankers were very unhappy and very unpopular.

In the election, Roosevelt received 27.4 percent and Taft 23.2 percent of the popular vote. Together, Roosevelt and Taft had won a clear majority—but they weren't together, but in opposition. The vote-splitting project worked like a charm. Socialist candidate Eugene Debs received 6.0 percent. After the election, the results of which he accepted without protest, Taft spent the World War I years as a professor at Yale University, where as a student he had been a member of the elite Skull and Bones. He got his reward in 1921, when President Warren Harding named him Chief Justice of the United States.

After Wilson was inaugurated as the 28th president of the U.S. on March 4, 1913, he had little to say about banking reform. Instead, Democratic Congressman Carter Glass of Virginia, Chairman of the House Committee on Banking and Currency, took up the issue. After some murky exchanges with Wilson, Glass got the message that the bill he would be proposing had better be very close to the Aldrich Plan that the bankers held in such high regard. In fact, he had been told as much by the Currency Committee of the American Bankers Association that now entered the fray and that ultimately got their way. Carter Glass also got his reward by later becoming Wilson's treasury secretary.

The Federal Reserve Act, a bill with immense significance for the future of money creation in the U.S. and the world, sailed into passage with little debate. It was passed on December 23, 1913 by the House of Representatives, then by the Senate the following day, Christmas Eve. It was signed immediately by President Woodrow Wilson. The Federal Reserve Board took office in August 1914.

The Sixteenth Amendment to the Constitution, authorizing a federal income tax, resolved decades of controversy about such a tax. This had taken place prior to ratification of the Federal Reserve Act. Between the income tax amendment and the Federal Reserve Act, the stage was set just in time to enable the massive explosion of American spending on World War I and the huge profits the financial system would make in handling all that money. Some researchers believe that another reason European bankers like Paul Warburg and the Rothschilds were instrumental in instigating the Federal Reserve was so that Britain and France could use it to finance their project in fomenting a major war against Germany. If this was the plan, it worked.

But some spoke up in opposition. Congressman Charles Lindbergh, Sr., said:

> This Act establishes the most gigantic trust on earth. When the President signs this bill, the invisible government by the Monetary Power will be legalized.... The worst legislative crime of the age is perpetrated by this banking and currency bill. The caucus of the party bosses have again operated and prevented the people from getting the benefits of their own government.

Lindbergh spoke of the "invisible government." President Wilson was well aware of this also, having written in *The New Freedom* in 1913:

> Since I entered politics, I have chiefly had men's views confided to me privately. Some of the biggest men in the United States, in the field of commerce and manufacture, are afraid of something. They know that there is a power somewhere so organized, so subtle, so watchful, so interlocked, so complete, so pervasive, that they better not speak above their breath when they speak in condemnation of it.[14]

The Federal Reserve was created by this "something," this "power." Wilson knew of it but did nothing to stop it or even slow it down, becoming among those who would not speak above their breath "in condemnation of it."

Through the Federal Reserve Act of 1913, the U.S. Congress ceded its Constitutional power to create the nation's currency and regulate its value. It ceded that power to the private banking industry. So whoever controls the banking industry controls the U.S. It was then that the U.S. ceased to be a sovereign republic. From then until today it's the Money Trust that rules.

ENDNOTES FOR CHAPTER 9

1 Ferguson, *Empire*, 227.
2 Ibid., 224.
3 Confederated Salish & Kootenai Tribes of the Flathead Reservation. https://csktribes.org
4 Nomi Prins, *All the President's Bankers: The Hidden Alliances that Drive American Power* (Nation Books, 2014), 52.
5 In 1915 the J.P. Morgan interests bought control of the top twenty-five U.S. newspapers. Editors were hand-picked. The focus was news favorable to big finance along with pro-British war propaganda. See Ed Whitney, *The Controllers: Secret Rulers of the World,* (American Free Press, 2015), 54. Media consolidation and control by big finance continues to this day, particularly after President Bill Clinton signed the Telecommunications Act of 1996.
6 Prins, 5.
7 Studenski and Kroos, 254.
8 Ibid., 255.
9 Prins, 20.
10 Eustace Mullins, *Secrets of the Federal Reserve* (privately printed, 1952), 41.
11 Ibid., 83. The question arises as to why Senator Aldrich thought the Federal Reserve would shift world financial power toward the U.S. when he was involving European bankers in writing the legislation. Briefly, his plan did not work. The U.S. did not gain world financial dominance until World War II. This was why the Bank of England was able to engineer the Great Depression in 1929.
12 William Howard Taft had been a longtime Roosevelt ally. For example, when he was U.S. solicitor general, Taft had urged President William McKinley to appoint Roosevelt Assistant Secretary of the Navy. Later, Roosevelt made Taft first civil governor of the Philippine colony. He then oversaw the Philippine genocide.
13 Prins, 30.
14 *The New Freedom* Quotes, Goodreads. https://www.goodreads.com/work/quotes/2528363-the-new-freedom-a-call-for-the-emancipation-of-the-generous-energies-of

CHAPTER 10

"The War to End All Wars"

World War I Begins

By the 20th century, the European system of states built on the balance of power that had been so revered by the British during the four hundred years they had spent constructing their empire had reached the brink of catastrophe. It was one thing to found, grow, and protect this empire using wooden sailing ships, trade goods, cannons, and muskets. But with the industrial age having arrived, with its steel warships, refined fuel, machine guns, barbed wire strung across trenches, poison gas, and submarines—with tanks and airplanes being readied for battle—the world had a new potential for a living nightmare.

In the America of 1914, with the income tax amendment enacted and the Federal Reserve System coming on-line, the U.S. powers-that-be in charge of managing the explosion of credit needed to finance the coming war looked to the future with sanguinity. If the wars leading up to the great conflagration—the Boer War and the Russo-Japanese War—were any indication, the cost of the great European war, with or without the U.S. being a combatant, would be immense. The money to be made—unimaginable.

J.P. Morgan, the banker with most to look forward to, had passed away in 1913, but his son, J.P. "Jack" Morgan, Jr., heir to his bank and fortune, took over without missing a beat. As Jack Morgan wrote to President Woodrow Wilson in a letter dated September 14, 1914, "The war should be a tremendous opportunity for America."[1]

Two armed camps were fixed in place—the Central Powers of the German and Austrian Empires vs. the Triple Entente of Britain, France, and Russia. Later, other nations would pick sides. The Ottoman Turks would join Germany/Austria as would Bulgaria. Italy would side with the Entente.

But why did Britain, the main manipulator in any balance of power contest, take sides with France and Russia and not with Germany? During the latter part of the 19th century, it had seemed that a German alliance would provide Britain a valuable counterweight to its greater imperial rivals—France in the Middle East and Africa and Russia with respect to control of the Black Sea-Mediterranean naval route and the Russian push through Central Asia towards India.

What may have been key, however, was money and finance. France was solvent, and like Britain, a lender to the world. And the interests of the Bank of England and the Banque de France were joined in controlling the world's gold supply. Nothing could matter more in a world based on the gold standard, where nations literally sent gold bullion back and forth on steamships, depending on the exigencies of trade and lending. A nation whose gold ran out was "kaput." Historians tend to cite Britain and France as exemplars of "democracy," but that's propagandistic nonsense. Both nations were plutocracies, as the U.S. was on its way to becoming. And rich people tend to hang together, especially when they are raking in fortunes by financing businesses such as the Maxim-Nordenfelt Guns and Ammunition Company, a Rothschild flagship.

As far as Germany and Austria-Hungary were concerned, neither was sitting on top of the world financially or politically. In fact, both nations were politically volatile, potential power kegs. Germany, under Bismarck, had imposed viciously repressive policies in outlawing the Socialist Party. German workers were not impressed by Prussian militarism and were not keeping up with living expenses in a country that lacked a lucrative empire to funnel profits into the home economy, as Britain and France possessed. Germany's would-be African empire never got off the ground—not much in the way of diamonds and gold. Austria-Hungary was an outdated hodgepodge of divergent nationalities, languages, and religions that was not that far advanced economically over its medieval past. Did the German and Austrian imperial elites go to war to divert the masses? Maybe. And maybe so did Russia which had its own powder keg of dissidents.

The alignment in Europe was almost the worst imaginable from the standpoint of potential for carnage. The foregoing might tempt us to believe that if only Europe had a functioning financial system, instead of a system where bankers financed both sides and reaped bonanzas on the manufacture and utilization of armaments, World War I might not have happened. At the same time, deeper spiritual forces may have brought world civilization to the point of either hitting the wall or rising to the many challenges a new age had wrought. Mechanistic analysis may take us only so far in understanding events.

Meanwhile, Marxists and Socialists thought worker solidarity would prevent the outbreak of war. Eugene Debs, head of the Socialist Party of

America, ran for president of the U.S. five times, the last time in 1920, when he ran from a prison cell in which he had been confined for alleged violation of the Sedition Act of 1918. There were many who believed that a modern industrial war would be so horrible that this fact alone made it impossible. And yet…war had always been a part of things in Europe, still viewed as the path to glory, power, and wealth.

On June 28, 1914, the Austrian heir to the throne, Archduke Franz Ferdinand and his wife were assassinated by Serbian nationalists. The Austro-Hungarian Empire was breaking apart, with Serbia, friendly to Russia, seeking autonomy. When Austria delivered an ultimatum to Serbia to essentially give up its sovereignty in an investigation of the assassination, then declared war when Serbia balked, Russia mobilized. Then Germany mobilized. Then Britain and France did the same. Mobilization in 1914 had no reverse gear.

Germany knew that a two-front war against Russia in the east and Britain and France in the west could be fatal. So Germany activated the Schlieffen Plan by sweeping through Belgium to outflank the French. Belgium leaned toward France and Britain but was technically neutral. Britain and France feigned outrage. The lie that they acted in order to preserve the honor of poor Belgium is still mouthed in some circles. It's similar to today's idiocy that the U.S. finagled its proxy war against Russia in defense of poor Ukraine or that Custer attacked the Sioux because they refused to become "civilized."

The disaster had begun: the Christian nations of Europe, the dominant world powers, began destroying each other and themselves. Though their governments claimed they could knock out the enemy by Christmas, no one believed it. On the Western front, the British and German troops came out of their trenches to set up Christmas trees and celebrate the holiday together with toasts. But it would be the last time. The inexorable process that would lead to the death of millions had begun and would not stop for four years. Maybe Europe's leadership wasn't so "Christian" after all.

The people of the U.S. did not want to go to war. The largest number of U.S. immigrants, after those from the British Isles, had been German. There were places in the U.S. where German was still spoken. So the U.S. government under Woodrow Wilson feigned neutrality. U.S. bankers began to lend to both sides. But the British and French were by far the favorites, particularly for the house of Morgan with its long-standing ties to the bankers of London—the Barings, the Rothschilds, the Cassels, and others.

In Britain, with David Lloyd George's Liberal coalition and Round Table head Alfred Lord Milner embedded as a permanent member of the War Cabinet, it may have seemed that this was the time that Cecil Rhodes' dream would be achieved—"recovering America for the Empire"—indeed, while rescuing the British Empire from ignominy and dissolution.

Germany failed in its initial thrust to achieve a decisive victory by the aforementioned strategy of outflanking France through Belgium in the autumn of 1914. Then it became Total War: thirty-one million combatants killed on the two sides and at least eight million civilians.

The two-front war that Germany faced saw Russia go down first. In the East, Russia's armies collapsed before the Czar was overthrown and that nation swallowed up in the Bolshevik Revolution, leading to a separate peace between Russia and Germany through the Treaty of Brest-Litovsk.

The Western Front is a thing of legend, with a generation of young European men destroyed by the insanity of war. Less remembered, at least in Europe and America, is the Middle Eastern front. Russia begged Great Britain to attack the Ottoman Empire and force an opening through the Dardanelles to the Mediterranean Sea. Some say that Britain deliberately lost the Gallipoli campaign to assure that Russia would be blocked from that longstanding objective.

Meanwhile, Britain succeeded in breaking up the Ottoman Empire to the south and eventually gaining control of Palestine, Transjordan, and Iraq, as well as influence over Arabia and Persia to ensure perpetual access to the oil resources required to fuel its fleet.

U.S. Dollar Neutrality

On August 19, 1914, President Woodrow Wilson announced the U.S. policy of neutrality. The House of Morgan, now being run by Jack Morgan, jumped in as the acknowledged leader of U.S. finance and as being totally and publicly on the side of Great Britain and France. His stance came close to costing him his life when a would-be assassin—a German nationalist—entered his Long Island mansion and shot him twice in the groin. He recovered, while the shooter killed himself.

Despite proclaimed U.S. neutrality, Wilson's entire economic program was designed to benefit the Allies. As soon as Wilson was inaugurated on March 4, 1913, a major tariff reduction was passed by Congress and signed into law by Wilson, making its tariffs the lowest since 1857. Tariff reduction by the U.S. and adherence to free trade principles had long been the goal of Britain in influencing U.S. policy. To compensate for lost customs revenues, Congress enacted the first national income tax under the new 16th amendment, with a top rate of six percent for incomes over $500,000.[2]

Financing the war cost the U.S. ten times as much money as the Civil War, with income tax rates much higher. The government steadily inflated the money supply by selling bonds through the commercial banks. But with the U.S. no longer dependent on British or European money markets, there was no danger of drawing down gold deposits. The U.S. had become the allies' supply depot.

When World War I began, European investors panicked and sold off securities at such a pace that a mild recession took hold in the U.S. But soon the favorable U.S. balance of trade grew each year from 1914 to 1917, reaching over $3 billion. Unemployment disappeared, with the gross national product increasing by twenty percent, as New York sought the financial leadership of the world. Such was the goal the Money Trust had pursued through the creation of the Federal Reserve.

While Britain aimed to use borrowing from U.S. banks and bond sales to investors to finance its war, a power struggle between British and U.S. bankers was underway as the world's gold began to weigh down the ships steaming to New York, with the U.S. accumulating the largest gold stocks of any nation in history. Nothing demonstrated more the absurdity of the gold standard than these awkward and well-guarded movements of bullion.

Within the domestic U.S. economy, the availability of cash caused interest rates to plummet. Bank loans increased by thirty-eight percent and bank deposits by fifty percent. The Federal Reserve was scarcely needed during this flush period. In order to put the growing economy to use for the war effort, Congress raised taxes again, and the Treasury began to sell its own certificates of indebtedness. In spite of neutrality, appropriations for the U.S. Army and Navy shot up.[3]

To pay for U.S. military expenditures and federal loans to the Allies, Secretary of the Treasury William Gibbs McAdoo, Jr., proposed that fifty percent be covered by taxes. The financiers lobbied for lower taxes that would create less of a bite on their own and their investors' incomes, with Jack Morgan lobbying for twenty percent. A compromise was reached at 33-1/3 percent. The rest would be borrowed, with the banks the prime beneficiary due to their commissions.

The war could not have been financed without the Wall Street bankers. Early on, Morgan and his partners formed a syndicate to manage a $500 million bond issue for the British and French governments.[4] Other U.S. banks followed suit as the war dragged on. U.S. banks also became the depository of choice for foreign securities. This influx of capital greatly increased the reserves against which the banks could lend globally.[5]

Another source of U.S. war funding was through Liberty Loan bonds sold to private citizens in a fashion similar to Jay Cooke's bond sales during the Civil War. Seventeen billion dollars was raised by the government through a succession of Liberty Loan issues. Many purchasers paid by borrowing from banks at low interest rates, anticipating the practice of borrowing "on margin" that would help trigger the stock market crash of 1929.

With Treasury now following a cheap money policy, the Federal Reserve began to make reserves easily available to banks for expansion of their lending. The reserve ratio for national banks was set at a rock-bottom level of 18 percent. Never in U.S. history had so much money created "out of thin air" been in circulation. Interest rates to consumers ranged from 3.4 percent to 5.9 percent. With the continuing availability of cash, the government raised income taxes again to a top rate of over 60 percent for high earners. Liquor and tobacco were both taxed heavily, bringing in almost $700 million annually by 1919.

During the war, the federal government largely took over the nation's economy, nationalizing railroads, directing the flow of products, controlling prices, and even limiting profiteering by the armaments industries. The U.S. economy and government performed surprisingly well, even as millions in Europe perished under fire.

The U.S. Enters the War

It was Germany's reliance on submarine warfare to break Britain's maritime blockade that supposedly brought the U.S. into the war. On May 7, 1915, a German U-boat torpedoed the British luxury liner *Lusitania* off the coast of Ireland, with 1,200 passengers, including 124 U.S. citizens, losing their lives. The *Lusitania* had departed for Liverpool from New York—in fact, the German government had taken out newspaper ads in New York, warning that the shipping lanes around Britain were now a war zone and that ships were "liable to destruction." The British and American governments knew that German U-boats were active in the area.

Besides passengers, the *Lusitania* was carrying munitions being sent by U.S. manufacturers to supplement the British war effort—170 tons of rifle ammunition and 1,250 cases of artillery shells, as well as fifty barrels each of flammable aluminum and bronze powder. Today we know that Britain's First Lord of the Admiralty, Winston Churchill, likely had operatives tracking the course of the *Lusitania* who nonetheless failed to either route the ship around the north of Ireland or provide an escort as it approached shore. Either step might have prevented the disaster.

Between the sinking of the *Lusitania* and the U.S. entry into the war on April 6, 1917, incidents of U-boat attacks on both military and civilian shipping continued, as Britain's blockade of Germany tightened in an effort to starve the German population. The U.S. maintained back-channel communications with Germany, which pledged to cease attacks against Americans, but unrestricted submarine warfare prevailed. The U.S. also intercepted a note from Germany to its embassy in Mexico—the "Zimmerman Telegram"—that purported to offer Mexico recovery of its territories lost in the Mexican War if it entered the war on Germany's side. The American press reacted with outrage.

Colonel Edward House, a Texas magnate viewed by history as President Wilson's controller, made several trips to England and France during the years preceding the U.S. declaration of war. He was known to have been close to Alfred Milner,[6] a member of Lloyd George's War Cabinet and leader of Cecil Rhodes' Round Table. House returned to the U.S. in time to run Wilson's 1916 election campaign based on the slogan: "He kept us out of war." The slogan won Wilson votes in the Western states and probably swung what was a close election his way. It was not until votes were counted in California that Wilson was declared the winner over Republican Supreme Court Justice Charles Evans Hughes.

Wilson's sloganeering was a lie. Wilson, House, Morgan, and the rest of the U.S. elite had every intention of entering the war as soon as possible after the election. Former President Theodore Roosevelt and others of the East Coast upper crust were cheerleading vociferously for American entry.

During the summer of 1918, American soldiers under General of the Armies John Pershing, Commander-in-Chief of the American Expeditionary Force (AEF), were arriving in Europe at the rate of 10,000 men a day. During the war, the U.S. mobilized over four million military personnel and experienced the loss in battle of 65,000 soldiers.

Both of my grandfathers served in the war. My paternal grandfather, Frederick Steele Fitts Cook, was drafted into the Army in 1918 but never got overseas. My maternal grandfather, Carlton William Peilow, sailed on the naval convoys out of the Brooklyn Navy Yard that transported troops to Europe. While in Brooklyn, he met my grandmother, Ethel Brown. My mother, Marjorie Virginia Peilow, their only child, was born in Brooklyn in 1922.

Wartime Propaganda

How is it that one day we look at a situation without emotion and the next, once war has been declared, we hate and want to kill? Are human beings really so lacking a moral center? Surely there has to be some unconscious force that triggers such enflamed emotions.

In past American wars, the combatants and their supporters certainly had a direct emotional commitment. In the American Revolution, those favoring independence were willing to stake their property and even their lives on winning. The Continental Congress that voted for independence did so on the basis of a document—the Declaration of Independence—that provided much detail about the transgressions of King George III. Later wars were preceded by speeches and written statements by presidents and other officials on the reasons for declarations of war that were printed in newspapers and distributed in written flyers.

But it was really "Remember the *Maine*" at the start of the Spanish-American War that shifted the conversation in the direction of slogans and propaganda. And with World War I, propaganda in the generation of public support for the war came into its own. Now it was "Remember the *Lusitania!*" The mass media propaganda machine quickly became an industry. Similar to the proxy Ukraine war against Russia which is enflaming instant hatreds through modern cybercommunications, World War I propaganda was able to mold the very fabric of human consciousness, leading even highly educated individuals to become pro-war zombies.

A thorough discussion of the madness of World War I may be found in Oliver Stone and Peter Kuznick's *The Untold History of the United States* and will not be repeated here. They describe the activities of the newly-minted Committee on Public Information, the government's official propaganda bureau, the firing of university faculty members who spoke out against the war, and the jailing of dissidents under the Espionage Act of 1917 and the Sedition Act of 1918. "Hundreds of people were jailed for criticizing the war, including IWW leader 'Big Bill' Haywood and Socialist Eugene Debs."[7]

There was little in U.S. history that could explain the extent and vehemence of the suppression of free speech by the U.S. government during World War I. It was as though an alien force had taken over.[8] This appears to be the same force of nihilism and societal self-destruction that has continued to erupt time and again over the past century, a force depicted by George Orwell in *Nineteen Eighty-Four.* Today that force seems to be on the move again, possibly leading us into the final nuclear conflagration.

Zionism and the Balfour Declaration

On November 2, 1917, British Foreign Secretary Arthur Balfour sent a letter to Lord Lionel Rothschild, a leading British banker. This letter contains what is known as the Balfour Declaration and includes the following statement:

> His Majesty's Government view with favour the establishment in Palestine of a national home for the Jewish people, and will use their best endeavours to facilitate the achievement of this object, it being clearly understood that nothing shall be done which may prejudice the civil and religious rights of existing non-Jewish communities in Palestine, or the rights and political status enjoyed by Jews in any other country.

The original of the letter is in the British Library with a notation listing Walter Rothschild as the author, along with Arthur Balfour, Leo Amery, and Lord Milner. Walter Rothschild was the son and heir of Nathaniel Rothschild, who died in 1915. Leo Amery was a member of Parliament and a close associate of Milner, whose exploits were addressed earlier. Given this authorship, it is evident that the Zionist project that resulted in the creation of the modern state of Israel was in fact a joint undertaking of the Rothschild family, the British government, and the Round Table secret society founded by Cecil Rhodes of which Milner was the head.

It was once possible, but today less so, to discuss the merits or provenance of Zionism, since the U.S. establishment is on a hair-trigger alert to condemn anyone they perceive as critical of the Jewish people, Zionism, or the modern state of Israel. Those who do so face the dreaded epithet "anti-Semitic," to the point where it might be concluded that Israel and the Jews can "do no wrong," though it is universally agreed that no human on earth can "do no wrong."

Zionism was originally a movement that aimed at establishing a "homeland" somewhere on the planet for the Jews. At different times, the British government considered East Africa, the Sudan, and Mozambique, but those options were rejected by the Zionists and their supporters who insisted on Palestine.

Why did people in Europe who identified as Jews want their own homeland in Palestine—in addition to the obvious religious associations? Could it have had anything to do with the desire on the part of Britain to control the Middle East and its oil?

The Russian Revolution

Discussions concerning the history of the Russian Revolution, the attempted erasure of the Christian roots of Russian society and culture, and the massive suffering of the Russian people are complex and still rife with controversy. With the Russian army suffering massive casualties on the front against Germany and Austria, the government of Czar Nicholas II had begun to consider a separate peace. The British were outraged and favored the creation of the Provisional Government under Kerensky that vowed to continue fighting.

Some state that what enabled Lenin, Trotsky, and the Bolsheviks to succeed in overthrowing Kerensky and the Provisional Government in November 1917 was that the Russian people, sick of the bloodletting, backed them in their determination to take Russia out of World War I—that this was the policy that formed the catalyst and core of their support rather than the struggle for communism per se. It was the Germans who allowed Lenin to journey from Zurich to Russia on the "sealed train," understanding that the Bolsheviks would sue for peace.

It was the Jews of Russia and the world that had been most strongly opposed to Czarist Russia. Millions of Russian and Eastern European Jews had already migrated to America. New York banker Jacob Schiff of Kuhn, Loeb & Company provided substantial amounts of financial aid to millions of Jews leaving Russia before the war. Schiff is also known to have financed the return of Leon Trotsky from New York to Russia to help steer the Bolshevik Revolution. Trotsky became head of the Red Army that crushed the Whites in the Russian Civil War. Later, having broken with Stalin, Trotsky fled Russia, emigrated to Mexico, and was assassinated by one of Stalin's agents in 1940. Some say that Britain and the U.S. secretly favored the Bolsheviks, despite their token aid to the Whites in the civil war, because they feared a future alliance between the aristocrats of Germany and those of Russia.

With Russia out of the conflict, Germany was able to shift its army to the Western front. But this was not enough to counter the arrival of over a million fresh troops from the U.S. Britain always knew that U.S. "boots on the ground" were their "ace in the hole."

The War Ends

By November 11, 1918, Germany had enough and asked for a cease-fire, so the war ended with an armistice. According to some sources, the end was delayed because President Woodrow Wilson was trying to maneuver the combatants into an agreement to stop fighting on the basis of his Fourteen

Points. Germany assumed that the Allies would adhere to the Fourteen Points in the coming peace conference, and that the Allies would leave Germany as a functioning nation. They did not, so Germany felt betrayed.

One thing is certain about World War I: it destroyed four empires: the German, the Austro-Hungarian, the Ottoman, and the Russian. At war's end, one empire stood victorious: the British. But as the BBC acknowledges, "The expense of World War I destroyed British global pre-eminence. Territorially the British empire was larger than ever."[9] In fact, Great Britain still carries debt from World War I—to the tune of about two billion pounds ($3.2 billion).[10]

With regard to the defeated empires, there was no clear model as to what kind of government would replace the ones that were shattered. Germany, Austria, Hungary, and Turkey became republics, after a fashion, and Russia became the Union of Soviet Socialist Republics. But what any of this meant, no one was certain. The collection of nations that existed after the war did not bring the world stability. The condition of the German Weimar Republic was particularly rocky.

One of the most disturbing things about World War I, beyond the extensive loss of life, was the widespread use of chemical weapons. These were used by both sides. After the war, claims were heard that if the war had not ended as it did in November 1918, the Allies were ready to use new chemical weapons in their planned 1919 spring offensive that would be so deadly the Germans would surrender immediately.

The U.S. never joined the League of Nations because the U.S. Senate refused to ratify the Treaty of Versailles. The main reason senators opposed the League of Nations was the provision in Article X of the League Covenant that would have required members to come to the defense of any other member that was attacked. This provision was seen as violating Congress's war making powers. Article X was similar to the future Article V of today's NATO pact. The League of Nations was also viewed by the European powers as their vehicle to enforce the gold standard, which the U.S. Senate also opposed.

The Treaty of Versailles placed a much more onerous burden on Germany than the Fourteen Points. Germany was required to pay reparations for damages in the amount of $33 billion in addition to returning Alsace-Lorraine to France, territories annexed by Germany after the Franco-Prussian War of 1870. Britain was given Germany's colonies in Africa and mandates over the former Ottoman regions of Palestine, Transjordan, and Iraq. France got mandates over Syria and the part of the Levant which became Lebanon.

The reparations were twice what Germany expected to pay but less than the $40 billion which the House of Morgan demanded.[11] Germany also had to accept the "War Guilt Clause," which made them accept sole responsibility for the war. The grudge the Germans bore at being "betrayed" at Versailles became a factor in the buildup to the next world war.

The territorial division in the Middle East put Britain in control of Middle Eastern oil. U.S. Senator William Borah, R-ID, called the Treaty of Versailles "a cruel, destructive, brutal document" that resulted in "a league to guarantee the integrity of the British Empire."[12] He was right. According to historian Niall Ferguson, "The Secretary of State for India, Edwin Montagu, commented that he would like to hear some arguments against Britain's annexing the whole world."[13]

The Treaty of Versailles didn't come close to addressing the underlying cause of the war. As Congressman Charles A. Lindbergh, Sr., put it, the cause was failure of an international system based on the control of national economies by the financial elite who extracted unbridled profits from lending money at interest, with the borrowing undertaken by the various national governments they effectively owned.

The subsequent debt drove the nations of Europe to colonize the world in order to pay it off. They needed a flow of value from their colonies to the master nations, but not every colonial power could continue their extractive policies at the same time. Inevitably, there were conflicts, competition, and clashes.

While the nations of Europe licked their wounds, Germany had been driven into a corner and Russia collapsed in revolution and civil war. Though Britain had been hurt badly and was no longer the financial master of the world, it had its colonies, still controlled much of the resources of the undeveloped world, especially oil, and would attempt a comeback.

The U.S., now unburdened of any further need to rescue the Western Allies from their own stupidity, was becoming the lender of choice, particularly in Latin America. During the 1920s, until the stock market crash of 1929, U.S. banks moved aggressively to invest in Central and South America and made many of these nations *de facto* financial colonies, including Cuba, Nicaragua, and Venezuela, where the big U.S. petroleum companies moved in to exploit the oil fields.

The U.S. continued to attract the Western world's gold reserves as backing for the financial boom of the Roaring Twenties, while Britain was threatened with eclipse.

The Unlikelihood of an End to War

There was a time when people really believed that World War I would be the last war, including British author H.G. Wells in his 1914 propaganda pamphlet, *The War That Will End War*. Very soon the world saw that the next war was inevitable. But what Ludwig Dehio calls the two "flanking powers"—Russia and the U.S. (now replacing Britain)—were temporarily absent from the world scene.

The attention of Russia was consumed by the communist revolution and its attendant horrors. The U.S. had turned its back on Europe by rejecting the Treaty of Versailles and the League of Nations. French nationalists took charge at Versailles, seeking to so load down Germany with constraints and reparations that it could not arise again, though Britain seemed to wish to keep some semblance of German functionality, now as a buffer against former ally Russia. So, on the Continent it was the same old power politics. Europe had learned nothing.

With Britain's resurgence through the acquisition of new colonies, the Empire seemed it might emerge triumphant. Yet the Round Table knew that Britain would only be safe if it gradually began to integrate its colonies into partnership under the Crown. Thus came into existence the British Commonwealth of Nations, declared at the 1926 Imperial Conference and codified through the Statute of Westminster in 1931. The "commonwealth" was assented to by eight heads of state (including Newfoundland). It was still the British Crown in charge.

Meanwhile, American and British elites took steps to keep the two nations marching toward the future together. The result in Britain was creation of the Royal Institute of International Affairs, also known as "Chatham House." The Royal Institute was closely aligned with the Round Table and the Milner influence coming down from Cecil Rhodes and Nathaniel Rothschild. A key figure in defining the mission was the English historian Arnold Toynbee.

In his book *The Anglo-American Establishment*, Carroll Quigley writes as follows:

> Among the ideas of Toynbee which influenced the Milner Group we should mention three: (a) a conviction that the history of the British Empire represents the unfolding of a great moral idea—the idea of freedom—and that the unity of the Empire could best be preserved by the cement of this idea; (b) a conviction that the first call on the attention of any man should be a sense of duty and obligation to serve the state; and (c) a feeling of the necessity to

do social service work (especially educational work) among the working classes of English society. These ideas were accepted by most of the men whose names we have already mentioned and became dominant principles of the Milner Group later. Toynbee can also be regarded as the founder of the method used by the Group later, especially in the Round Table Groups and in the Royal Institute of International Affairs.[14]

What Quigley fails to mention is that all of these fine-sounding words rest on one foundation: the vast financial power that propped up the British state, including the Crown, and that derives in a significant sense from the extraction of profits and resources from the colonies. This power was now threatened by British indebtedness to the U.S. and possession by the U.S. of much of the world's gold.

In the U.S., a parallel institution to the Royal Institute was created: the Council on Foreign Relations. The Council grew out of joint meetings between British and American diplomats, scholars, and other interested parties at the Hotel Majestic in Paris in May of 1919, just before the signing of the Treaty of Versailles. While the British went home to set up the Royal Institute, the Americans were more circumspect. Almost a year earlier, a set of secret meetings had taken place in New York City, headed by corporate lawyer Elihu Root, who had been secretary of state under President Theodore Roosevelt. The meetings were attended by 108 high-ranking officers of banking, manufacturing, trading and finance companies, together with many lawyers. The Council on Foreign Relations was the eventual result.

President Woodrow Wilson was succeeded by Republican Warren Harding, who won the 1920 presidential election with his slogan of "Return to Normalcy." That meant, principally, that big banking and big finance were unleashed, and the Council on Foreign Relations, which was legally chartered on July 29, 1921, would be the instrument of those interests.

In 1922, the first issue of the Council's *Foreign Affairs* was published, using money donated by those whom they called the "thousand richest Americans." The Rockefeller fortune was and remains instrumental in the operations of the Council on Foreign Relations throughout its history as the premier U.S. instrument of international financial control. This New York City institution run by and for the elite would be at center stage in the U.S. decision to enter World War II and in setting what the nation's war aims would be.

ENDNOTES FOR CHAPTER 10

1 Prins, *All the President's Bankers*, 40.
2 Studenski and Kroos, 273. Also see Switzer.
3 Wartime statistics from Studenski and Kroos.
4 Prins, 44.
5 Ibid., 47.
6 Now Alfred "Lord" Milner.
7 Oliver Stone and Peter Kuznick, *The Untold History of the United States* (Gallery Books, 2012), 14.
8 An outstanding account of the extreme development of pro-war propaganda as well as the financial machinations by Wall Street during World War I may be found in F. William Engdahl, *Gods of Money: Wall Street and the Death of the American Century.* Some believe that the early Zionist forces made a deal with the British that if they could bring the U.S. into the war then they would receive British support on setting up a homeland in Palestine. See "Who Wrote the Balfour Declaration and Why: The World War I Connection," *IMEMC News,* October 25, 2017. https://imemc.org/article/who-wrote-the-balfour-declaration-and-why-the-world-war-i-connection/. Also: Shlomo Avineri, "Britain's True Motivation Behind the Balfour Declaration" (opinion), *Haaretz,* November 2, 2017. https://www.haaretz.com/opinion/2017-11-02/ty-article/.premium/britains-true-motivation-behind-the-balfour-declaration/0000017f-dc3d-d3ff-a7ff-fdbdc5ed0000
9 Rebecca Fraser, "Overview: Britain, 1918–1945," *History,* BBC, February 17, 2011. https://www.bbc.co.uk/history/british/britain_wwtwo/overview_britain_1918_1945_01.shtml>
10 "The NYC and London Banksters who financed the Bolshevik Revolution," *The Millennium Report,* July 17, 2018. https://themillenniumreport.com/2018/07/the-nyc-and-london-banker-who-financed-the-bolshevik-revolution/
11 Stone and Kuznick, 36.
12 Ibid., 36.
13 Ferguson, *Empire*, 315.
14 Carroll Quigley, *The Anglo-American Establishment* (GSG and Associates, 1981), 42.

CHAPTER 11

The Roaring Twenties and the Depression

"Good Times"

A Republican president, Warren Harding, was elected in 1920, though he served only two years before dying of a heart attack. Both he and his successor, Calvin Coolidge of Massachusetts were pro-business and in favor of letting go of control over the U.S. economy that had been dictated by the war. Accordingly, they cut taxes deeply and sought to reduce the national debt. By 1926, taxes on millionaires had been reduced by two-thirds. Also cut was the federal budget, so impact on the national debt was negligible.

On the surface, the "Roaring Twenties" was a period of remarkable economic vitality, with eight years of "good times" after a post war recession. From mid-1921 to 1929 the U.S. gross national product grew from $69.9 billion to $99.4 billion, and productivity increased three percent annually. Remarkably, there was virtually no inflation. Much of the economic growth was enabled by consumers taking on debt by buying on the installment plan, but no one worried about that.

By this time, the Federal Reserve had adopted its now-familiar role of lending to the banks so they could lend in turn, though at a higher rate, to business and consumer customers. The flush times were made even more so by low Fed discount (interest) rates, resulting in an easy money policy. The Federal Reserve Act of 1913 led to the implementation of a uniform interest rate for the entire country, a rate set by the Federal Reserve Bank of New York, owned by member banks on Wall Street. This policy destroyed the ability of local or regional banks to adapt their lending policies to local economic conditions. The Money Trust now reigned supreme and still does today.

Then, as at present, the federal government favored low interest rates as it would reduce the interest it had to pay on the national debt. Even at low rates, T-bonds were in demand as a safe parking place for wealth. Their use as a basis for fractional reserve bank lending was an early form of "quantitative easing."

The Federal Reserve's expansive monetary policies were encouraged by the Bank of England, whose higher interest rates were luring back the gold it had relinquished to the U.S. during its massive World War I borrowing. London was also sucking up gold from India and Germany, among the causes of the infamous German hyperinflation of 1922–1924.

The Bank of England and the New York Federal Reserve Bank had been maneuvering to deal with adverse conditions in Europe, where Germany was defaulting on its war reparations to France and Britain, causing France in particular to renege on its debts to American banks. When Germany defaulted on its reparations in 1923, the French and Belgians sent in troops to occupy Germany's Ruhr industrial region. But after the hyperinflation was done wreaking havoc, Germany's reparations problems seemed to be fixed by the U.S. Dawes Plan.

The Dawes Plan set up a circular flow of scaled-down payments from Germany to France, with money then reverting to the U.S. for recycling back to Europe in new loans, then back along the same path, with the banks always raking in commissions. More than one observer called the whole process "absurd," but for the time it seemed to work. And the Ruhr was evacuated.

Much of the loose money now floating around within the U.S. economy went into stock market and real estate speculation. Starting in 1924, U.S. stocks and bonds also began to attract large amounts of capital from Europe. This included French and German investors seeking a safe haven, with many well-off Germans having already transferred capital abroad before the German hyperinflation.

A lot of U.S. money went into bank deposits, which increased bank lending capabilities and enabled a spree of overseas lending, as well as domestic loans to allow individuals to buy stock "on margin." In fact, a majority of the money going into the stock market had been borrowed from U.S. banks.

With so much foreign lending, the dollar was beginning to compete with the British pound as a world reserve currency, a battle the dollar would one day win—but not yet. U.S. corporations were meanwhile expanding overseas, with Standard Oil starting to invest in the oilfields of Iraq and Saudi Arabia.

Nineteen-twenty-four was the most prosperous year in history for the New York banks. The 1920s were even spoken of as a second "Gilded Age." When Herbert Hoover won election as president in 1928, he said, "We in America are today nearer to the final triumph over poverty than ever before in the history of every land."[1] The stock market reached record highs by the fall of 1929. But some sensed trouble, with insiders beginning to make money "by shorting the market"; i.e., betting on a decline.

Millionaire Secretary of the Treasury Andrew Mellon was in the middle of everything, and his main interest was in cutting income taxes. Anticipating President Ronald Reagan's "supply-side" tax cuts in the 1980s, Mellon argued that lower taxes would mean more disposable income and greater economic growth. For a while he seemed to be right. *The New York Times*, always the house organ for the U.S. establishment and the New York City power elite, cheered it all on. Even after Black Thursday, the *Times* assured Americans that the "Big Six" banks had everything under control.

It wasn't that no one saw what was going on. Even President Coolidge himself said, "The whole country from the national government down has been living on borrowed money."[2] Not long after the 1929 crash, a third of the nation's banks closed and unemployment hit twenty-five percent.

Rebuilding the Italian and German Economies

Meanwhile, the Morgan Bank, under top Morgan executive Thomas Lamont, also with Rockefeller connections, became the principal lender to Italy's Prime Minister Benito Mussolini and for the next decade provided the money for the rise of Italian fascism. Later, when Lamont became Morgan Bank chairman, he continued the bank's relationship with Mussolini almost until World War II.

Germany too began to rebuild its economy with credit supplied by American banks. In late 1923, Britain and the U.S. installed German banker Hjalmar Schacht as the Weimer Republic's new Commissioner on the National Currency.[3] Schacht's program was based on forming giant German industrial cartels that would issue their own bonds, primarily to American investors. This would enable the building of a new German currency based on the Reichsmark, now under British control through the gold standard system run out of London. Schacht next became governor of the German Central Bank.[4]

Behind the scenes, Britain and the U.S. had decided to encourage Germany to rearm as a bulwark against the Soviet Union.[5] Ever since the short-lived 19th century League of the Three Emperors linking Germany, Russia, and Austria-Hungary, it was an axiom of British foreign policy *never* to allow Russia and the German-speaking nations to form an alliance.[6] Such a step would have violated Britain's longstanding policy of *always* opposing a hegemonic Continental force. "Divide and conquer" was ever Britain's rule. That rule enforced another foundational axiom: that Britain and other nations were in a perpetual state of hot or cold war, always egged on by the hyena-like British press.

Between the two world wars, the Soviet Union itself was rebuilding its military machine, using Western money obtained from selling Czarist gold. The impoverished Germany, however, could only rebuild by taking out British and American loans. By 1930, $28 billion flowed into Germany, half from the U.S.[7] This included loans by Morgan & Co., the Rockefeller's Chase National, Dillon & Reed, V.A. Harriman, and Brown Brothers to such German armaments firms as I.G. Farben, the Vereinigte Stahlwerke coal and steelworks, and AEG, Germany's General Electric. Thus, the ground for World War II was being deliberately enabled by the Anglo-American financial elite, in anticipation that the German military would strike against their emergent class enemy, the Soviet Union.

Prohibition and Organized Crime

Organized crime has held the U.S. in its grip for decades and does so still today, especially with the emergence within the U.S. of murderous Mexican drug cartels encouraged by the Biden administration's open border policies. The frolics of the 1920s produced enough surplus cash within the U.S. to fuel a drastic rise in organized crime, but at this time, as described by Stephen Fox in *Blood and Power: Organized Crime in Twentieth-Century America*, it was the prohibition of alcohol consumption that enabled organized crime to spread its tentacles across the entirety of the U.S.

Prohibition lasted from 1920 to 1933. The U.S. had an excise tax on liquor since 1791 which had long been a big money-maker for the federal government. Ever since, there had been an undercurrent of tax evasion by the manufacture and sale of bootlegged liquor. But the evasion became overwhelming with Prohibition. This failed attempt to outlaw all alcoholic beverages resulted in the creation of vast fortunes from criminal enterprise and turned a majority of the U.S. population into scofflaws.

Prohibition was among the worst public policy failures of modern history. The gangsters who controlled the alcohol trade combined it with profiteering in every vice, including gambling, prostitution, extortion, bribery, human trafficking, and later illicit drugs. Alcoholism did see a slight decline, but at a huge cost. Rates of liver cirrhosis, alcoholic psychosis, and infant mortality also dropped. But the net result of all this was a profound decline in private and public morality induced within the entire U.S. population, with effects that persist today.

Organized crime remains a gigantic problem in the U.S. Illegal drug trafficking has had savage effects on public health that the U.S. government has never effectively dealt with. It's hardly surprising, when agencies of the U.S. government such as the CIA have also been involved. Indeed, the agency itself has admitted to drug involvement in reports from its own inspector-general that would become established over time through the drug trade in Southeast Asia, Latin America, and Afghanistan.

The FBI under J. Edgar Hoover failed to take organized crime seriously. In fact, the FBI and CIA made an alliance with organized crime where it carried out various unsavory projects for these agencies, including assassinations.

While the Roaring Twenties, vividly characterized in F. Scott Fitzgerald's *The Great Gatsby,* seemed to portray the era as one in which everyone was having fun—though Gatsby himself was shot to death in a swimming pool—the need for loose money was ever present to enable the U.S. to keep up with its explosive economic growth and all the new consumer industries, particularly automobiles, fancy new homes, and bank buildings that looked like Greek and Roman temples.

The Bankers' Catalyst Sparks the Great Depression

Surprisingly in hindsight, almost no one in America was aware of what was going on in Britain and Europe that might have foreshadowed disaster; the bubble of the Roaring 20s was that all-encompassing. But explosions happen fast. There were signs as early as 1924:

> Chase's chief economist, Benjamin Anderson, expressed concern about a dangerous speculative bubble caused by "the present glut in the money markets, with excessively cheap money and its attendant evils and dangers to the credit structure of the country.... Both incoming gold and Federal Reserve Bank investments are reflected almost entirely in an increase of member bank balances with immediate and even violent effect upon the money market. The situation is abnormal and dangerous.[8]

In 1926, Swiss banker Felix Somary also warned of a stock market bubble in America where, if any big investors pulled out, markets would crash.[9] A large part of the money being lent by U.S. banks to foreign governments was being spent on the purchase of goods manufactured in America, but only the interest on these loans was being repaid, not the principal. So debt stayed on the books.

The pro-business Republican administrations of Warren Harding, Calvin Coolidge, and Herbert Hoover had brought prosperity, but by Black

Thursday, October 24, 1929, the world was heading in a different direction. On that day, the U.S. stock market crashed, following the long indulgence in stock inflation and speculation engineered by U.S. banks and the Federal Reserve. Within a month, stocks valued at $80 billion were worth $50 billion. The Dow Jones Industrial index fell from 381 in September 1929 to 50 by May 1932.

The Federal Reserve failed to respond and seemed to be a passive player. The monetary manipulations that were starting to lead to the rearming of both the Soviet Union and Germany were directed by the Bank of England, with the Federal Reserve going along with it.[10]

Here is how the Depression impacted the U.S.: National income dropped from $83.3 billion in 1929 to $68.9 billion in 1930 with a further descent to $40.0 billion by 1932. Unemployment was estimated at 3.5 million in 1930 and 15 million by 1933. The wholesale price index fell from 95.3 in 1929 to 86.4 in 1930 and 64.8 by 1932.

It was clear that people wanted to work and were able to do so. The natural resources required for production were available, as had been shown by the prosperity of the 1920s. The infrastructure of railroads, highways, motorized vehicles, water and sewage systems, electrical grids, telephones and telegraphs, international shipping, housing for families and workers—all these were up and running.

What was missing was an effective, functioning medium of exchange: money. But the U.S., along with other Western nations, had long since turned over their monetary systems to the privately-owned banking industry and their gold standard. Even if that industry provided money, it had a string attached to it. That string was "interest." With any contraction of the money supply, the additional money needed to cover the interest payments—which would require an expansion, not a contraction of the money supply (and hence inflation)—was just not there. Without it, the economy was kaput.

It's no mistake that the people in charge of banking are the wealthiest on the planet. Presumably their wealth is the recompense they receive for making modern industrial civilization possible through the extension of credit, yet there was a fatal flaw (interest) in that system, and accordingly the civilization that it had enabled had now broken down. The system that the bankers had promoted had failed miserably at keeping their end of the bargain, as it inevitably would.

Even the industrialists were subservient to the financiers, not to mention the people who did the hard, physical work. And by 1929, the international financial system was largely in the hands of the Bank of England. When the U.S. banks raised their interest rates in response to the Bank of England's call on gold which offered greater returns to investors by shifting funds to the British markets, the smart money bailed out of the U.S. markets. Some used insider information to bail before the crowd did so. Then, "crash."

Looking deeper, the monetary dynamics now reversed from the 1920s when it was the U.S. that was the leading holder of gold. This had produced so much instability in Europe that the heads of the Banks of England, France, and Germany had demanded that the U.S. raise its own interest rates to reduce inflation and move gold back to Europe. The decision to do so was reportedly taken at a meeting over lunch. Eustace Mullins writes:

> The secret meeting between the Governors of the Federal Reserve Board and the heads of the European central banks...was held to discuss the best way of getting the gold held in the United States...back to Europe to get the nations of that continent back on the gold standard.... The movement of that gold out of the United States caused the deflation of the stock boom, the end of the business prosperity of the 1920s, and the Great Depression of 1929–1931.... The [American] bankers knew what would happen when that $500 million worth of gold was sent to Europe. They [i.e., the bankers] wanted the Depression because it put the business and finance of the United States completely in their hands.[11]

Germany was also hit hard by the Depression. Unemployment began to decimate the country. On paper, there were over three million jobless individuals by 1930. Despair would lead to many of them committing suicide.[12]

Other Germans began to support the Nazis, which had been struggling to gain adherents. On January 30, 1933, German President General Paul von Hindenburg named Adolf Hitler Chancellor. This was another result of the Great Depression.

Roosevelt and the New Deal

The U.S. banking industry had failed so badly that in February 1933, after Franklin D. Roosevelt had been elected president over incumbent Herbert Hoover in 1932 and just before he was inaugurated in March, the banking system collapsed altogether. It happened due to runs on the banks by depositors trying to withdraw their savings in gold:

> During the week ending March 1, over $200 million in gold was withdrawn from the Federal Reserve banks, and during the next

few days another $200 million was withdrawn, bringing the gold reserve down almost to the legal minimum.[13]

With the runs pulling money out of the banks, the banks simply shut down. It was called a "bank holiday." On March 4, Roosevelt was inaugurated. Six days later, by Executive Order, he took the U.S. off the gold standard. Now, no one, other than a Federal Reserve bank, would even be allowed to hold gold or gold certificates except for up to $100 for use in the arts or collectibles.[14] Those who say the U.S. never defaulted on its debt are wrong in that in 1933 and again in 1971 the U.S. suspended gold convertibility of cash assets held by depositors in the U.S. and abroad. Creditors could get paid by paper or credit transfers but not in hard currency.

The elimination of gold-based currency was another step in depriving the U.S. public of purchasing power not dependent on borrowing from the banking system. The Federal Reserve then pumped $200 million into the system through new loans to the banks, and the run stopped. Roosevelt told the public in his first "Fireside Chat":

> ...there is an element in the readjustment of our financial system more important than currency, more important than gold, and that is the confidence of our people. Confidence and courage are the essentials of success in carrying out our plan. You people must have faith; you must not be stampeded by rumors or guesses.[15]

In his 1932 presidential campaign, Roosevelt:

> ...identified those responsible for the current dismal state of affairs: "The money changers have fled from their high seats in the temple of our civilization. We may now restore the temple to the ancient truths. The measure of the restoration lies in the extent to which we apply social values more noble than mere monetary profit." He called for "strict supervision of all banking and credits and investments" and "an end to speculation with other people's money."[16]

Never mind that Roosevelt was of the same social class and had long-standing business, personal, and social relationships with many of the guilty bankers. He attacked the Depression with an alphabet soup of solutions, many quite effective, but others less so. For instance, the Agriculture Adjustment Administration was criticized for supporting farm prices by ordering the destruction of crops and livestock at a time when rural populations were short on food. Many farmers, including black sharecroppers in the South, were put out of business.

But the point, in part, was simply to spend money which the government raised by borrowing. Roosevelt's policies were influenced by a new financial guru, British economist John Maynard Keynes, who argued that deficit spending and high income taxes would unlock a nation's wealth, boost employment, and generate enough economic growth to pay for the cost of financing an activist government.

This was not entirely new. The U.S. had done something similar during the Civil War and World War I. What was new was to put a nation's economy on a wartime footing when there was not a war. Of course, there was an impact on prices. While the Great Depression was deflationary, pulling out of it generated inflation. So the race was on, and has been ever since, between price inflation and economic growth, necessitating that growth prevail. If growth faltered, even for two quarters, it was called a "recession."

This is but another variation on the "business cycle" that fluctuated between expansion and panics throughout U.S. history. The Federal Reserve was created to stop these panics. But in 1932, by pegging interest rates higher than it should have in a deflationary period, the Federal Reserve utterly failed. The Depression even worsened.

But President Roosevelt was not relying solely on the Federal Reserve or the business community. His response to the Great Depression was marked by federal government action based on several worthwhile principles. One was the creation of government lending agencies such as the Reconstruction Finance Corporation, which recognized the concept of *credit as a public utility*.

The creation of the Federal Deposit Insurance Corporation acknowledged the government's responsibility to protect the public's savings; it also was essential to protect member banks against failure. The creation of the Civilian Conservation Corps and Works Progress Administration saw the government as the employer of last resort in roles more constructive than drafting young men without means into the military. Nevertheless, it took the World War II draft to create a full-employment economy.

Via Social Security insurance and Civil Service retirement the government took on the role of a savings bank, creating Trust Funds in ways that had been foreshadowed by earlier experiments in postal savings. Next, federal funding began aiding infrastructure development that formerly devolved mainly on state and local governments. New projects began such as the expansion of the U.S. highway network, later becoming the interstate highway system, and infrastructure projects such as the Tennessee Valley Authority or Hoover Dam.

These government programs conceived of public spending for the public good. State and local governments had been doing this for decades, but now it was being done at the federal level. This was where Roosevelt differed from his Republican predecessors, which opened him to charges of being a socialist or even a communist.

Of course, private companies serve the public as well by producing goods and services, but they also make a profit that goes to managers who guide the enterprise, the owners who risk their capital, and banks that provide loans. But through private sector banking, a major segment of the economy had broken away from emphasis on producing goods and services of value by giving priority to *just making money*.

The innovations of the Roosevelt administration obviously required funding. No matter what government does, at any level, the question arises: where is the money going to come from? The common answer is through taxes or credit. With the New Deal, taxes shot up again, reaching a marginal rate for high incomes of 90 percent. And the borrowing was massive.

Roosevelt's New Deal was accomplished through a compliant Congress that was determined to regain at least some of the Constitutional power of money creation by its commandeering of credit through taxation and borrowing. But the Roosevelt administration never went as far as to have the government itself create the nation's money as the Lincoln government had done with the Greenbacks. Congress now authorized such spending, but Roosevelt didn't use the authority.

Today, "printing money" is viewed as one of the worst things a government can do, a practice that is inflationary to a catastrophic extreme. But this is a misnomer. In actuality, as it presently stands, the U.S. government isn't really "printing money"; it's really still borrowing from a privately-owned central bank that uses the government's credit as reserves.

The Greenbacks are still the only true money printing option ever exercised by the federal government. But again, as Ford and Edison once asked with respect to the 1920s Muscle Shoals project, why shouldn't the government create its own money instead of borrowing from the banks who were creating their own money "out of thin air"? How to do so was answered by Congressman Dennis Kucinich's proposed 2011 NEED Act.[17] This question should be asked again today—along with the question: why is this not being done, and who/what is preventing it?

U.S. Banks' and Investors' Role in the German Recovery

No one can say when the Great Depression was over, as unemployment both in the U.S. and Great Britain remained above ten percent through to the end of the 1930s. It is clear that Roosevelt's New Deal programs aided recovery, that the principle of credit as a public utility was affirmed, and that Social Security began to supply a "social safety net" for older men and women.

It is also clear that by reforming it, Roosevelt's government saved the banking system. Even though bank deposits were no longer redeemable in gold, the yellow metal was still used to settle international bank transfers, and U.S. banks benefitted by being regarded as safe havens by foreign investors uneasy about the rising threat of renewed warfare in Europe. What followed was called the "gold avalanche." As with World War I, so was World War II "money in the bank" for the U.S.

U.S. banks and investors continued to make huge amounts of money through financing German rearmament. In addition to loans from Morgan and Chase—both now Rockefeller-controlled banks—Germany profited from cash infusions and partnership agreements with IBM, General Motors, the Ford Motor Company, Du Pont, and Standard Oil. According to a 2001 research report by the Ford Motor Company, at the start of World War II, 250 American companies owned more than $450 million worth of German assets. Among these companies were Standard Oil, Woolworth, IT&T, Singer, International Harvester, Eastman Kodak, Gillette, Coca-Cola, Kraft, Westinghouse, United Fruit, and Ford and GM.

There are still people today who sing the praises of the Nazi "economic miracle" in bringing Germany back from the Great Depression and providing full employment. But if you look at American loans and industrial investment you get a clearer idea of where this "miracle" came from.

The Role of Germany's Central Bank

But the biggest boost for the German economy came from the German central bank's lending money against assets that had been foreclosed due to the Depression then lent at low rates to the German government, which in turn funded economic enterprise, particularly infrastructure and armaments. Germany *de facto* went off the gold standard and lowered the prime lending rate from 8 to 2.31 percent. Lowering interest rates *always* enhances economic activity.

Germany also fixed consumer prices and left wages relatively low. This allowed money to be diverted into the arms industries. From 1933–1936,

Germany's GNP increased by 9.5 percent annually. Half of government outlays were going to the military.[18] But Germany was out of the Depression.

Unlike Germany, when World War II broke out in 1939, the number of American unemployed was still high, numbering eight million. A huge amount of unused factory capacity was also available. Bank lending abroad continued to be facilitated through a restoration of U.S. bank savings. But what finally pulled the U.S. out of the Depression was lending to Great Britain and France under President Roosevelt's Lend-Lease Program, which then enabled the rise of full employment here while allowing Britain and France to fight World War II on credit.

Family Matters

John C. Hill

Back in Chapter 7, I told the story of my great-grandfather John Clark Hill, who was born in Illinois, entered the newspaper business, worked with the cattlemen of Caldwell, Kansas, moved to Kingfisher, Oklahoma, and rode in the 1892 land run.

J.C. operated a general store and a farm, joined the Masonic order of Knights Templar, married my great-grandmother Edna Hubbard, and with her raised five children. Their daughter Carolyn was my paternal grandmother. The time came when the store went bankrupt, due to unpaid bills by customers, so J.C. and Edna sold the farm and moved to Oklahoma City. The records contain a testimonial to J.C. from his minister published in July 1944 upon his retirement from the church council. He was then 78:

> Salute the Living: One of the elders that retired this year, by reason of the rotating system adopted by the Congregation for its officers, was J.C. Hill. He has been a faithful member of our board of elders, since the church was organized thirteen years ago last April. Before that time he was a member of the First Church. He was an "89er"—but didn't stake a claim until 1892 when the Cheyenne strip was opened. It is a real treat to hear him talk of Pioneer days. He owned and edited the *Kingfisher Free Press* for several years. When lead poison forced him to relinquish that job, he went into the mercantile business at Kingfisher. When the Hills came to Oklahoma City in 1922 he was in the Agriculture Department of the State for a while, and then helped organize The Mutual Home Savings and Loan Company of this city and served as its secretary for several years.

His has been, and is, an interesting life—and how he enjoys life! Life has no anticlimax for him. There are many interesting things about this man worth recording, but I want to salute him for two qualities in particular. One is his devotion to his lifelong companion. When the Hills celebrated their Golden Wedding he glowed with enthusiasm for his friends but his greatest thrill was Mrs. Hill. She came up and spoke to us when we were having a little side-chat. When she left us he said, "Isn't she a good looker?" Really, such devotion and loyalty through fifty years of marriage is unusual and admirable. Nothing could be a better indicator of Christian character than that.

The other quality of Mr. Hill that I salute is his devout spirit. The elders of our church meet every Sunday morning in the preacher's office for a prayer session before the church service. I find their prayers to be very helpful, especially so when Brother Hill prays for the preacher. His loyalty to me as his preacher and to the church give not a little boost to my soul. So I salute this good man.

John Clark Hill lived until 1950 to the age of 83. At one point he said of his wife Edna, "There are few her equal and none her superior."

The Peilow Family

My maternal grandfather Carlton William "Bill" Peilow was born in Gladstone in the Upper Peninsula of Michigan in 1894. His family was French-Canadian. His father, Joseph, who was born in Ontario, died in a farming accident, so Bill went to live with family in Butte, Montana.

When World War I began, Bill joined the U.S. Navy and sailed out of the Brooklyn Navy Yard on transport convoys carrying American troops to France. He met my grandmother, Ethel Brown, who was living in Brooklyn and working as a secretary in an insurance firm. Ethel was the granddaughter of William Forster, who had left Ireland in 1849 during the Great Famine and served in the Union Army during the Civil War.

Bill and Ethel were married in Marion, North Carolina, where his mother, who was married to a fundamentalist preacher, was living. My mother, Marjorie Virginia Peilow, was born to Bill and Ethel in Brooklyn on December 30, 1922. She was baptized in the Christ Church Episcopal congregation in Bay Ridge. Four years later, when she was four years old, she waved a little flag in the direction of Charles A. Lindbergh, Jr. This was when "Lindy" returned to New York City after his transatlantic flight from New York to Paris for a ticker tape parade on June 14, 1927. It is almost unimaginable today how much excitement Lindbergh's flight evoked. My mom was among the estimated four million people who turned out to cheer Lindbergh's arrival back in New York.

At some point, my grandparents left Brooklyn to travel west. They owned a Model-A Ford which they loaded with their belongings, then drove to Atlanta, where Bill's mother Lizzie was living. After a visit, Bill and Ethel, with little Marjorie, then drove the 2,200 miles from Atlanta to Missoula, Montana. They used a geography textbook to plot their route. They likely followed motor routes along the old Oregon and Bozeman Trails. Bill had lived in Butte, Montana, as a boy and had a cousin named Alberta Driscoll in Missoula. Bert's husband Jerry was an engineer for the Northern Pacific Railroad.

The 1930 census shows the Peilow family living in Missoula. The stock market had crashed, the Depression was underway, and there were no jobs back east, so Bill got a job working for the Anaconda Forest Products Company. The company had formed in 1898, when Anaconda founder Marcus Daly purchased the lumber mill at Bonner on the Blackfoot River east of Missoula. The mill provided mine timbers and lumber for Daly's Butte and Anaconda mining and smelting operations. The mill also provided railroad ties for the Anaconda's mining and timbering enterprises.

Gradually the timbering moved up the Blackfoot River in the direction of the Continental Divide. Thirty-two miles upriver at Clearwater, the Swan Valley begins between the Mission and Swan mountain ranges and runs north toward what is now Glacier National Park. Fifteen miles north of Clearwater is Seeley Lake, with a town of the same name, which had become a hub for lumbering in the Swan Valley. The main source of timber was the huge tamarack, or larch, trees—the largest in the world—that grow abundantly in the region.

In 1934, lumbering operations moved to Woodworth, just south of Seeley Lake. The camp had both permanent and portable buildings, including bunk houses, a mess hall and kitchen, an office and community building, a school, a library, a car repair and machine shop, and family housing. The Peilows lived there for a time before moving down to Seeley Lake a couple of years later. My grandfather learned the job of a scaler, whose task is to measure standing timber and estimate the number of board feet of finished lumber it would produce. Ethel earned money by taking in laundry for the workers at the lumber camp.

The lumber was cut from national forest land and was purchased at auction. The national forests around Missoula were among the first to be part of the National Forest Service system, with Missoula becoming the location of a regional office. Timbering was the main economic engine of the Missoula area, along with ranching. The Army's Fort Missoula was established in 1877, and the University of Montana opened in 1895.

Life around Seeley Lake was hard for my grandparents and my mother, but the beauty and wildness of the surrounding mountain landscape was also a source of joy, and they were part of a small, closely-knit community. The family lived in a log cabin without indoor plumbing, but there was a stream behind the house that served as a water source and icebox. My grandmother had a large garden, and they fished for trout in Seeley Lake. She canned their food and had plenty of fish in Mason jars for the winter. My mom would go cross-country skiing. In milder weather she would take her rowboat with her rifle and her dog and putter around on the lake. She said that some of the local boys were awfully mean bullies. She slapped one of them hard in the face at a community dance.

But out in the Montana wilderness, nothing could have been compared to the hardship of trying to scrape by in a big eastern city during the Great Depression, which would have been the case if they had stayed in Brooklyn.

Often Ethel and my mom would be alone while Bill was traveling in the area working for the Forest Service. This was common for the lumbermen as a supplement to their income. Forest fires were a constant threat in the national forests and still are. Early on, the Forest Service established a practice of one hundred percent fire containment, which meant that whenever a fire was spotted, the firefighters would rush to put it out. Today, controlled burning like the Indians used to practice is starting to be done.

There were times when men putting out the fires were trapped and burned to death. Often, for the bigger fires, the Forest Service would dispatch a mule pack train to bring the men food and supplies. Work building roads and trails for the Forest Service was also available. There were also fire lookout towers, where my dad worked one summer. He said he did a lot of reading.

After the family moved to Seeley Lake, my mother attended the one room schoolhouse until she reached fourteen. Most of the children in the community stopped going to school at that age and went to work, either on family farms, in town, or in the logging business. My mother was lucky in that her Aunt Bert in Missoula took her in. She worked at a drug store in Missoula to meet expenses and graduated from Missoula County High School in 1940. During vacations she would return to Seeley Lake. That was where she met my dad—at one of the Seeley Lake dance halls. Also at some point, my grandfather's mother Lizzie moved to Missoula where she made a living rehabbing houses.

After graduation from high school, my mom spent a year at Montana State College in Bozeman, where she studied art history. But her money ran out, so she withdrew. Early in World War II, the Army set up a program where young women could enroll to become nurses, with the agreement that when they finished their studies they would go overseas to the war zones. My mom enrolled and became a nurse through the program at Missoula's St. Patrick's hospital. But the war ended before she could be sent abroad.

The film *A River Runs Through It* produced by Robert Redford gives a good depiction of life in and around Missoula at the time my mother lived at Seeley Lake in the 1930s. The fishing scenes were shot on location along the Blackfoot River. Author Norman McLean (1902–1990), who wrote the original autobiographical story, had a cabin on Seeley Lake in his later years. He also spent a couple of summers working for the Anaconda lumber camps around the time my grandfather arrived on the scene. Maybe they ran into each other.

The museum at the Seeley Lake Historical Society has a room devoted to Norman McLean and the film. It also has exhibits on the logging industry, including posters of the lumbermen's diet at the camps and details on how the scalers did their work. In fact, it was at Seeley Lake that the techniques of scaling were developed.

If you want to read some history on what life at Seeley Lake was like in those days, take a look at *Cabin Fever: A Centennial Collection of Stories About the Seeley Lake Area.* Edited by Suzanne Vernon, the book was compiled by the Seeley Lake Writers Club. In a cover blurb, Robert Redford writes:

> We used *Cabin Fever* as a model for our research and casting of *A River Runs Through It.* It is a wonderful book and I highly recommend it to anyone interested in the history of this magical part of Montana.

I can only add that I remember my grandfather, Bill Peilow, the sailor/logger/firefighter/trail builder from Montana, as the kindest, funniest, warmest man I ever met.

The Cook Family

My paternal grandfather Frederick Steele Cook was born on November 17, 1896, in Tioga, Texas, a crossroads just north of Fort Worth, about thirty miles south of the Oklahoma state line. His father was William Damarcus Fitts, about whom nothing is known except that he was from Mississippi and died in Tioga in April 1904 at the age of thirty-five, when Fred was only seven years old.

Fred's mother, Ida Florence Steele, was born on September 14, 1874, also in Tioga. According to the Tioga website, this was before the first general store was built. She lived to be sixty-one and died in California in 1936. According to my dad, Mrs. Steele did have a "steely" personality. All of the forebears of the Steeles and the Fittses were from the South, except for one of Fred's great grandfathers, who was from the UK.

After William Damarcus Fitts died, Ida married William Cook, who adopted Fred and gave him his last name, "Cook." So Fred was now Frederick Steele Cook or Frederick Steele Fitts Cook.

William Cook was a gambler, whether professional or not, we don't know. My dad told this story: One day someone ran into the house shouting, "Grandpa killed a man." William Cook had been accused of cheating at cards. The accuser had rushed him and fell onto Cook's knife. Cook was either not charged or acquitted.

My paternal grandmother was Carolyn Hill, born on November 30, 1899, in Kingfisher, Oklahoma. She was the daughter of John C. and Edna Hill and the granddaughter of Col. George Hubbard and Caroline Lasher. On November 4, 1917, she and Fred Cook were married at Kingfisher. According to the wedding notice in the newspaper, the service was officiated by Rev. H.E. Stubbs at the Hill home. Since my grandmother had not yet turned eighteen, she needed permission from her father to marry.

At the time of the wedding, Carolyn was working as a saleslady at Logan-Ames & Co. No occupation was given for Fred. The notice says, "The groom is an exemplary young man." The notice states that his parents, Mr. and Mrs. W.E. Cook, recently relocated to Kingfisher and were in the auto business. The notice also states, evidently referring to Fred, that, "He has accepted a position with Mr. Sandefur." Mr. Sandefur seems to have been a local businessman.

Fred's career, as was the case with so many young men, was interrupted by the war. He registered for the draft, as required, and entered the Army in June 1918. He did not go overseas but was discharged in November 1918. By then Carolyn had given birth to an infant son, who died when the family was in El Dorado, Arkansas. In September 1919, a daughter, Bobbie, was born when they were back in Oklahoma City. My dad, Richard Edward Cook, was born in 1924, also in Oklahoma City.

For a time, Fred Cook worked as a salesman for Hardeman-King, a feed company. My dad once wrote that Fred had a bright future there. He also wrote that "I remember asking passers-by who they were going to vote for, Al Smith or Hoover." This was in the 1928 presidential election. "We were for Al Smith, and I was disappointed that so many people were going to vote for Hoover." There was a clear divide then between the political parties. The Republicans were for business, while the Democrats were for working people.

Though the sequence is not entirely clear, at some point, Fred decided to follow his stepfather, William Cook, into gambling and for the next thirty years plied the gambling trade. His specialty was "dealing a roulette wheel." Fred first went to Mexicali, Mexico, the capital of Baja California, which was right across the U.S. border from Calixico, California. From there the family went to Las Vegas, where casino gambling had been legalized in 1931. My dad attended fourth grade there.

My dad wrote, "To be a house dealer in a legal gambling place is to be a working stiff." So one day Fred and my grandmother Carolyn put Bobbie and Dick in the back seat of their car and drove to Los Angeles.

My dad wrote of the move to California, "Fred was going to deal a wheel in the Colony Club." This was in Palm Springs. He wrote:

> This was big time illegal gambling run by tough guys for rich and famous people, in this case, Hollywood people. It could take place because the rewards were so great you could pay off the right people to stay open and the dealers had to be the best. The ones that made sure the players lost most of the time.

He continued:

> In 1937 he went to dealing the wheel in the Dunes, a high class club in Palm Springs, for Al Wertheimer. If you've ever read or ever will read about crime in California in the late 30s you will come across the name of Al Wertheimer. He was considered a big shot gambler, and I guess he was. We had two nice years there.... The Dunes closed in 1939 after Al had a terrible car accident.

Al Wertheimer was associated with the Purple Gang out of Detroit. One time Fred was busted at the Dunes and got his name in the newspaper, though the movie stars who were rounded up didn't.

Fred made enough money to take his family all over the West. My dad said there were times when Fred had been drinking and got the whole family to hop in the car for a ride to Yosemite. Fred would keep his money in $100 bills under the wallboards in the garage. His mother, Ida Steele, had followed him to California and often came to visit.

My dad says that during this time they summered in Montana for two years at Seeley Lake. On a map in the book *Cabin Fever* there is a little square eight miles to the north on Lake Inez that says "Cook Summer Home." Whether that was Fred's house, I don't know. Judging from the family photos, Fred became a great fly fisherman, as did my other grandfather and my mom. Back in California, my Aunt Bobbie attended West Hollywood High School with Judy Garland and Mickey Rooney.

But then something happened, and I'm not sure exactly what. As goes the family legend, Fred lost his house and the family had to huddle at a bus station to catch the next bus out of town. They went to Missoula. Fred tried to get a gambling operation going in Montana, but it didn't work out.

My dad went to high school in Missoula for a while. He had met my mom at a dance in Seeley Lake. He went back and forth between Montana and Oklahoma but ended up with Fred in Kalispell, Montana, where he got his driver's license and registered for the draft. In 1943, my dad was drafted into the Seabees, and I'll pick up more of the story later. By then Fred and Carolyn had divorced; she took the kids and moved back to Oklahoma. Fred stayed out west. Eventually he turned up in Arizona, where he lived until October 15, 1976.

My dad described Fred as very private and somewhat sinister. The two of them never talked about anything serious. That was reserved for talks between my dad and his mom. He quoted Fred as saying, "When I meet a man, he doesn't know if I've got a million dollars or if I'm flat broke."

I am certain that my grandfather deeply compromised himself with his gambling career and that his poor life choices affected subsequent generations. But he was a man of his time.

Flathead Reservation Life—1910 to 1940

By 1910, the Salish, Kootenai, and Pend d'Oreilles Indians living on the Flathead Reservation had been given land allotments of 80–160 acres as a result of the Dawes Act of 1887. Most Indians selected land on the periphery of the reservation near the Mission Mountains to the east and the Salish

Mountains to the west, thinking they would find more big game for hunting, especially elk and deer. Then in 1910, the "surplus" land was opened to white homesteaders. About 600 plots of 160 acres each were distributed by lottery.

Prior to the homestead lottery allotments to whites, the Indians were able to graze their cattle and horses on the open range. With the buffalo gone and the annual buffalo hunts beyond the Continental Divide eliminated, open grazing of cattle and horses had allowed a semblance of the old way of life. Now that too had been destroyed, and not only on the Flathead Reservation. Wherever allotment took place, the integrity of the U.S. reservation system was shattered. The federal government once again had betrayed the Native Americans.

The result was a downward plunge in the economic status of reservation Indians, with many sinking into poverty, as living off the land was now much harder. Some Indians did take up residence on their allotments and began to operate subsistence farms. Others sold or leased their allotments and either lived from the proceeds or left the reservation. For the Flathead tribes, the main destinations for off-reservation migration were the city of Seattle, Washington, almost 500 miles to the west, and California.

The Indians that had already been farming were in the habit of hand-digging irrigation ditches to their plots from local streams. This was if they chose to irrigate, as the lack of rainfall in the arid Rocky Mountain ecosystem could not always sustain farm fields, gardens, or orchards. Irrigation had already begun around the St. Ignatius Mission in the late 19th century, where the waters of Mission Creek were diverted for farming and to operate the mission's saw and grist mills.

To pave the way for white homesteading, Congress passed two separate acts that authorized construction of the "Flathead Project." The act of April 23, 1904, provided for the distribution and irrigation of Indian allotments, and the sale of "surplus" lands. Then on April 30, 1908, Congress authorized $50,000 for the construction of irrigation systems on the reservation.

Almost immediately in 1908, the federal Bureau of Reclamation began work on the Flathead Reservation irrigation system. After 1924, the project was taken over by the Bureau of Indian Affairs and continued until today. In 2021, a revitalization of the system was approved by Congress. But the project managers complained early on that their irrigation work produced far more water than there were farmers to utilize it.

It appears that the primary impetus for the massive amount of work that was done was to benefit the contractors and the politicians and bureaucrats who enacted and oversaw the program. Some Indians lost their allotments altogether when they were unable to pay the assessments charged by the government to finance the project.

The Seli'š Ksanka Qlispe' Dam, previously known as the Kerr Dam, was completed in 1938 by the Montana Power Company, with financial help from the federal government. The dam was built at the south end of Flathead Lake, where the Flathead River flows out of the lake into a canyon on its way south to join the Clark Fork River. The system is part of the Columbia River watershed. Tribal members helped build the dam over an eight-year period, along with workers from the Civilian Conservation Corps and the Works Progress Administration. When the dam was designed, its size was increased in order to provide sufficient electricity to the Anaconda Copper Company, 177 miles away near Butte. The Confederated Salish and Kootenai Tribes operated the dam jointly with successive electric companies before purchasing the dam outright in 2015. They became the first Indian nation to own and operate a hydroelectric facility.

During this period, the St. Ignatius Mission, founded by the Jesuits in 1854, continued to operate, with the boarding school under the Sisters of Providence and Ursuline Sisters. By 1929, the federal government had cut off all financial aid to religious schools on Indian reservations. But there were there were still 230 pupils at the mission school—170 girls and 60 boys.

After white settlement on the reservation began, so did intermarriage with the Indians, so gradually children of mixed race increased in number. As time went on, these seemed less amenable to Catholic education. By now, the government had also begun to open its own public schools on the reservation with no religious instruction.

Relations between the mission and the BIA reservation agency were generally good, except during the Teddy Roosevelt administration.[19] The Indian boys usually dropped out of school when they reached puberty. Attempts by the priests and the Indian chiefs to force them into line caused resentments that linger to this day; charges of brutality are still levied by descendants. Another setback came when some of the school and mission buildings burned down. But the Indian school continued to operate into the 1970s. Today St. Ignatius is a parish church with an active congregation of both Indians and whites.

Meanwhile, life went on for those Indians who had stayed on the reservation, with or without their own allotments. Pick-up work was often available at the government agency or on farms and ranches or in nearby Montana towns. Gambling and alcohol were always problems, though alcohol was forbidden on the reservation. But the whites would bring it in, particularly those associated with the railroad that ran through the southern end of the reservation, as did non-reservation Indians from elsewhere.

Since 1898, the reservation has celebrated an annual gathering at the town of Arlee which encompasses five days of activity, visiting, and traditional dancing. The modern name is "powwow," though some Indians don't like the term. It takes place on the Arlee Powwow grounds around the Fourth of July, with people staying in tents, campers, or traditional teepees.

When my father was a teenager, he made his own teepee and attended the powwow because, he said, he liked to hear the old men chant. In recent times, the Arlee Powwow has become a well-attended event, with anyone welcome, though it went into abeyance for two years due to the COVID shutdowns.

During the period under discussion, 1910–1940, there was a rich life among the Indians on the reservation that can be grasped by reading a book on the life of Duncan McDonald. Duncan was the son of the original Hudson's Bay Company trader, Angus McDonald, and his Nez Perce mother. The book is *Duncan McDonald, Flathead Indian Reservation: Leader and Cultural Broker, 1849–1937* by Robert Bigart and Joseph McDonald. Duncan received his fee patent for his allotment in 1910 and became a U.S. citizen.

The Indian Reorganization Act passed in 1934 was part of the "Indian New Deal" promulgated by the Franklin D. Roosevelt administration. The IRA was the initiative of John Collier, Commissioner of the Bureau of Indian Affairs from 1933 to 1945. Collier characterized contemporary American society as "physically, religiously, socially, and aesthetically shattered, dismembered, directionless" and saw traditional Indian society as morally superior to that of whites.[20]

The Flathead tribes were the first Indian nation in the U.S. to organize themselves under the Act. In 1935, what is now the Confederated Salish and Kootenai Tribes ratified a tribal constitution and created an elected government of ten tribal council representatives. Charlie McDonald, a member of the first Tribal Council from Hot Springs, Chief Martin Charlo, and Chief Koostahtah were named as life council members and members of all committees. The first committees to be established were Land, Finance, Law and Order, Health, Labor, and Education. As one of its first actions, the Tribal Council designated a section of the Mission Mountains as a tribal wilderness area.

Back in "Civilization" Disaster Looms

Meanwhile, Britain was playing the same imperialistic games as always. The primary game, whatever happens, is to look for the strongest Continental power and bring them down. In 1914 it was Germany. By the 1920s and 1930s, that power had shifted east to the Soviet Union. In order to stop communism, Britain had invested in a German upstart named Adolph Hitler. But Hitler would soon be the enemy and the Soviets, allies. As I've said before, it was another old game: "Let's you and him fight." You and him now being Hitler and Stalin.

But the complacent Americans, who had been giving "isolationism" a try by rejecting the Treaty of Versailles and League of Nations, were increasingly the real power behind the British throne. The U.S. would soon become the enforcer and in many ways the prime beneficiary of Anglo-American hegemony. Stay tuned. The "Good War" is about to begin.

ENDNOTES FOR CHAPTER 11

1 Prins, op. cit., 88.
2 Ibid., 79.
3 Ibid., 161.
4 Ibid., 170.
5 These machinations were concealed from France, which feared a German military resurgence.
6 The policy continues to this day. Britain is the biggest cheerleader for the U.S./NATO proxy war against Russia in Ukraine. A main objective of this war is to prevent German industry from being supplied by Russian gas and oil.
7 Prins, 166.
8 Ibid., 86.
9 Guido Giacomo Preperata, *Conjuring Hitler: How Britain and America Made the Third Riech* (Pluto Press, 2005), 175. A new edition of this book is titled *Conjuring Hitler: How Great Britain and America Made the Third Reich and Destroyed Europe* (Citta' di Castello, Hemlock, NY: Ad Triarios Press, 2023).
10 Ibid., 180.
11 Eustace Mullins, *The Federal Reserve Conspiracy* (Mansfield Centre, Conn.: Martino Publishing, 1954), 89. Also see Mullins, *The Secrets of the Federal Reserve*.
12 Ibid., 192.
13 Studenski & Krooss, 381.
14 It was Roosevelt's discarding of the gold standard that appears to have been the immediate trigger for the attempt by Wall Street plotters to persuade famed Marine Corps Major General Smedley Butler to carry out a coup against Roosevelt in 1934. Butler, the most decorated military officer of the day, reported the plot to a congressional committee, which, acting on Roosevelt's advice, took no action against the conspirators. Among the named plotters was banker Prescott Bush, later a U.S.

senator and father of future President George H.W. Bush and grandfather of President George W. Bush. Another was Thomas Lamont, head and future chairman of J.P. Morgan & Co. and with Rockefeller affiliations.

15 Ibid., 384.

16 Stone and Kuznick, 46.

17 Kucinich's NEED Act is discussed in the Appendix to this book.

18 Preparata, 222.

19 Gerald Lee Kelly, *History of St. Ignatius Mission, Montana* (University of Montana, 1954), 67.

20 John Collier, "Does the Government Welcome the Indian Arts?" *The American Magazine of Art. Anniversary Supplement,* Vol. 27, No. 9, Part 2 (1934), 10–13.

CHAPTER 12

The National Security State

The Good War

World War II was the "Good War"—right? We all know that the good guys—us—kicked the bad guys' butts—them. We know who the bad guys were—the Germans and the Japanese. Thousands of war movies have told us that. And we taught the bastards a lesson, didn't we? It was mainly the British who firebombed the major German cities, reduced them to smoking rubble, but we joined in. We did the same to the citizens of Tokyo, and we dropped nukes on Hiroshima and Nagasaki.

Why then, if it was such a "good war," have we been fighting more wars ever since? Korea was a "good war" too, right? Vietnam was also a "good war," I guess. Then we fought Desert Storm against Iraq. Then we bombed and dismembered Yugoslavia. Both "good wars." So was the "War on Terror," with the destruction wreaked on Afghanistan, Iraq again in 2003 on the pretext that they had WMDs (which they didn't), and Libya. We still have forces in Syria.

Now we're conducting another "good war," our proxy war against Russia over Ukraine, a Ukraine whose government we *created* in an illegal coup in 2014 and have armed to the teeth and egged on ever since. And America's president and our incredibly sophisticated propaganda media are once again telling us that it's *all* the other guy's fault. It's always *"unprovoked aggression." Always,* just like in the killing of the Sioux Indians.

This doesn't even include the governments we've attacked, overthrown, and/or subverted in smaller-scale conflicts over the last seventy-five years, the foreign leaders we've assassinated, places to which we've sent troops, countries we've ransacked with economic sanctions, the "color revolutions" we've instigated using the resources of the CIA and/or the National Endowment for Democracy. Yes, we've really done a great job of creating "open societies" with our weapons, our propaganda, our pressure, and our manipulations. And we have military bases in over eighty nations around the world to be sure we keep up the good work.

But back to World War II. What if it wasn't a "Good War"? What if conservative commentator, Reagan speechwriter, and presidential candidate Patrick Buchanan was right? What if both of these world wars were "unnecessary wars"? What if, as Buchanan says, these wars were "hideous and suicidal," that they "advanced the death of our civilization"?[1]

Britain Always Has a Plan

Great Britain, in its centuries-long quest to build an empire, has attacked any competing power on the European Continent that threatened to establish its own hegemony. As a result, Britain has been at war, or poised for war, for its entire modern history, including the conflicts it has waged for control of the various imperial components in North America, Africa, the Middle East, India, and East Asia. While Britain may often pay other nations to do the fighting, Britain itself, with its "royals" atop the heap, has been the chief imperial power of modern history.

As Britain neared the end of the 19th century, it had already waged successful war against the Spain of the Habsburgs; the Dutch, with whom it merged institutionally through the Glorious Revolution of 1688; then with the France of Louis XIV and later with the French republic and Napoleon. Each time, those threatened attempted to launch an invasion of the British homeland, and each time failed. The closest instance was by the Dutch, who succeeded in getting their fleet up the Thames estuary and won a major naval battle in 1667 but did not disembark land forces.

By the first years of the 20th century, it was clear that another war of continental proportions loomed. Every statesman in Europe knew it. But it was not clear who Britain's foe would be. I have tried to explain in a previous chapter why the Central Powers of Germany and Austria-Hungary were finally designated as the World War I opponents, rather than the other two imperial rivals, France and Russia.

In the process, the most prescient minds within the British establishment were also likely wondering who they could engage as allies, because whenever Britain fought a major war, it always did so through a coalition. Often Britain's partners could simply be paid to fight. On occasion, Britain would send its own forces to the Continent, as it did against Napoleon. But faced with a project of the scope of World War I, who would be its partner in the coming conflagration?

Of course it would be the U.S. By the end of the 19th century, the British and their American cousins were joined at the hip financially, even if the bulk of America's population had no intention of going to war in Europe on anyone's side. No one in the U.S. government bothered to read the section of the Monroe Doctrine that pledged the U.S. would stay out of internal European political affairs.

The British could be so persuasive, especially if, as did Cecil Rhodes and Lord Nathaniel Rothschild, they had the fantastic wealth of South African diamonds and gold at their disposal. And especially if Rhodes then bequeathed his wealth toward the formation of a secret society aimed, in his words, at "recovering" America for the British Empire. As explained previously, this secret society was the Round Table. Rhodes's successor in the enterprise was Alfred, Lord Milner.

It took a while, but Britain was evidently able to stage the sinking of the *Lusitania*, and America finally got itself in gear. Germany and its allies tried to call a truce, even as the Bolshevik Revolution plunged Russia into chaos. In fact, as Guido Preparata argues in *Conjuring Hitler: How Britain and America Made the Third Reich*, Britain—and the U.S.—*allowed* the Bolsheviks to defeat the White army in the five-year Russian Civil War. Russian Admiral Kolchak himself, leader of the White armies, said he was betrayed by the Western powers. The point being, says Preparata, that a future combination between a royalist/aristocratic Russia with a resurgent Germany would be a lethal threat to Britain's imperial future. *Britain's greatest fear was an alliance between Germany and Russia fighting on the same side.*

After World War I ended, Germany would soon be able to fight again. Britain and the U.S. had assured this with massive investments in German heavy industry in the 1930s. But rearming Germany only made sense if Britain could direct Germany's newly found might against the Soviet Union, in the hope that the two might destroy each other, and that the British Empire would be spared.

To prepare for the next war, the British and American elites would encourage a reactionary movement within Germany that would view the Soviets as Germany's mortal enemy. The charismatic leader, one Adolph Hitler, was identified very early on—in fact, by 1919. He was groomed, flattered, financed, dressed up in military glory, and gotten ready to act as Pied Piper to the nation's future destruction, all carefully prepared by British lords and diplomats well practiced at this sort of thing. Behind the scenes schemed the Round Table and other assorted British "clubs," as Preparata calls them.

Hitler loved the British and saw the Soviet Union—Russia—as the enemy. He had written in *Mein Kampf*:

The National Security State 233

> If land was desired in Europe, it could be obtained by and large only at the expense of Russia.... For such a policy there was but one ally: England.... No sacrifice should have been too great for winning England's willingness.... Only an absolutely clear orientation could lead to such a goal: renunciation of world trade and colonies.... Concentration of all the State's instruments of power on the land army.[2]

But then, neither the U.S. nor Britain neglected the Soviet Union in the rearming process. Standard Oil and Ford built installations, and other companies signed on for gold and oil extraction. The great Dnipro dam on the Dnieper River, for instance, was built from 1927 to 1932 with U.S. money and British engineering skill.[3] The British also looked the other way as the Stalinist Terror killed a million people, including Leninist and Trotskyite partisans and the top military echelons.

At any time during Hitler's dictatorship, which was consolidated after the false-flag burning of the Reichstag in 1933 and the purge of the Nazi Party's left-wing in 1934—the "Night of the Long Knives"—the military power of Britain, France, and the Soviet Union could have squashed the still-rebuilding German *Wehrmacht*. Britain chose not to, France was appalled at Britain's inaction, and the Soviet build-up continued.

First Russia, then France, then Germany, had felt the British stab in the back. And the Soviet Union knew well that their turn would be next. So Germany and the Soviet Union shocked the world by signing a non-aggression pact on August 23, 1939, "the Ribbentrop-Molotov Pact."

The Ribbentrop-Molotov Pact guaranteed peace between the parties and made the commitment that neither nation would aid or ally itself with an enemy of the other. But there was also a Secret Protocol which defined the borders of Soviet and German spheres of influence across Poland, Lithuania, Latvia, Estonia and Finland.

Britain may have realized its mistake in standing by idly. But by then it was too late: war now loomed. The German plan was to bring Europe under control by subordinating all industry to ownership and coordination by the German national banking system with the Reichsmark the reserve currency, much as the dollar would become for the world after U.S. victory in World War II. The Germans also discarded the gold standard as being a compromise of sovereignty. The Germans foresaw a Eurasian economic union that would include the Soviet Union and Japan, a prospect that was anathema to Britain and the U.S.[4]

On September 1, 1939, only days after the signing of the Ribbentrop-Molotov Pact, Germany invaded Poland. Two days later, on September 3, 1939, Britain and France declared war on Germany. On September 16, 1939, the USSR invaded Poland. Implementation of the Soviet-German Secret Protocol was underway.

The British set out to wean Stalin's Soviet Union away from the German alliance. They succeeded. The Soviets had been cooperating with Britain for a decade in allowing the Nazi war machine to attain its present prowess.[5] Its pact with Germany couldn't hold; Stalin had little choice but to play along. Nonetheless, it wasn't until 1941, after Hitler broke his pact with the USSR and invaded that, on July 12, 1941, the USSR then signed a military alliance with Britain.

Even through the Battle of Britain, fought in the skies over England, Hitler appeared to nurture the hope that he and the British might one day share a common future, so long as America remained on the sidelines. His vision was always for Britain to rule the sea while Germany controlled the continental land mass. This was why he felt compelled to neutralize the Soviets. He hoped in vain.

Looking back, the only fly in the ointment was that if Britain had to count on the U.S. to provide the muscle once the next phase of the war began, the Americans might feel entitled to take charge of the entire Anglo-Saxon enterprise themselves. But the Americans, though mighty and rich, were not practiced at this sort of thing and might be easy to steer in another direction.

Today, not without reason as we shall see, the British sometimes refer to themselves as the "tugboat" to the American "destroyer." And Americans have always been easy to dupe with a "fistful of dollars" waving in their faces.

America Opts for Global Military Dominance

Like an engine that was running out of gas, the New Deal was sputtering as economic recovery slowed in the mid-to-late 1930s. Though it has never been proven conclusively that President Franklin D. Roosevelt deliberately helped provoke World War II against Germany and Japan in order to rescue the U.S. economy, such calculations were likely part of the thinking of the time. Everyone knew that war was a potent economic stimulus. The banking fraternity, in particular, had been growing rich off war for a *very* long time.

It's a little-known fact that even before the U.S. entered the war on the side of Great Britain, a decision had been made in America's highest official circles that the long-term objective of the U.S. was to become the world's *dominant military power*. The fact of planned American global military dominance has been documented in extensive detail in an impeccably precise book published in 2020, *Tomorrow the World: The Birth of U.S. Global Supremacy* by Stephen Wertheim, a senior fellow at the Carnegie Institute for International Peace. Another corroborating source is F. William Engdahl's *Gods of Money: Wall Street and the Death of the American Century.*

Both Wertheim and Engdahl relate that as war clouds began to gather over Europe in the late 1930s, consultations were taking place within the Roosevelt White House, the U.S. State Department, and the War Department—all still minuscule by today's standard of bloated executive bureaucracies and intelligence agencies—on what would be the policy objectives of the U.S. once war in Europe broke out.

When Germany attacked Poland in September 1939, causing Britain and France to declare war, the debate within the U.S. became more urgent. When the Germans occupied Denmark and Norway, followed by its *blitzkrieg* against the Low Countries and France in 1940, the U.S. faced two contingencies. Either Britain would also be defeated, leaving all Europe in German hands—except for the Soviet Union—or Britain would hold out until the German victories could be rolled back. Meanwhile, in the Far East, Japan had invaded China following its earlier conquest of Manchuria in the mid-1930s.

Whether or not Britain would stand or fall, the war was obviously an excellent business opportunity for the U.S. It was President Calvin Coolidge who had said in 1925 that, "The business of America is business," and it was Roosevelt's implementation first, of Cash-and-Carry on September 21, 1939, and then the Lend-Lease Act, that made the U.S. the "arsenal of democracy." This ended the Great Depression and set America on a course of staggering economic prosperity lasting until the 1960s.

Still, the U.S. government's assumption at the time was that mainland Europe was going to be controlled by the two authoritarian states of Nazi Germany and the Stalinist Soviet Union, at least in the near term. So where would the U.S. draw the line that it would defend at all costs? A consensus was forming that the U.S. would be able to secure control of the Western Hemisphere, but possibly not much else.

The debate was fierce, with a more radical party emerging which believed that long-term U.S. economic power could not be assured unless the goal were established for *total global military dominance.*

The Council on Foreign Relations Moves In

The studies delivered to President Roosevelt, along with the State and War Departments, were drafted by the Council on Foreign Relations (CFR), established in New York following World War I, with funds supplied largely by the Morgans and Rockefellers.

By 1939, the Rockefeller dynasty was under the control of the founder's son, John D. Rockefeller, Jr. The Rockefellers had been intimately involved, personally and by marriage, with the growth of the U.S. banking industry after the creation of the Federal Reserve, with David Rockefeller, one of John D. Jr.'s sons, eventually becoming head of the Chase Manhattan Bank and the figure at the center of the global financial spider's web until his death at the age of 101 in 2017.

The Council on Foreign Relations never had any official standing with the U.S. government. It was rather an elite instrument giving voice to the Rockefellers' global ambitions, in league with the big New York banks, and was intimately linked with the parallel imperial and financial interests of Great Britain. At the head of these interests stood the Federal Reserve Bank of New York and the Bank of England. The Council on Foreign Relations is the U.S. equivalent of the Royal Institute of International Affairs which, according to American scholar Carroll Quigley, was founded by Cecil Rhodes' secret society, the Round Table, becoming a "fief" of Rhodes's successor in influence, Alfred, Lord Milner.[6]

The CFR lobbied President Franklin Roosevelt and his administration to adopt policies in the prosecution of World War II that would not only defeat Germany, Italy, and Japan, but that would also set the stage for long-term competition with the Soviet Union, and eventually transform the U.S. into a juggernaut of multi-spectrum warfare against any country that stood in its way.

Within two weeks of the German invasion of Poland in September 1939, the U.S. State Department turned to the Council on Foreign Affairs for advice on what to do.[7] Despite the fact that Britain and France had declared war against Germany, it appeared that few Americans wanted the U.S. to do the same. A poll in late 1939 identified only seventeen percent of Americans as wanting to enter the war. In fact, Congress had passed a series of Neutrality Acts in 1935 that banned the export of weapons, granting of loans, and travel of citizens to nations at war. But in 1937, the Acts were modified to allow the president to discriminate between "aggressors" and "victims."[8]

On September 12, 1939, Hamilton Fish Armstrong, a founder of the CFR and the editor of its *Foreign Affairs* journal, along with CFR director Walter Mallory, met with Undersecretary of State Sumner Welles and several aides, and told them that the war that just began was a "grand opportunity" for the U.S. to become "the premier power in the world." They offered to undertake planning for the post-war peace. Welles agreed, provided that Armstrong and Mallory kept it quiet.[9]

The CFR created an *ad hoc* organization called War and Peace Studies that ended up sending 682 memoranda to U.S. government policymakers. Head of the project was Prof. Isaiah Bowman, president of Johns Hopkins University and CFR director.[10] The Rockefeller Foundation funded the entire project's cost of $350,000. The team's Armaments Group was led by future CIA director Allen Dulles, who would one day be fired over the Cuban missile fiasco by President John F. Kennedy. Dulles would then go on to sit on the Warren Commission that investigated Kennedy's assassination.

While the CFR was working in the shadows, the Roosevelt administration followed the same path of being an observer as President Woodrow Wilson had done early in World War I. In February 1940, Roosevelt sent Undersecretary Welles to meet with the conflict's leaders in Berlin, London, Paris, and Rome to secure a peace agreement that would include military disarmament. While Welles talked, Allen Dulles's CFR Armaments Group was examining military expansion no matter which way the European war went.[11]

As Germany began to sweep across Western Europe in 1940, invading Norway and Denmark in April, then Belgium, Luxembourg, the Netherlands, and France in May-June, several organizations in the U.S. were lobbying for continued American neutrality, including the Keep America Out of War Congress, the American Peace Society, and the America First Committee, where aviator Charles A. Lindbergh, Jr., became a spokesman. The America First Committee advocated defending the Western Hemisphere but going no further. Journalist Walter Lippmann began to derisively label opponents of entering the war as "isolationists."

By the autumn of 1940, the U.S. had begun its largest military buildup ever, far greater than in the run-up to World War I. This included a decision to build the world's largest navy, exceeding Britain's. With the enactment of the military draft, the first in American peacetime history, unemployment ended.

In 1940, the U.S. was still more than a year away from entry into the war. The common assumption is that the U.S. slept until Pearl Harbor, then suddenly awoke with its world on fire. This is far from the truth. As the country was building its military forces, the CFR and government insiders were already envisaging a much-expanded role of the U.S. in future world affairs, as it was increasingly clear that the British Empire no longer could control the world. But in the autumn of 1941, President Roosevelt had a problem. The American people still didn't want war.

Meanwhile, Britain had terminated its alliance with Japan in 1923. Japan had been fighting a war of conquest in China since 1931 and had been allied with Germany and Italy since 1940. Now, with the U.S. anchored in the Philippines, and the British arc of Hong Kong, Singapore, Burma, and India in place, a Japanese clash with the Anglo-Americans seemed likely, particularly after September 1940, when Japan invaded and occupied French Indochina.

A modern consensus has grown that President Roosevelt deliberately provoked Japan to attack the U.S. fleet at Pearl Harbor on December 7, 1941. I would refer you to the article "Pearl Harbor: Hawaii Was Surprised; FDR Was Not" by investigative journalist James Perloff. This article appears on-line and contains references to additional books and articles.[12]

The provocations were these: freezing Japanese assets being held in U.S. banks, a move that Japan viewed as an act of war; a far-reaching plan to cut Japan off from overseas petroleum supplies through an embargo, a ban on exporting steel to Japan; repeatedly sending U.S. warships into Japanese territorial waters; keeping the U.S. fleet stationed at Pearl Harbor late into 1941, rather than anchoring safely at its home Pacific Coast ports, especially San Diego. Instead, Roosevelt left the fleet anchored at Pearl Harbor, where he was told by his highest-ranking officers it was a sitting duck. All these circumstances are well-known to today's military establishment.[13]

The U.S.-British Divergence

The U.S. was finally in the war, and Britain rejoiced. But not only Britain. The U.S. banks that handled the sale of $150 billion in war bonds were charging the government 12–13 percent of every dollar in service charges.[14]

But the U.S. and Britain soon diverged in a manner that would have profound effects on the post-war world.

Once the U.S. entered the war in December 1941 after the Japanese attack on Pearl Harbor, Germany declared war against the U.S. immediately thereafter. The question facing the U.S. and Britain—with Germany having launched Operation Barbarossa against the Soviet Union six months earlier—was when would the two allies attack Germany on the European continent? Or would they let Germany and the Soviets fight it out and risk a German victory?

General Dwight D. Eisenhower, then stationed in London, was the leader of the U.S. armies in Europe. He favored an early assault across the English Channel on German-occupied France, but British Prime Minister Winston Churchill said this was premature. U.S. Chief of Staff of the Army

George C. Marshall, along with President Roosevelt, went along instead with a plan to attack the Axis first in North Africa, then Sicily, then mainland Italy.

It's believed today that Churchill delayed opening a Western front against Germany for three years until the Russians had routed the German armies in the east. The turning point was the largest battle in history at Stalingrad, where four million soldiers fought and over 1.2 million died. The Battle of Stalingrad ended in February 1943 with the Soviets pushing Germany back along the entire front. The Allies landed in Sicily in July, crossing to Italy in September.

Eisenhower and the British were now planning for the cross-Channel attack through France sometime in the following year. To the south, as the Americans led the drive through Italy, Churchill wanted them to turn east into the Balkans, where Tito and the Yugoslav communists were waging a bitter guerrilla war against the Germans. Despite Britain's July 1941 alliance with Stalin, Churchill was already thinking of a move to dominate a region that the Soviets would soon be eyeing as their own—yet another British stab in the back. But Roosevelt refused to comply with the British plan. He did not want to risk a clash with the Soviets in Eastern Europe.

On June 6, 1944, D-Day, the Allies crossed the Channel to Normandy in France. As they pushed the Germans back across France, Churchill began advocating for a British flying assault to take Berlin before the Soviets got there.[15] But Roosevelt disagreed, concerned that once the Soviets had pushed the Germans back to the Russian border, they would stop, thereby allowing Germany to move its forces from their eastern front to face the Allies on the Rhine. Marshall, Roosevelt, and the Americans also wanted their army in Italy to attack through southern France to protect Eisenhower's southern flank during his push toward the German homeland.

Churchill was stymied. In the end, the Americans prevailed, with Eisenhower telling the Soviets that the Americans would not attempt to take Berlin but would meet the Russians further west at the Elbe River. This became the boundary between East and West Germany. Churchill never got the British army to Berlin. He also had to watch while Tito and the communists set up a national state in Yugoslavia allied with Stalin.

All this was epochal. The British plan for the Nazis and Soviets to destroy each other had failed. It failed, at least in part, because Roosevelt had reached an understanding with Stalin and was willing to allow the Soviet Union to create a corridor of communist governments extending from what became East Germany and a reconstituted Poland in the north, down through Czechoslovakia, Hungary, Romania, and Bulgaria, with a southern anchor in Yugoslavia. The Baltic Republics of Latvia, Estonia, and Lithuania had already been incorporated into the Soviet Union.

So instead of a Europe under joint British-American control, the continent was now drastically divided, with a powerful Soviet Union controlling much of it. And France, now under control of a determined French nationalist, General Charles de Gaulle, could in no way be called a British/American satellite. True, the British oversaw a zone within Germany, but that was all. Even Italy had reconstituted itself as a self-directed republic, and Greece, later occupied by the British but evacuated, was collapsing into civil war with a strong communist influence.

It's possible to read the history of World War II as a constant juggling for position among the purported allies—Britain's Churchill, the U.S.'s Roosevelt, and the USSR's Stalin—to determine who would control the world at the end of the most devastating war in history. *The U.S. and the Soviet Union were the winners of World War II*, with the U.S. winning postwar dominance in the West, while the Soviets had won the worst of the actual fighting. The losers were Germany, Italy, Japan—and Britain. Roosevelt had no intention of allowing Britain to reconstitute its empire, especially now with India on the brink of independence.

The result of the war was a stalemate between two superpowers—the U.S. and the Soviet Union—with Germany reduced to rubble by Allied bombing and Japan devastated by American aerial attacks and the A-bomb.

Far from Britain "recovering" the U.S. for the British Empire, Britain had been relegated to second-power status. Britain's days as a U.S. "poodle" had begun.

Bretton Woods

Before the war ended, the U.S. convened an international conference at Bretton Woods, New Hampshire, on July 1–22, 1944, which established the supremacy of the American dollar as the basis for post-war international trade and commerce. Bretton Woods was a big step in the direction of American global hegemony, because where the dollar went, military force would follow. The Soviet Union, while it attended the conference, did not join the Bretton Woods system.

The U.S. had the power to create the Bretton Woods system because U.S. allies had sent most of their gold to America to pay for weapons purchases. The U.S. now controlled *seventy percent* of the world's monetary gold.[16]

Rejecting John Maynard Keynes' idea for a new global currency—the *bancor*—the Bretton Woods system established the U.S. dollar as the benchmark for international monetary transactions. Now the British pound as the world's reserve currency was *kaput*. The U.S. would be in the driver's seat from this point on—or so it thought until Britain wormed its way back in by becoming the center of world Eurodollar trading by 1970.

Since every nation would be expected to hold a substantial dollar reserve, they could get dollars through trade or by borrowing from U.S. banks. Borrowing could be done directly from the banks or through a new intergovernmental organization, the International Monetary Fund (IMF).

The IMF mission was to monitor exchange rates and lend reserve currencies to nations with deficits in their balance of payments trade ledgers. A major purpose of the system was to prevent nations from unilaterally devaluing their currencies in order to improve their trade postures with other countries, though the British eventually did just that. The system also kept the gold standard for settlement of international trade balances.

The IMF would become one of the key instruments of international political control, a role it continues to play until today. The IMF has lent to dozens of nations around the world, especially those from the Global South with financial problems. Whenever a nation seeks its help, the IMF insists it undertake "free market reforms," which in actuality means selling off its publicly owned utilities and industries to the big U.S. and international banks and corporations while its population languishes in destitution.

The IMF required the borrowing nations to allow Western banks and corporations to take over and exploit that nation's mining, agriculture, and industry so that they could earn enough income from selling their products abroad to repay the IMF loans. Countries under IMF control gradually lost the ability to create and manage a sustainable economy and policies promoting domestic wellbeing. It was neocolonialism—American style, making the IMF the world's greatest loan shark.

With the dollar triumphant, the U.S. had no need to borrow from the IMF or anyone else. Instead, Federal Reserve interest rates would determine the amount of money available for trade, created as always through fractional reserve lending. Later, this would be used to finance the growing U.S. trade deficits through sale of Treasury bonds to foreign nations. It would also pay for the growth of the U.S. military machine's hundreds of foreign bases.

Eventually, Federal Reserve "money printing" would backfire. In actuality it was a hidden method of devaluing the U.S. dollar. Long-term inflation has been the bane of the U.S. and world economy ever since, especially after Nixon's abandonment of the dollar peg for international exchange in 1971. Now, in 2023, with Russia, China, and other nations leading the charge against the dollar as a reserve currency, the entire system is poised to blow up with hyperinflation. The U.S. population can scarcely afford to buy a home, a car, or food.[17]

Neophyte Truman's Outsized Role in U.S. History

The start of the Cold War was presided over by President Harry S Truman after Roosevelt died in office on April 12, 1945. Truman had few qualifications, but he knew how to take orders from powerful people.

Following World War II, Truman acquiesced in *the largest military expansion in American peacetime history*. Truman was determined to fight against communism, or any other progressive movement, though U.S. belligerence was temporarily halted by its near catastrophe in Korea. The military industrial complex also organized the rearmament of Germany and Western Europe, with bank and corporate profits soaring into the stratosphere. Latin America meanwhile became a *de facto* U.S. colony.

The measures put in front of Truman for his signature were epoch-making. Truman knew *almost nothing* of what had transpired in the decision-making process presided over by Roosevelt for the past twelve years. This included the intricate discussions among Roosevelt, Churchill, Stalin, and their staffs and diplomats during the war.

Truman approved the August 6 and 9, 1945, dropping of A-bombs on Hiroshima and Nagasaki.[18] When Roosevelt died the previous April, Truman had *not even known* of the existence of the Manhattan Project and the development of nuclear weapons. On the other side of the world in Moscow, even Stalin had known what was going on. It is certain that the Soviets had been given nuclear secrets held by the U.S. and Britain by spies and informants. This allowed them to quickly match the U.S. and to explode an A-bomb in 1949 and an H-bomb in 1953, less than a year after the U.S. had detonated theirs. What Truman was told was a U.S. monopoly on nukes was a myth.

In a presidential directive of January 22, 1946, Truman created the Central Intelligence Group led by a Director of Central Intelligence. The National Security Act of 1947 changed the name to the Central Intelligence Agency, and in 1952 Truman approved the National Security Agency. The CIA was the outgrowth of the Office of Strategic Services that ran covert operations in Europe during World War II.

The National Security Act of 1947 defined "covert action," as:

> ...an activity of the U.S. government to influence political, economic, or military conditions abroad, where it is intended that the

role of the U.S. government will not be apparent or acknowledged publicly.

The act specified that mass propaganda, paramilitary operations, and lethal force could be carried out against *anyone* deemed a threat. Soon, the policy of "plausible deniability" would become standard operating procedure; that is, our proud Constitutional government created by Washington, Franklin, Hamilton, Jefferson, and other patriots would become, and still is, a clique of trained professional liars. Sadly, when then-CIA director Mike Pompeo admitted to a Texas A&M University audience in 2018 that "I was the CIA director. We lied, we cheated, we stole. We had entire training courses. It reminds you of the glory of the American experiment," there was no protest from the audience; it joined him in laughter.

Just as Congress earlier gave away its Constitutional prerogative to create money, it now, through the National Security Act of 1947, gave away its prerogative to declare war. Conflict between the executive and legislative branches of government over war powers now became endemic, with the executive and its wall of secrecy almost always winning the battle.

On the recommendation of Britain, Truman signed into law the Economic Cooperation Act of 1948 that established the Marshall Plan, named after Secretary of State George C. Marshall. Funding would eventually rise to over $12 billion for the rebuilding of Western Europe. Besides Britain and France, the beneficiaries of the Marshall Plan would include erstwhile enemies Italy and Germany. The plan overrode the advice of some U.S. policy makers, including Secretary of the Treasury Henry Morgenthau, who wanted Germany to be turned into a huge but harmless tract of farmland.

The proponents of the Marshall Plan, including the British who suggested it, reasoned that the West needed a strong industrialized bulwark against future Soviet expansion, as long as Germany was demilitarized and strictly controlled by the Western powers. Again, Germany was to be raised up as a deterrent to Russia. For the U.S., flooding Europe with American dollars, even as loans, not grants, as the Marshall Plan would do, served to amplify the role of the dollar as the world's reserve currency. Much of the money lent would come back to the U.S. as the Europeans purchased American machinery and factory equipment for their reconstruction.

The "Truman Doctrine" was announced in a speech by Truman to Congress on March 12, 1947. Most immediately, Truman pledged to contain communist uprisings in Greece and Turkey—purely domestic struggles that were *not* being supported by the Soviet Union. The U.S. had pulled most of its armies out of Europe except for a small force in Germany, so now

Congress appropriated financial aid to support the economies and militaries of Greece and Turkey rather than send troops. U.S. action was forced by the British pull-out of their own troops.

The Truman Doctrine marked the start of American support for other nations it claimed were threatened by the Soviets or by internal communist or progressive upheavals. As a result, Greece became ruled by a corrupt monarchy, and Turkey was in need of no military support at the time of Truman's speech. This new orientation suited many other nations as well. "All the Greek government, or any other dictatorship, had to do to get American aid was to claim that its opponents were communist."[19] This pattern has persisted throughout post-World War II history. The real beneficiaries of the Truman Doctrine have been right-wing oligarchs and the ruling classes of ostensibly pro-U.S. nations—and of the U.S. military industrial complex.

The Truman Doctrine was a step toward the formation in 1949 of the North American Treaty Organization—NATO. Historians often use Truman's speech of 1947 to indicate the start of the Cold War, though tensions with the Soviet Union had started to run high even before the end of the war. NATO now was to be the answer. Once again, NATO started as the brainchild of the British, specifically British Foreign Secretary Ernest Bevin. In fact, Bevin conceded that American independence had allowed it to attain the strength needed to save Britain. NATO was also a British ploy to maintain its relevance by furthering the narrative of the U.S. and Soviet Union being arch enemies. Stalin wanted the Soviet Union to be part of NATO but was rebuffed. Later, in 1955, the Soviets created the Warsaw Pact as a defensive alliance among the Soviet Union and its Eastern European satellites.

June 1950 brought the U.S. into the Korean War, when North Korea was enticed to invade South Korea following clashes along the border. North Korea was supported by China, which by now had been taken over by Mao Tse Tung's communists, and by the Soviet Union, while South Korea was supported by the newly-formed United Nations—effectively, the U.S. and Britain, since the Soviets were boycotting the vote in protest at the UN's recognition of the Republic of China (Taiwan) as China.

Called a "police action" by the U.S., the Korean War registered over two million combatants as well as two to three million civilians killed, wounded, and missing. The fighting ended with an armistice on July 27, 1953, after Truman had left office, with the original borders between North and South Korea being confirmed.

According to F. William Engdahl, the U.S. had provoked North Korea into invading in order to create outrage and fear among the U.S. population with regard to the communist "menace" and to mobilize public opinion against the Soviet Union, which would now become the scapegoat justifying continued U.S. military and economic dominance worldwide.[20]

Churchill's "Iron Curtain" Speech

On March 5, 1946, Winston Churchill gave his famous speech at Westminster College in Fulton, Missouri, where he announced that the Soviet Union had caused an "Iron Curtain" to fall across Europe.[21] Churchill had just been voted out of office as Prime Minister by a British electorate weary of war. But he was a master of war propaganda. More than that, it was yet another manifestation of the British maneuvering the U.S.

Churchill introduced the notion of a "special relationship" between Britain and the U.S. He spoke of U.S. possession of the atom bomb and the unity of the "English-speaking peoples" in terms that Stalin called "warmongering" and imperialistic "racism." In the speech, Churchill advocated the "common use" between the two nations of all their military facilities. Churchill also gave voice to the idea of "common citizenship" between the British and Americans.[22] Today this sounds like a desperate gambit to maintain British influence after its imperial power had been irretrievably lost.

Handing Palestine to the Zionists

At midnight on May 14, 1948, Zionist leaders in Palestine proclaimed a new state of Israel. On the same day, President Truman recognized the provisional Jewish government as the *de facto* authority of the Jewish state. Formal U.S. recognition was extended on January 31, 1949. The British Balfour Declaration of 1917 had authorized the creation of a Jewish homeland in Palestine. With Britain now diminished, the chief protector of Israel in a hostile Middle Eastern environment became the U.S.

In supporting the creation of Israel, Truman reneged on promises President Roosevelt had given to Saudi Arabia at the end of World War II that the U.S. did not support large-scale Jewish emigration to Palestine. But the Zionist movement won out, with massive financial support from wealthy Jews in the U.S., including organized crime kingpin Meyer Lansky. Over the next generation, American Zionists would swell into the Neocon faction that began seizing control of the U.S. national security state during the Reagan, Bush I, and Clinton administrations and that rules U.S. foreign policy today.

A Seabee at War—My Dad

Let me address how World War II appeared to one particular person in uniform caught up in it all: my father, Richard Edward Cook, born in Oklahoma City, Oklahoma, on January 7, 1924.

After my grandfather Fred Cook and my dad's mom, Carolyn, were divorced, she moved back to Oklahoma City, where my dad graduated from high school in 1943. Now without an income, my grandmother Carolyn worked in a grocery store, where she became assistant manager, and sold real estate. The whole family were card players, so she hosted bridge tournaments at her house. Her parents, John C. and Edna Hill, lived nearby, so they enjoyed good family times. I've heard J.C. was proud of the pecan trees in his yard. My dad's older sister Bobbie and her two kids also lived with Carolyn, her mom and their grandmother.

My dad gravitated back to Montana, where my own mom-to-be, Marjorie Peilow, was now in nursing school at St. Patrick's Hospital in Missoula. My dad moved up to Kalispell, just north of the Flathead Indian Reservation, and Fred joined him there. My dad obtained a Montana driver's license and registered for the draft in Kalispell. Oddly, Fred registered for the draft also, under a law that made veterans of World War I subject to being drafted if they were under forty-five. But Fred was not called up.

In late 1943, my dad was drafted into the Seabees, the branch of the Navy responsible for engineering and construction projects. The Seabees were trained at Camp Peary, Virginia, on the York River, where a small settlement of black residents had been displaced, with the population relocated a few miles away. The government constructed a training center and barracks on the ten-square-mile Camp Peary property.

Ironically, when my dad was sent to Camp Peary, he trained at a place only a couple of miles from where we would later reside in Williamsburg, Virginia, after he was transferred by Dow Chemical to a factory near the colonial capital in 1960. During World War II, Camp Peary was also used as a German POW camp. By the early 1950s, Camp Peary had become the CIA's premier training center for covert operations. It was called "The Farm," because the military had set up a hog farm on the premises to help provide food for the base.

My dad said that as soon as he stepped off the bus at Camp Peary for basic training they gave him a pack of cigarettes. This started a lifelong addiction, due to which he would one day suffer from emphysema and COPD. Then they shaved his head. From Camp Peary, the Seabees sent my dad to San Francisco, where he embarked for one of the most remote and inhospitable places on the planet—Attu Island. He arrived on September 25, 1943.

About 35 miles long and 25 miles across, Attu is the last island in the long sweep of the Aleutian island chain in the north Pacific, 1100 miles from mainland Alaska. It's the westernmost point in the U.S. In fact, it's so far out in the Pacific Ocean that they had to adjust the International Date Line so that it bends around the island, leaving it part of an Alaskan time zone.

In June 1942, the Japanese occupied Attu. About forty-five native Aleuts were removed to a Japanese prison camp. Earlier, the U.S. had removed 880 Aleut residents to an internment camp on the Alaskan panhandle. During their assault, the Japanese killed an American radio operator, Charles Jones, and sent his wife Etta as a prisoner to Japan. She was released at the end of the war and lived in Florida until age eighty-six.

On May 11, 1943, the U.S. Army had launched an attack to retake Attu. After some of the most bitter fighting of the Pacific war, almost the entire Japanese garrison of 2,900 were killed or committed suicide, capture being dishonorable. Only twenty-nine Japanese had survived when the fighting was over on May 30. On the American side, there were 549 killed, 1,148 wounded, and 1,814 sick or dead from disease.

Soon the Seabees arrived, my dad with them. Their job was to build airstrips, with barracks, service buildings, and control towers. The purpose was to prepare the site for use by the Navy and the Army's Air Force in flying patrols over the North Pacific and in the bombing of Japan.

The work was grueling and the weather conditions horrendous. Equally so, the Japanese body parts that my dad said were lying all over. The men lived in tents and worked sixty hours a week. My dad sent letters home to his mom, describing the ordeal of spending long days in knee-deep mud, working past the point of exhaustion, including days spent operating a jackhammer in the rocky soil. He also spent time as a "grease monkey," lubricating trucks and jeeps, and he worked in the camp laundry.

When he was able to return to the mainland for a thirty-day furlough, he spent it in Montana visiting his now-girlfriend Marge and with his own family in Oklahoma. He wrote to his dad, Fred, now estranged from the rest of the family and living in Idaho Falls and mentions receiving Fred's gifts of candy and piñon nuts. My dad's mom and his sister Bobbie sent him letters, books, and photos. He wrote to them a couple of times a week. He also wrote to his grandparents, J.C. Hill and Edna.

The first airfield built by the Seabees was called the Alexi Point Army Airfield. The first attack run against Japanese territory used the Japanese-built airstrip and left Attu for the Kurile Islands on July 10, 1943. Subsequent attacks on the Japanese mainland were carried out from the Seabees-constructed runways.

The Seabees built runways and a seaplane base for patrol bombers and flying boats. For the duration of the war, naval aircraft taking off from Attu patrolled the North Pacific. As at Alexi Field, the Navy's two runways were first covered with steel mats. By 1944, however, the Seabees had laid asphalt, and the Navy made the runways available to Army planes as well as its own. The 11th Air Force also established maintenance facilities.

My dad stayed on Attu Island for the duration of his service. As time went on, with the war eventually winding down, he realized that this would be his only assignment. They had a regular movie night, with one of the movies being *The Song of Bernadette*. Now and then the USO would come in with a live show. He said one time there were four women in the show, which seemed to be a big deal. They also sometimes heard propaganda broadcasts from the Japanese, with appearances by men who supposedly were American POWs. He liked listening to Fred Waring's choir sing *The Battle Hymn of the Republic*.

Life on Attu was boring, and my dad was often homesick. But he wrote that, "If I can just not be too selfish and be thankful for what good things I have instead of being bitter over the bad, then I feel a lot better." One of his blessings was that "I'll come back in one piece when I do come home though." There were church services, and he liked the ones on Easter and Christmas. He often asked in his letters how his nephew Butchie and niece Judy were doing. Early on in his tour he got some letters from Herman Iselin, Bobbie's husband, who was destined to fly in Europe as a pilot, surviving that but then dying in a plane crash after the war.

My dad often commented on the weather. He wrote home, "The Aleutian weather is just what you read about only much worse." In February 1945 he wrote that the snow would soon be done, and that he could look forward to an Aleutian spring of rain and slush. Sometimes the rain would leak through the tent onto the men's bunks, making it hard to sleep. Sometimes the weather was too bad for the men to work. Once he wrote that the men were off a whole day because, "There is a beautiful blizzard and the tent is so warm and cozy."

One of the many books he read was *A Tree Grows in Brooklyn*. Another was *The Princess and the Pirate*. He also read mysteries by Agatha Christie and Ellery Queen. He read Hemingway's *For Whom the Bell Tolls,* then after seeing the move, said that it kept close to the book and remarked that, "Ingrid Bergman was more beautiful than a woman should be."

The men on Attu heard about the D-Day Allied landing in Normandy eight days after it took place. He later said that, "I guess we are on our way now towards the winning of the war, but I predict it won't end until late '45 or early '46." The war actually ended with the surrender of Japan on September 2, 1945.

When the job on Attu Island was finally done, my dad shipped back to the naval base in Bremerton, Washington, and spent the remainder of the war doing office work. He and my mom were married in San Francisco in September 1945. He says he was pleased at being able to type his own discharge papers in the base office.

At one point my dad had written home saying he wanted to buy a ranch someday. But it was not to be. He also wrote that, "Maybe in the next war I will be one of the high shots, and I'll start a new order." But he was also philosophical, writing "It is better to be a good little man than a poor big one."

After discharge, he enrolled at the University of Montana in Missoula. He and my mom were able to rent a small house on campus at the foot of Mount Sentinel, where the university's famous mountainside "M" is located. He went to college at the government's expense on the GI Bill, majoring in chemistry.

I was born on October 20, 1946. Two years later my sister Barbara was born. Unfortunately, she was a "blue baby." Not much was known back then about this syndrome. My sister was retarded and spent the rest of her life in a Montana institution.

In 1950, when my dad graduated from the university, we moved briefly to Norman, Oklahoma, where he planned to pursue graduate school. But soon afterwards he was offered a job as a chemist with Dow Chemical in Midland, Michigan, in the Saran Wrap Division. That's where we spent the 1950s, with my sister Sandy being born in 1952 and my sister Christine in 1956. In 1960 Dow transferred him to a plant near Williamsburg and Camp Peary.

ENDNOTES FOR CHAPTER 12

1 Patrick J. Buchanan, *Churchill, Hitler, and the Unnecessary War: How Britain Lost Its Empire and the West Lost the World* (Crown Publishers, 2008), x.

2 Preparata, 135. Otto Strasser, an early Hitler collaborator, commented that during those early Munich days, Hitler was always flush with cash from unexplained sources, while his mates could barely outfit themselves with a decent set of clothes. The implication was that even then Hitler was on someone's payroll.

3 Ibid., 244.

4 Engdahl, 183.

5 Ibid., 249.

6 Quigley, 3ff.

7 Stephen Wertheim, *Tomorrow the World: The Birth of U.S. Global Supremacy* (The Belknap Press of Harvard University Press, 2020), 8.

8 Ibid., 32.

9 Ibid., 37. Maybe the U.S. could take over the world without anyone noticing?

10 Engdahl, 137.

11 Ibid., 42.

12 The New American Magazine, "Pearl Harbor: Hawaii Was Surprised; FDR Was Not," *Four States News*, December 7, 2015. https://www.fourstatesnews.us/2015/12/07/pearl-harbor-hawaii-surprised-fdr-not/

13 *Douglas MacGregor – We are co Belligerents,* video, 29:33, January 28, 2023. https://youtu.be/t8FG34KX0h4?si=4IU2xYmBYnP9P8Am F. William Engdahl also recounts the U.S. government's actions in enticing a Japanese attack in *Gods of Money.*

14 Prins, 159–160.

15 Stephen E. Ambrose and Douglas G. Brinkley, *Rise to Globalism: American Foreign Policy Since 1938,* eighth revised edition (Penguin Books, 1997), 28. While I don't agree with all their interpretations, these two mainstream historians present a competent chronology of events.

16 Engdahl, 212.

17 This is the essence of "Bidenomics" as characterized by Robert Barnes on *The Duran,* August 21, 2023.

18 See Eustace Mullins, *The Secret History of the Atomic Bomb: Why Hiroshima Was Destroyed, The Untold Story* (June 1968).

19 Ambrose and Brinkley, 82.

20 Engdahl, 205–206.

21 Even as World War II was raging, with Britain and the Soviet Union as allies, the British government had begun making plans for future war against the Soviets in Europe. An iteration of this plan was called Operation Unthinkable.

22 "Sinews of Peace" (Iron Curtain Speech), March 5, 1946, National Churchill Museum. https://www.nationalchurchillmuseum.org/sinews-of-peace-iron-curtain-speech.html

CHAPTER 13

Nightmare

Postwar Finance Merges with Covert Operations

With World War II over and President Harry S. Truman now in office, U.S. banking moved aggressively to control the devastated economies of Western Europe. The Marshall Plan, in force from 1948 to 1952, had not been a disinterested act of charity. The $15 billion provided to Europe through grants and loans was money then returned to the U.S. through purchase of American products. The greatest beneficiary was the Rockefellers' Standard Oil, which supplied Europe with huge quantities of petroleum at marked-up prices.[1]

The newly founded CIA received more than five percent of the funds allocated under the Marshall Plan, which it then used to establish "front" businesses in European countries. Another focus was to deflect elections in Italy away from leftist parties.[2] Notably, the CIA at this early date tried but failed to create a fifth column of spies and saboteurs in Ukraine.

The Soviet Union had outraged the Roosevelt administration by refusing to sign the Bretton Woods agreements mandating the primacy of the U.S. dollar. It has been argued that the main reason Truman nuked Japan, when the war against Japan had already been won, was to frighten the Soviets, whom the Rockefeller-controlled U.S. establishment, along with the British, had designated the enemy of the future. It was war, first, last, and always for which the gods of money lusted.

The term "Cold War" was first used by Bernard Baruch, Wall Street magnate and adviser to presidents, who said in a speech on April 16, 1947:

Let us not be deceived. We are today in the midst of a Cold War. Our enemies are to be found abroad and at home. Let us never forget this: Our unrest is the heart of their success.[3]

Note the words "at home." A Red Scare was coming, with Senator Joe McCarthy doing the dirty work. But *the Cold War was started by U.S. financiers* as part of their mission of global conquest.

The Soviets were still reeling from the war but did not remain idle as the world began to split into two armed camps. In 1948, a communist coup took place in Czechoslovakia. Faced with the Marshall Plan, the Soviet Union launched its own Council of Mutual Economic Assistance (Comecon) in 1949 followed by Cominform—the Communist Information Bureau. Cominform was meant to balance the longstanding British/American propaganda operation that used psychological methods pioneered by the Tavistock Institute. As much as they were able, the Soviets supported anti-colonialist movements around the world and opened their universities to students from Third World countries.

Meanwhile, the Rockefellers' Chase National Bank had become the biggest U.S. lender to Europe. In 1946, David Rockefeller, a son of John D. Rockefeller, Jr., had joined the staff of Chase Bank in New York as assistant manager in the international division. Chase became the first U.S. bank to establish branches in Germany and Japan. Chase was also the key lender to Latin American governments and was engaged in establishing as many branch banks as possible. David's brother, Nelson Rockefeller, managed the family's South American banking empire, forging alliances with the families of oligarchs and army generals in Venezuela, Brazil, Argentina, Peru, and Chile. It was Rockefeller money and credit that forged the backbone of South American fascism.

And it was the CIA that would become the backstage enforcer of U.S. financial hegemony—the "jackals,"[4] charged with overthrowing progressive governments and assassinating their leaders.

Creation of the CIA

The CIA was the successor agency to the World War II Office of Strategic Services (OSS) headed by "Wild Bill" Donovan. Prior to the OSS, the U.S. did not have a secret intelligence service. Britain, of course, has had one for centuries, going back to Elizabethan days.

The OSS learned much of its craft from the British, which had formally institutionalized its intelligence in MI6, chartered in 1909.[5] In fact, when Roosevelt named Donovan head of the predecessor to the OSS, the Office of the Coordinator of Information, he created the unit on the advice of Winston Churchill. Roosevelt authorized the unit on July 11, 1941, five months before Pearl Harbor.

With the OSS, U.S. foreign policy took its first steps toward permanent linkage to covert violence. The fact that the OSS was continued in peacetime as the CIA meant that covert violence would continue. Given the regular turnover of U.S. presidential administrations, political oversight and control of the "spooks" became impossible. This was a major step toward today's ubiquitous "Deep State."

Truman had issued an executive order on September 20, 1945, terminating the OSS. He and his advisers were aware that U.S. wartime intelligence had been built on cryptologic success, not covert subversion, and there are suggestions that Truman feared that a peacetime agency operating without tight control would be a potential menace to Americans citizens.[6] Truman had also received a report originally written for President Roosevelt by Col. Richard Park, Jr., stating that the OSS had "all the earmarks of a Gestapo system."[7]

Truman and his advisers saw no need for a continuation of covert activities, while intelligence gathering could be handled by the military services and the State Department. With no war, against whom would a covert action agency be fighting?

At least in this instance, Truman had good instincts. But now John McCloy stepped into the breach. McCloy was a long-time Rockefeller man soon to become one of the main U.S. figures in rebuilding Western Europe as a market for U.S. bank loans. In 1945, McCloy was Assistant Secretary of War.

McCloy was instrumental in keeping some of the key elements of the OSS intact, including, with ominous portents for the future, the 101st Detachment in Burma and Southeast Asia and the Jedburgh operation in Burma in charge of covert action against the Japanese and later the Chinese. These units continued to be active in Burma, Laos, and Vietnam, where the CIA became instrumental in guiding the U.S. into the war in Vietnam and where it oversaw opium production by Laotian army officials.

These OSS/CIA units were never disbanded and were the origin of CIA involvement in the international drug trade that later branched out into South and Central America, Afghanistan, and U.S. cities.

Truman caved in to McCloy's pressure. In a presidential directive of January 22, 1946, Truman created the Central Intelligence Group headed by a Director of Central Intelligence. The first DCI was Rear Admiral Sidney Souers, Assistant Chief of Naval Intelligence. Souers had an installment ceremony of sorts. The diary of President Truman's chief military adviser, Fleet Admiral William D. Leahy, records the event of January 24, 1946:

At lunch today in the White House, with only members of the Staff present, RAdm. Sidney Souers and I were presented [by President Truman] with black cloaks, black hats, and wooden daggers, and the President read an amusing directive to us outlining some of our duties in the Central Intelligence Agency "Cloak and Dagger Group of Snoopers."[8]

This amusing skit marked the start of almost eight decades of CIA mayhem, still ongoing.

The National Security Act of 1947 changed the name of the Central Intelligence Group to the Central Intelligence Agency, and in 1952 Truman approved the National Security Agency "for special intelligence gathering." As stated in Chapter Twelve, the Act specified that mass propaganda, paramilitary operations, and lethal force could be carried out against *anyone* deemed a threat.

CIA secrecy was codified in the Central Intelligence Act of 1949 that exempted the CIA from disclosing the functions, names, titles, salaries, and even the number of its employees. It was also given the power to spend money "without regard to the provisions of law and regulations relating to the expenditure of government funds."[9]

The CIA and the Rockefellers

So, was the CIA under the direction of *nobody*? Not exactly.

David Rockefeller's ties to the CIA became legendary. During World War II, he served in military intelligence in North Africa and France. At the end of the war, he served as an assistant military attaché at the American Embassy in Paris, where the Rockefeller family's Chase Bank had operated unmolested throughout the German occupation.

Later ensconced in his Manhattan office at Rockefeller Center in New York City, David Rockefeller was regularly briefed on covert intelligence operations by CIA division chiefs.[10] In 1949 he became a director of the Council on Foreign Relations. As we have seen, the CFR did the legwork during World War II in advising the government on future U.S. global military dominance. During the 1950s, David Rockefeller, John J. McCloy, and Henry Kissinger were referred to as the "Triumperate" in describing backroom influence over U.S. foreign policy.

David Rockefeller's brother Nelson Rockefeller, future New York governor and vice-president of the U.S. under Gerald Ford, was also heavily involved in the early days of covert operations. In 1954, President Eisenhower appointed Rockefeller as Special Assistant for Cold War Planning. Part of his portfolio was monitoring and approving covert CIA activity.

Henry Kissinger was part of covert operations early on. In 1938, when Kissinger was fifteen years old, he and his family fled Germany as Nazi persecution of Jews increased. Drafted while in college, Kissinger became a U.S. citizen, joined U.S. military intelligence, and served in the Army's Counterintelligence Corps. After the war he enrolled in Harvard, where he earned his bachelor's, MA, and doctorate degrees. Seymour Hersh reports:

> In 1952, Kissinger was named a consultant to the director of the Psychological Strategy Board, an operating arm of the National Security Council for covert psychological and paramilitary operations.... In 1955, Kissinger, already known to insiders for his closeness to [Nelson] Rockefeller and for Rockefeller's reliance on him, was named a consultant to the Operations Coordinating Board, the highest policy-making board for implementing clandestine activity against foreign governments.[11]

Kissinger would retain his close working relationship with the Rockefellers as he moved toward becoming President Richard Nixon's National Security Advisor and Secretary of State.

Thus the Rockefeller financial empire and the CIA were two sides of the same coin and the same power center—one side operating more or less openly and the other in the shadows. Both had the same objective: global U.S. control. But it could only be done with propaganda leverage. The leverage was to use the U.S. mainstream media to generate sufficient fear and hatred of the Soviet Union. The leader in propaganda was *The New York Times,* which covertly employs numerous CIA operatives.

Where was all this headed? Later, David Rockefeller wrote in his memoirs:

> For more than a century, ideological extremists at either end of the political spectrum have seized upon well-publicized incidents such as my encounter with Castro to attack the Rockefeller family for the inordinate influence they claim we wield over American political and economic institutions. Some even believe we are part of a secret cabal working against the best interests of the United States, characterizing my family and me as "internationalists" and of conspiring with others around the world to build a more

integrated global, political, and economic structure—one world, if you will. If that's the charge, I stand guilty, and I am proud of it.

He continued:

We are grateful to *The Washington Post*, *The New York Times*, *Time* Magazine and other great publications whose directors have attended our meetings and respected their promises of discretion for almost forty years......It would have been impossible for us to develop *our plan for the world* [emphasis added] if we had been subjected to the lights of publicity during those years. But the world is more sophisticated and prepared to march towards a world government. The supernational sovereignty of an intellectual elite and world bankers is surely preferable to the national auto-determination practiced in past centuries.[12]

The Rockefellers and the CIA were the *keepers and enforcers* of "our plan." And the U.S. media were fully cooperating.

Eisenhower is Elected as the Cold War Accelerates

War hero Dwight D. Eisenhower, Supreme Allied Commander for D-Day, first NATO commander, lately president of Columbia University, was recruited by both the Democratic and Republican parties to be their 1952 presidential candidate. Nelson Rockefeller persuaded Ike to go Republican.

Eisenhower's main competitor at the Republican National Convention was U.S. Senator Robert Taft of Ohio. Taft had opposed U.S. entry into World War II prior to Pearl Harbor and was opposed to the creation of NATO and involvement of the U.S. in post-World War II entangling alliances. Taft represented a long-standing Republican opposition to global entanglement that had prohibited U.S. entry into the League of Nations.

Taft was anathema to the Rockefeller globalists, who were now pushing the "American Century."[13] Eisenhower and Taft ran a close one-two on the first convention ballot, though Eisenhower did not have a majority among five different candidates. But he won on a revised count.

The Republicans were desperate to have a president after being shut out of the White House for twenty years. While Truman had not neglected the big banks and Wall Street, Eisenhower was close pals with the financial big shots, down to the choice of his favorite golfing partner in Senator Prescott Bush[14] of Brown Brothers Harriman, a bank that had been cited for collaboration during wartime with the Nazis.

At a time of Senator Joe McCarthy and the Red Scare *redux*,[15] Eisenhower was an anti-communist icon. Still, the president was criticized because the peacetime growth of what he would later call the military industrial complex actually *declined* from Truman years. Eschewing all-out total war against the Soviets, Eisenhower had embraced the "containment" strategy espoused by State Department adviser George Kennan. But even with containment instead of open warfare, the Soviets were *the enemy*.

The Soviets still had the largest land-based military force on the planet, the Red Army, which had crushed the Nazis in an awesome display of force. True, Stalin had kept the Red Army mostly at home in Russia during the postwar years, but there was no way the U.S. could send another army to fight in Europe on the scale of the two world wars. The public wouldn't stand for it, and Congress wouldn't pay for it. But what about the Europeans?

Et voila, NATO. From the start the North Atlantic Treaty Organization was a British-instigated but U.S.-directed military alliance intended to utilize mainly European national armies as ground forces if another war broke out. Insofar as it would obviously be France and Britain that would bear the burden, and neither gave any indication of wanting to fight the Red Army on European soil, the solution was to rely on the U.S. Air Force armed with nuclear weapons. Western Europe, especially West Germany, became, in essence, a U.S. air base.

But saber-rattling was still allowed and made for good press. It was U.S. Secretary of State John Foster Dulles, a long-time Rockefeller crony, who now invented the term "brinksmanship," a byword for threats to carry out massive retaliation if the Soviet Union overstepped its bounds. Dulles denounced neutrality as immoral. It was "us" vs. "them." But this quickly became a standoff, as the Soviets acquired their own nuclear arsenal. Thus, the potential of military conflict shifted away from Europe to Asia and the so-called Third World, where, post colonialism, the non-aligned movement gained adherents among nations wishing to keep its distance from U.S. financial power.

No one in power within the U.S., even for a second, thought of coming to some kind of compromise or accommodation with the Soviets. They were the communists; we were the "Free World."

But Eisenhower, with his "New Look," reminded people of the need for a "balance between minimum requirements in the costly implements of war and the health of our economy."[16] In a speech on April 16, 1953, he told the American Society of Newspaper Editors, "Every gun that is made, every warship launched, every rocket fired signifies, in the final sense, a theft from those who hunger and are not fed, those who are cold and are not clothed."[17]

Eisenhower's Crises

The popular image of Eisenhower is of an old gent playing golf and lounging around his Gettysburg farm. But his two terms were filled with crises. Among them:

1953

With Eisenhower's approval, the CIA overthrew the elected government of Iran. The new government brought the Shah of Iran back from exile and divided up Iran's oil assets: The British and the U.S. each got forty percent; the Dutch got fourteen percent; and the French six percent. The Shah himself was later overthrown in the Islamic Revolution of 1978–1979. The CIA has been engaged in covert warfare against the government of Iran ever since.

1954

Again with Eisenhower's approval, the CIA overthrew the government of Guatemala, with the CIA dropping napalm on civilian areas with its own aircraft. The U.S. was heavily criticized by UN Secretary-General Dag Hammarskjold, who later was killed when a terrorist bomb blew up his airplane in Africa in 1961.[18]

1954

The French stronghold at Dien Bien Phu in French Indochina fell to North Vietnamese insurgents. The president of North Vietnam, Ho Chi Minh, had attended the Versailles Conference in 1919 to ask for Western recognition of a free Vietnam. He was ignored. Now he led an insurrection against the French colonizers. The U.S. Air Force wanted to drop nukes on the Viet Minh, the North Vietnamese guerrilla force, while Eisenhower wanted conventional bombing. At the subsequent peace conference at Geneva, the U.S. walked out, with the attendees dividing Vietnam at the seventeenth parallel. U.S. military advisers soon started training the South Vietnamese army.

1959

On January 8, 1959, Fidel Castro ended the eight-year Cuban Revolution by driving dictator Fulgencio Batista out of Cuba. The U.S. owned eight percent of Cuba's utilities, forty percent of its sugar, ninety percent of its mining wealth, and occupied Guantanamo Bay. CIA Director Allen Dulles told Eisenhower that, "Communists and other extreme radicals appear to have penetrated the Castro government."[19] In 1961, Eisenhower severed diplomatic relations with Cuba and gave the CIA clearance to plan for an invasion and train Cuban exiles to carry it out. This would result in the Bay of Pigs fiasco under President John F. Kennedy's watch.

1960

Soviet Premier Nikita Khrushchev announced that Russia had shot down an American U-2 spy plane inside its territory, an incontrovertible fact since pilot Francis Gary Powers, Jr. was captured alive. This was just before a Paris summit with the USSR. Eisenhower refused to apologize, and the summit was called off.

Nonetheless, Eisenhower said in his Farewell Address:

> The conjunction of an immense military establishment and a large arms industry...new in American experience, exercises a total influence...felt in every city, every state house, every office of the federal government.... In the councils of government, we must guard against the acquisition of unwarranted influence, whether sought or unsought, by the military industrial complex.

Eisenhower's warning failed. The military industrial complex, owned by big finance, and served by CIA global subversion, rules America today.

The CIA Agenda Justified

To head off a threatened congressional investigation proposed by Senator Mike Mansfield in response to the CIA's overthrowing the governments of Iran and Guatemala, President Eisenhower enlisted World War II aviator General James Doolittle to produce his own classified report on the CIA's clandestine activities. Doolittle wrote:

> It is now clear that we are facing an implacable enemy whose avowed objective is world domination by whatever means and at whatever cost. There are no rules in such a game. Hitherto acceptable norms of human conduct do not apply. If the United States is to survive, long-standing American concepts of "fair play" must be reconsidered. We must develop effective espionage and counter-espionage services and must learn to subvert, sabotage, and destroy our enemies by more clever, more sophisticated, and more effective methods than those used against us. It may become more necessary that the American people be made acquainted

with, understand, and support this fundamentally repugnant philosophy.[20]

The top-secret report identified the Soviet Union as the "implacable enemy" but was not released to the public or Congress until 1976.[21] Though Eisenhower never asked Congress for a declaration of war, the report can be read as a call for an open-ended conflict using what might even be called state-sponsored terrorism, with no defined objectives or end point. The CIA was asserting its own rules and agenda, let the results for American foreign policy be what they may.

Flathead Reservation "Termination"

We now return to Montana, with the 1.25 million acres of the Flathead Indian Reservation containing some of the most beautiful landscapes in North America. The reservation begins in the south where U.S. 93 cuts through the Lolo National Forest just outside Missoula and continues north about 90 miles to include the lower half of Flathead Lake. This is the largest freshwater body of water west of the Mississippi River and among the cleanest lakes in the world. East to west the lower Flathead River valley extends from the line of the Mission Mountains across wetlands, low hills, and prairie to the Salish Mountains beyond the remote town of Hot Springs, where there is a native-owned mineral bath.

Driving north on U.S. 93 in the spring, we see the snow-capped Mission Mountains rising to the right in a jagged line to almost 10,000 feet, an unforgettable sight. The mountains were sculpted by the same glaciation events that dug out Flathead Lake. Living in the mountains among the profusion of wildlife are moose, grizzlies, elk, mountain lions, eagles, and many other inspirational species. The tribal-run CSKT[22] Bison Range, enclosed with a herd of over 400 animals, including the annual crop of newborns, lies in the south of the reservation.

As mentioned previously, the most numerous tribe on the Flathead Reservation is the Salish:

> Although never a large tribe, the Salish had a reputation for bravery, honesty, and general high character and for their friendly disposition towards the whites. When first known, about the beginning of the [19th] century, they subsisted chiefly by hunting and the gathering of wild roots, particularly camas, dwelt in skin tipis or mat-covered lodges, and were at peace with all tribes,

excepting their hereditary enemies, the powerful Blackfeet who lived across the Continental Divide on the Great Plains.[23]

The Salish language is spoken and taught today. The Salish and the related Pend d'Oreilles and Kootenais, had custody of the Flathead Reservation confirmed through the 1855 Hellgate Treaty. Owing to the Dawes Act of 1887, after "allotment" of portions of the reservation to tribal members, the "surplus" was opened to white homesteaders in 1910.

The allotment policy ended with the Indian Reorganization Act of 1934 and the "Indian New Deal." Since then, the first tribal government to be established in the U.S. has operated under a Tribal Council and constitution. Today, according to 2020 Census Bureau data, 8,029 American Indians live on the reservation, plus an undetermined number of mixed race.[24] This includes non-CSKT Indians.

The Indians living on the reservation scraped by during the Depression, as did others in rural areas of the U.S., depending particularly on farming and gardening, annual gathering of camas roots, use of herbal medicines, hunting, and intermittent jobs with local and regional employers. Some traveled to Helena or Missoula for work or to neighboring states. Some "rode the rails."

During World War II and the Korean War, many Native Americans joined or were drafted into the military, with many being sent overseas. About 25,000 Native Americans served during World War II, most in the Army. Several hundred women served as nurses. More than 10,000 Native Americans served in the Korean War. Military service, while often a matter of honor or adventure, brought money to the reservations through pay and pensions.

In 1954, the St. Ignatius Mission celebrated its centennial. Even though the earlier generation of reservation residents was passing away, the Jesuit priests and Ursuline sisters continued to serve the community, with the boarding school still in operation. Among the paintings in the chapel were life-sized depictions of an Indian Chief Jesus and an Indian Madonna. The biggest festival for native churchgoers was Christmas, with large numbers also turning out for wakes and funerals. By this time, people were not living so much on scattered farmsteads, but had started to reside in neighborhoods, especially in the towns that were growing up along or near U.S. 93.

In 1953, the federal government tried to *abolish* the Indian reservations across the nation, with the Flathead Reservation being designated the first to go. A study of termination may be found in a book by Jaako Puisto: *"This Is My Reservation: I Belong Here: The Salish Kootenai Indian Struggle Against Termination."*[25] The movement toward termination began during the Truman administration and was portrayed as a benefit to Indians who would no longer suffer from "dependence" on the federal government. The hidden motive was similar to the days of allotment: a land grab by the whites. State and local governments, wanting to tax Indian lands, joined others with commercial interests. Indians lived on the reservations tax free, which, for many, was the only way they *could* live. It was also the era of McCarthyism, with right-wing Republicans alleging that the reservations were "un-American strongholds for foreign ideas and political systems" and even "hotbeds" of communism.[26]

The Flathead tribes and their supporters succeeded in driving home the point that the agreement which the Indians made with the government in the Hellgate Treaty of 1855 was sacred. Thus, the Flathead Reservation was preserved for tribal members and their descendants and survives today.

The Kennedy Presidency: Cuba

The most important geopolitical question in 1961, when John F. Kennedy was inaugurated, was what should be done about Cuba?

Three months later, on April 16, 1961, a weak and poorly armed CIA-trained force of about 1,400 Cuban exiles and mercenaries landed in Cuba at the Bay of Pigs and was smashed within seventy-two hours by Fidel Castro's army. This aforementioned Eisenhower project, foisted on him by a CIA that by now was out of control, nonetheless received Kennedy's approval for the attack.

With the invasion having failed, the CIA and military were furious that Kennedy did not send an American force, or at least U.S. bombers, to support the invaders. Equally angry, Kennedy asked for the resignation of Director Allen Dulles, believing that Dulles had lied to him about the attack. Dulles's replacement was John McCone, then serving as chairman of the U.S. Atomic Energy Commission. According to a *New York Times* article of April 25, 1966:

> President Kennedy, as the enormity of the Bay of Pigs disaster came home to him, said to one of the highest officials of his Administration that he "wanted to splinter the CIA in a thousand pieces and scatter it to the winds."

Kennedy now approved a plan to increase U.S. nuclear weapons capable of being delivered against the Soviet Union by Polaris submarines and long-range bombers. Fearing an American first-strike, the Soviets responded by expanding their own nuclear arsenal. A new arms race was underway.

In the summer of 1961, Kennedy met in Vienna with Soviet Premier Nikita Khrushchev, who argued that the U.S. should allow the existence of revolutionary movements in the Third World. Kennedy rejected the idea and ordered the U.S. military buildup to proceed. Khrushchev called Kennedy's stance "military hysteria."[27] Khrushchev's next move was to build the Berlin Wall.

Confirming Soviet fears, talk in the U.S. of a nuclear first strike now went public in the American press.

The escalation continued. The Soviets broke a three-year moratorium on nuclear testing by exploding a bomb three thousand times as powerful as the one used on Hiroshima. Kennedy also resumed U.S. atmospheric tests.

Events now ramped up in Cuba. On October 19, 1960, after Cuba nationalized U.S.-owned oil refineries, the U.S. placed an embargo on exports to Cuba, except for food and medicine. On December 2, 1961, Castro declared himself a Marxist-Leninist, and two months later, Kennedy extended the trade embargo to include almost all Cuban exports. The Soviet Union increased its trade with Cuba and now began to include military supplies. In August 1962, they began to build medium-range missile sites in Cuba. Kennedy now expressed the fear privately that if he failed to act, he would be impeached after the November congressional elections.[28]

Meanwhile, the CIA and military had come up with a proposal for a major false-flag incident similar to, but far beyond, the "Remember the Maine" campaign that preceded the U.S. invasion of Spanish-held Cuba at the start of the Spanish-American War. This was Operation Northwoods. The proposal was attached to a March 13, 1962, memo sent by Chairman of the Joint Chiefs of Staff Lyman L. Lemnitzer to Secretary of Defense Robert McNamara. The proposal called for the CIA to commit acts of terrorism *against American military and civilian targets*. The attacks would be blamed on the Cuban government and used to justify war against Cuba.

Kennedy rejected Operation Northwoods.[29] I am not the first to suggest that the proposed Operation Northwoods was of a scope to rival the 9/11 attacks almost four decades in the future, which was blamed on Osama bin Laden and then used to justify an attack on Afghanistan and Iraq.

On October 22, 1962, President Kennedy went on national TV to announce that the U.S. was placing "a strict quarantine on all offensive military equipment" being shipped from the Soviet Union to Cuba. Khrushchev responded with anger, accusing the U.S. of "degenerate imperialism" and stating the Soviet Union would not comply.

Khrushchev then sent a message stating he would remove the missiles from Cuba if the U.S. removed theirs from Turkey. At the time, this *quid pro quo* was not made known to the American public. The U.S. Joint Chiefs recommended an air strike against Cuba the next day. But JFK's brother, Robert F. Kennedy, had already met with Soviet Ambassador to the U.S. Anatoly Dobrynin, saying that the U.S. was in fact taking the missiles out of Turkey. This allowed Kennedy to agree with Khrushchev, after back-channel communications also agreeing that the U.S. would withdraw its naval blockade and would not invade Cuba. But as far as the American public was concerned, Kennedy had faced down the Soviets.

The crisis was over. It's been said that after the Cuban Missile Crisis, where the concept of brinkmanship had proven to be sheer folly, Kennedy's worldview began to move towards peaceful coexistence with the Soviet Union. His change of mind was undoubtedly hastened by a series of actions taken by French President Charles de Gaulle.

The de Gaulle Interlude

French General Charles de Gaulle had returned to the leadership of France by becoming president of the Fifth Republic on January 8, 1959. During World War II, de Gaulle frequently annoyed President Roosevelt and the American generals with his independent attitudes. He had insisted that France be given a seat on the UN Security Council and had been lukewarm towards NATO. With U.S. insistence on controlling NATO's nuclear weapons in Europe, de Gaulle announced that France would develop its own nuclear arsenal. Kennedy, who was adamantly opposed to nuclear proliferation (or otherwise put, to anyone else having them), opposed France's intentions but de Gaulle was adamant.

Reacting to the display of U.S. brinksmanship over Cuba, de Gaulle withdrew French naval forces from NATO and ordered NATO headquarters to leave France. Regarding the future of Europe, de Gaulle had a longstanding vision of a European federation "from Portugal to the Urals," a concept that would obviously include the Russians. By contrast, the founders of the European Union would later exclude Russia from their vision, one of many omissions that helped lay the groundwork for the current divide between the EU/NATO and Russia over the conflict in Ukraine.

The CIA has been credibly accused of participating in assassination plots against de Gaulle and in fomenting the riots that drove him to resign the French presidency in 1968. Little of his impact on a presently supine Europe remains.[30]

The Kennedy Presidency: Vietnam

It was in Vietnam that the U.S. would pioneer a new policy of sending Green Berets as advisers to build a counterinsurgency force that would stand up to the rebels, while American civil agencies and banking institutions would supposedly shore up the economy. The idea was that this program "would show the people that there was a liberal middle ground between colonialism and Communism."[31]

The trouble was that the North Vietnamese were determined to liberate all of Vietnam from foreign control. In September 1960, Hanoi recognized the South Vietnamese insurgents known as the Viet Cong, along with their political arm, the National Liberation Front. The Kennedy administration quickly invoked Eisenhower's "domino" theory, warning that if South Vietnam fell, so would the rest of Southeast Asia.

President Kennedy now began to send more American advisers. The result was the Strategic Hamlet program, where peasants would be herded into villages to prevent the Viet Cong from recruiting them. In 1961, Kennedy sent a delegation to Saigon headed by his adviser Walter Rostow and General Maxwell Taylor, who recommended that the U.S. send 10,000 American troops and start bombing supply routes from North Vietnam.

By November 1963, the U.S. force stood at 15,000, but Kennedy hadn't begun the bombing. South Vietnamese President Diem was proving corrupt, unpopular, and unreliable, and with CIA collusion, the South Vietnamese army assassinated Diem and his brother. Kennedy expressed dismay at the killings. Three weeks later, Kennedy himself was shot in Dallas.

One of the most persistent points of discussion with regard to the causal factors prompting Kennedy's assassination was whether he had intended to extricate the U.S. from Vietnam before the situation turned into a full-scale conflagration. It is a fact that Kennedy had committed the U.S. to opposing leftist wars of liberation. It is a fact that he had begun to send U.S. soldiers into Vietnam. But people argue to this day that he intended to get out of Vietnam before it devolved into a Korea-type situation.

In late 1962, Senator Mike Mansfield had advised Kennedy, after a visit to Vietnam, that the U.S. should get out. In April 1963, Kennedy told journalist Charles Bartlett, "We don't have a prayer of staying in Vietnam. We don't have a prayer of prevailing there. The people hate us."[32] Kennedy had also received messages from Europe that Vietnam would be a quagmire for the U.S. But European governments wanted the U.S. to "hold the fort" against Asian nationalism.

In what should put to rest the controversy over Kennedy's intentions, in May 1963, Kennedy approved a secret withdrawal plan developed by the

Joint Chiefs of Staff for Secretary of Defense Robert McNamara. The withdrawal would begin at the end of 1963 and be completed in two years.[33] But Kennedy did not want withdrawal to become an issue in the 1964 presidential campaign.

Meanwhile, Khrushchev was pushing for a new nuclear test ban treaty and sought a nonaggression pact between NATO and the Warsaw Pact. Both nations were also pushing ahead with their manned space programs. A Russian cosmonaut, Yuri Gagarin, got there first, but in February 1962, John Glenn orbited the earth. In June 1963, the Soviets sent the first woman to space, Valentina Tereshkova.

Kennedy expressed his hopes for peace in his American University commencement speech on June 10, 1963:

> I have...chosen this time and this place to discuss a topic on which ignorance too often abounds and the truth is too rarely perceived—yet it is the most important topic on earth: world peace. What kind of peace do I mean? What kind of peace do we seek? Not a *Pax America* enforced on the world by American weapons of war.... I am talking about genuine peace—the kind of peace that makes life on earth worth living—the kind that enables men and nations to grow and to hope and to build a better life for their children—not merely peace for Americans but peace for all men and women—not merely peace in our time but peace for all time.[34]

For the first time since before World War II, an American president took a stand against U.S. global military dominance, signaling a revolution in American foreign policy.

But by the fall, Kennedy would be dead and Lyndon B. Johnson, now president, would escalate in Vietnam beyond anyone's imagining.

Kennedy's Economics

John F. Kennedy's economics program was a modernization of New Deal thinking. When he took office in 1961, the nation was mired in a recession for which the Eisenhower administration had no answer. From 1941 onwards, the U.S. economy had been on a powerful wartime trajectory, but with Eisenhower's conservative budgetary policies, it had run out of steam.

Kennedy immediately took the initiative with tax policies favoring capital investment by U.S. industry, encouragement of low interest rates to promote consumer spending, and an emphasis on measures to enhance productivity over financial speculation. He also favored expanding low-cost credit for underserved areas, such as black communities. These measures were a frontal assault on banking profitability.[35]

Part of Kennedy's intended program to reform taxation was the elimination of the oil depletion allowance, which allowed investors in oil to deduct a significant proportion of their corporate income to compensate for the presumed depletion of oil deposits. By proposing to eliminate the oil depletion allowance, Kennedy made implacable enemies of a whole class of industrial and financial titans.

Kennedy also favored government deficit spending, even at times of economic growth. Upon Kennedy's death, the U.S. had experienced thirty-three consecutive months of GNP expansion. He also succeeded in beating back exploitative price increases by U.S. Steel, a controversy that earned him more enmity from big business and banking. This included criticism from David Rockefeller, whose letters to and from Kennedy appeared in *Life* in July 1962. Says Donald Gibson:

> Claiming to reflect the concerns of bankers in the U.S. and abroad, Rockefeller advised the president to make a "vigorous effort" to control government spending and to balance the budget. He also suggested to Kennedy that interest rates were being kept too low and too much money was being injected into the economy.[36]

Kennedy also believed that the government should regulate the flow of capital between the U.S. and overseas. This again placed him diametrically opposed to the big U.S. banks like David Rockefeller's Chase Manhattan Bank that were global powerhouses.

Finally, Kennedy was increasingly a friend to the civil rights movement, where Dr. Martin Luther King, Jr., and other leaders were beginning to advocate for economic fairness.

Kennedy's Assassination

Few historical figures have been so tragically prevented from pursuing their laudable objectives than the late President John F. Kennedy. Here is my own experience:

By the fall of 1960, my family had moved to Williamsburg, Virginia, when my dad was transferred by Dow Chemical in Michigan to a textile processing plant owned in partnership with the German Badische Corporation. That fall, I went around with a newly found friend handing out presidential campaign literature for U.S. Senator John F. Kennedy.

We had heard the candidate would be making a speech at the Williamsburg Court House. Instead, Robert Kennedy showed up and gave a stirring talk on his and his brother's hopes for our country. Within a couple of weeks, John F. Kennedy defeated Vice-President Richard M. Nixon in a close election.

A little over three years later, I was a senior at James Blair High School in Williamsburg, when a teacher told me I should go to the school office. They had a radio on and were listening to a broadcast saying that President Kennedy had been shot in Dallas, Texas. Later that day we heard he was dead.

That weekend I was working at a part-time job as an announcer at the local radio station WBCI, when a listener called to say that Lee Harvey Oswald had been shot. I hurried to the teletype machine, then back to the microphone, to announce that Oswald, Kennedy's presumed assassin, had been shot in the basement of the Dallas police station by a man named Jack Ruby. That week, I wrote an article for the school newspaper, the *Blarion*, decrying the terrible state of violence in our country and how badly the assassination of our young, brilliant president boded for the future.

Three years later, I was attending a talk in New Haven, Connecticut, being given by a law professor at the Yale University Law School on the Warren Commission Report on the assassination. The professor spoke strongly in favor of the Warren Commission and its "proof" that Oswald was the lone assassin.

I raised my hand and said that I had just finished reading Mark Lane's *Rush to Judgment* that questioned the "magic bullet" theory that the same bullet that killed Kennedy came out of his body and wounded Texas governor John Connally who was sitting on the other side of the limousine. Researchers to this day ridicule the absurd "magic bullet" theory.

The law professor then seemed to become hysterical. He talked at length about how the "magic bullet" theory was absolutely proven and should never be doubted by anyone. I went away unconvinced, but it was a long time before I felt I was getting a grasp on what actually took place on that terrible day in November 1963 and why it might have happened.

Today I feel I do have a grasp. One of the books I later read was Donald Gibson's book *The Kennedy Assassination Cover-Up*. In his book, Gibson writes that two of the institutions most instrumental in perpetrating the cover-up were the *Washington Post* and the Yale University Law School.

It is clear to me, and I think it has been proven more than circumstantially, that President John F. Kennedy was assassinated by a conspiracy coordinated by the CIA. That it was in fact a conspiracy, i.e., one involving at least two shooters, was established conclusively, based on acoustic evidence, by the House Select Committee on Assassinations, which submitted its report in 1979.

While they did not establish who the shooters were, some members of the committee believed it was done by figures from organized crime. The fact that Kennedy was shot from the front, not from the rear by Lee Harvey Oswald, as the Warren Commission claimed, has been proved by expert medical testimony that the frontal shot was an entrance, not an exit, wound.

But then, who coordinated the shooting, and why? I believe the evidence is persuasive that Kennedy may have been making a radical shift in post-World War II U.S. government foreign and domestic policy and that people working for the government habituated to killing their fellow human beings for "official" purposes may have done so in this case also.

The CIA had a litany of grievances against John and Robert Kennedy, starting with the Bay of Pigs fiasco. The mob was also bitterly angry over RFK's vendetta against them. They were accustomed to working in close partnership doing the dirty work of both the CIA and FBI.[37]

In fact, members of the CIA team were named as perpetrators by co-conspirator E. Howard Hunt in a death-bed confession to two of his sons.[38] Their names are David Morales, Frank Sturgis, Antonio Veciana, David Atlee Phillips, William K. Harvey, and Cord Meyer. I believe it likely that CIA director of counterintelligence James Jesus Angleton was one of the coordinators of the assassination and cover-up. There are indications that CIA involvement reached to the top of the agency,[39] whose director was then John McCone.

A vast quantity of journalistic research points to the likelihood of the FBI having been involved, along with organized crime, specifically the Chicago mob, the U.S. Secret Services, the U.S. military, Dallas police officials, Vice President Lyndon B. Johnson, various bankers and oil magnates, and possibly Israel.[40]

But despite the thousands of books and articles published, the fact remains that no one has yet identified the actual individuals who gave the orders to shoot the president. The same goes for the 1968 killing of Robert Kennedy.

Undoubtedly the action was not viewed with disfavor by the Council on Foreign Relations and the Rockefeller syndicate, whom Kennedy had antagonized by starting to disrupt their plans for global U.S. military dominance.

President Lyndon Johnson and the Vietnam War

During the election of 1964, President Lyndon Johnson had the audacity to portray himself as the *peace candidate* in comparison with Barry Goldwater's war mongering, citing Goldwater's threats to use nuclear force in Vietnam. Johnson's tack recalls Woodrow Wilson's "He kept us out of war," and Franklin Roosevelt's, "Your boys are not going to be sent into any foreign wars." Johnson said, "We are not going to send American boys nine or ten thousand miles away from home to do what Asian boys ought to be doing for themselves."[41] *All lies*, as events would prove.

On August 2–3, 1964, Johnson claimed to receive reports that American destroyers in the Gulf of Tonkin had been attacked by North Vietnamese torpedo boats. Later, Senator J. William Fulbright would disclose that allegations of the attack were a complete fabrication. Nevertheless, through the Gulf of Tonkin Resolution, Congress gave Johnson a blank check for expanding the conflict.

Meanwhile, in 1965–1966 in Indonesia, the CIA and the U.S. embassy were supporting the mass killing by the Indonesian military of approximately one million Indonesian civilians accused of leftist leanings, including most of the members of the Indonesian Communist Party. It was a political genocide.[42] The impact was to wrest Indonesia from the non-aligned movement and make it a military-controlled ally of the U.S.

In Vietnam, the North Vietnamese government was trying to negotiate a cease-fire, but President Johnson ignored the request, reasoning that the South Vietnamese and their American backers had to apply more military pressure to gain an advantage in any talks. After Johnson won the fall election and was secure in his only elected term, the U.S. began to bomb North Vietnam. Every Tuesday the president would pick the targets for the coming week over lunch with his top advisers.[43]

The bombing did not stop North Vietnam from supplying the Viet Cong insurgency. By 1970, the U.S. had dropped more bombs on Vietnam than on all other targets in human history to that point. In 1965, Johnson pledged, "We would not be defeated,"[44] and began to pour in more ground troops. By July 1965 there were 125,000 pairs of boots on the ground. Eventually there were more than 500,000. These troops also constituted a large and expanding market for consumption of the heroin that originated with growers under CIA control in Laos.[45]

In January 1968, another presidential election year, the Viet Cong and North Vietnamese launched the Tet offensive, driving the Americans and South Vietnamese out of the countryside and into the cities. The U.S. Army destroyed the ancient city of Hué to remove the insurgents there. After U.S.

General William Westmoreland asked for 200,000 more troops, Johnson announced a reduction in U.S. bombing then withdrew from the presidential race.

The American war effort had collapsed. Richard M. Nixon and his soon-to-be national security adviser Dr. Henry Kissinger were waiting in the wings. Protests against the war were filling the streets across the U.S. In April, Dr. Martin Luther King, Jr., was assassinated in Memphis.[46] On June 6, 1968, presidential candidate Robert Kennedy was assassinated in Los Angeles, after winning the Democrats' California primary.

On August 29, 1968, after a police riot in Chicago against demonstrators, the Democratic Party nominated Senator Hubert Humphrey over peace candidate Eugene McCarthy. President Johnson was insisting that Humphrey stick close to the pro-war narrative, so Humphrey lost any chance to capitalize on opposition to the war.

In November 1968, Nixon and his vice-presidential candidate from Maryland, Spiro Agnew, were elected. Nixon was able to turn Lyndon Johnson's support for civil rights against him by adopting the "Southern Strategy," used by the Republicans to shift the "Solid South" in their favor. The Republican Party was becoming the white person's party with an anti-communist twist.

Later, Nixon and Kissinger were accused of colluding with the North Vietnamese to forestall any plans by the Johnson White House to obtain a cease-fire prior to the 1968 presidential election. Nixon-era "dirty tricks" were underway. Nixon also claimed he had a "secret plan" to end the war. After his inauguration it turned out that his "secret plan" was more bombing, now secretly including Cambodia, followed by the "Vietnamization" program that resulted in the killing of hundreds of thousands of civilians under the CIA's Operation Phoenix.

The War and Me

The Vietnam War was accelerating when I graduated from James Blair High School in Williamsburg, Virginia, in June 1964. One of my strongest memories was of attending a talk at a local church by the chair of the College of William and Mary philosophy department, Dr. Frank McDonald. He had been asked to speak on "How can philosophy help us to fight communism?" He explained that philosophy might be able to help you think for yourself, but it wouldn't help fight communism. The people sponsoring the talk didn't know what to say.

Another recollection: I was at some event—it might have been "bridge night" at the nearby Fort Eustis Officers Club—when I happened to be talking to an army officer. I asked him why were we in Vietnam. The officer began to shout at me. "We are there because the North Vietnamese are ripping open pregnant women and killing their babies!"

Years later, when I heard U.S. claims that we were invading Iraq because Iraqi soldiers were in Kuwait tearing babies out of incubators, I thought of my encounter with the officer at Fort Eustis. I wondered, how stupid do these people think we are? Very stupid, I've come to realize.

Under no circumstances was I going to Vietnam to fight anyone, but I didn't have to. I had been accepted at Yale University, which was one of three universities our guidance counselor had me apply to. My parents drove me up to New Haven to enroll, but I quit after six weeks. I had thought that I wanted a career as a foreign service officer, but at Yale I realized there was nothing there that I desired. The classes were cold and regimented, and the atmosphere was elitist and conventional. The U.S. war machine was simply taken for granted. Plus Yale was known as the prime recruiting ground for the CIA. I was in the same class as George W. Bush, though I don't recall running into him.

Later that fall, while the campaign between President Lyndon Johnson and Senator Barry Goldwater was going on, I flew to Lima, Peru, to visit a girlfriend I'd gone to high school with. She was staying with her aunt and uncle at their villa. The villa was surrounded by a high cinder block wall lined on top with large shards of broken glass to keep intruders out. One day, the uncle took me for a drive around Lima's city streets which were bustling with the noisy life of a Latin American downtown. We kept the doors of the car locked and didn't get out, but it was still fun to see what life in Lima was like. I only stayed for a week, but before I flew back to the U.S., I offended my strait-laced hosts by telling their teenage son I kind of liked Fidel Castro.

I then returned to Virginia and enrolled at the College of William and Mary, which worked out well because the atmosphere was low-key, I was known and liked in the English Department, and was able to teach myself writing at my own pace.

The Vietnam War seemed not so far away. One time, Air Force recruiters parked a jet on the lawn of the Campus Center. It set off some excitement when students began to picket it with signs saying, "Get this war machine off our campus." Turns out that the organizers were mainly the kids of government workers from DC. I had long hair at the time, but the barbers at the local barber shop refused to cut it because to them long hair was unpatriotic.

When President Johnson came to Williamsburg to attend a church service at Bruton Parish Church in the historic area, the minister of the church, Rev. Cotesworth Pinckney Lewis, aroused controversy when he questioned

Johnson from the pulpit about why were we in Vietnam. Reportedly the President was fairly well-steamed. Two boys I knew from high school and the father of another died in Johnson's war.

Ironically, Bruton Parish Church was a place where Washington, Jefferson, and Patrick Henry worshipped while they were making a revolution. Johnson may have felt he was facing another insurrection.

ENDNOTES FOR CHAPTER 13

1 F. William Engdahl, *Gods of Money: Wall Street and the Death of the American Century* (Engdahl, 2009), 225.

2 "Marshall Plan," June 5, 2020, *History.com*.

3 Andrew Glass, "Bernard Baruch Coins Term 'Cold War,' April 16, 1947," *Politico*, April 16, 2010.

4 So-called by John Perkins, author of *Confessions of an Economic Hit Man* (Plume, 2005).

5 MI6 stands for "Military Intelligence Section Six." MI5, on the other hand, is the British equivalent of the FBI.

6 Richard Dunlop, *Donovan: America's Master Spy* (Rand McNally, 1982), 467–468.

7 Michael Warner, "The Creation of the Central Intelligence Group" (Center for the Study of Intelligence, Central Intelligence Agency, 1996), 112.

8 Ibid., 111.

9 Ambrose and Brinkley, 168. If the U.S. had a genuine Supreme Court, legislation such as this by which Congress effectively committed suicide with respect to its responsibilities would be declared unconstitutional. It is ironic that some figures engaged in antiwar protest regard Congress as a potential ally, given the requirement to achieve Congressional approval for U.S. wars.

10 As an example of David Rockefeller's access, see his December 8, 1958, letter to CIA director Allen Dulles on the subject of the "Russian economic warfare problem," Central Intelligence Agency, *Reading Room*, https://www.cia.gov/readingroom/docs/CIA-RDP80B01676R003800180005-9.pdf

11 Seymour M. Hersh, "The Price of Power: Kissinger, Nixon, and Chile," *The Atlantic*, December 1982.

12 "Author Quotes: David Rockefeller," *Goodreads*. https://www.goodreads.com/author/quotes/9951.David_Rockefeller

13 A phrase coined by famed American propagandist Henry R. Luce in a 1941 *Look* magazine editorial.

14 Father of future president George H.W. Bush and grandfather of George W. Bush.

15 The first "Red scare" came at the time of the Palmer Raids in 1919–1920.

16 Ambrose and Brinkley, 132.

17 Ibid., 132.

18 Emma Graham-Harrison, Andreas Rocksen and Mads Brügger, "RAF veteran 'admitted 1961 killing of UN secretary general,'" *The Guardian*, January 12,

2019. https://www.theguardian.com/world/2019/jan/12/raf-veteran-admitted-killing-un-secretary-general-dag-hammarskjold-in-1961

19 Ambrose and Brinkley, 168.

20 Ibid., 188.

21 James H. Doolittle, *Report on othe Covert Activities of the Central Intelligence Agency*, August 6, 1976. https://cryptome.org/cia-doolittle.pdf

22 Confederated Salish & Kootenai Tribes. https://csktribes.org/

23 "Salish Range (MT)," Summitpost.org. https://www.summitpost.org/salish-range-mt/501549>

24 United States Census Bureau, My Tribal Area: Flathead Reservation, MT. https://www.census.gov/tribal/?st=30&aianihh=1110

25 Jaako Puisto, *This Is My Reservation: I Belong Here: The Salish Kootenai Indian Struggle Against Termination* (Pablo, Montana: Salish Kootenai College Press, 2016).

26 Ibid., 24–27.

27 Ambrose and Brinkley, 178.

28 Ibid., 182.

29 U.S. Joint Chiefs of Staff, "Justification for U.S. Military Intervention in Cuba (TS)," U.S. Department of Defense, March 13, 1962.

30 De Gaulle's rhetoric on a European foreign policy independent of the U.S. is mirrored today by the decidedly more feeble efforts of French President Emmanuel Macron.

31 Ambrose and Brinkley, 91.

32 Stone and Kuznick, 315.

33 Ibid., 315.

34 Ibid., 316–317.

35 The best available account is Donald Gibson's *Battling Wall Street: The Kennedy Presidency* (Sheridan Square Press, 1994).

36 Ibid., 73.

37 See Seymour Hersh, "The Kennedys' Secret Sicilian Operation: What the CIA Didn't Tell the Warren Commission," Substack.com, March 29, 2023.

38 John Koerner, *Why the CIA Killed JFK and Malcolm X: The Secret Drug Trade in Laos* (Winchester UK and Washington DC: Chronos Books, 2014), 20–21.

39 The weekend after the killing, the CIA took custody of the famous Zapruder film. We now know that the film was altered by the CIA to obscure the frames that proved Kennedy had been shot in the head from the front. Another alteration removed evidence that Kennedy's limousine had stopped on the street when the shots were fired. See "The Zapruder Film Mystery" on *YouTube*.

40 See William Dankbaar, *Files on JFK: Interviews with Confessed Assassin James E. Files. YouTube* has an interview with James E. Files that describes his delivery of the fatal frontal shot on Kennedy from the Grassy Knoll. On the involvement of the CIA in the assassination cover-up, see *Mary's Mosaic: The CIA Conspiracy to Murder John F. Kennedy, Mary Pinchot Meyer, and Their Vision for World Peace* by Peter Janney. Regarding Israel's possible involvement, see "Did Israel Kill the Kennedys?" by Laurent Guyénot, *Unz Review*, June 3, 2018.

41 Ambrose and Brinkley, 199.

42 Vincent Bevins, "What the United States Did in Indonesia," *The Atlantic*, October 20, 2017.

43 Ambrose and Brinkley, 203.

44 Ibid., 204
45 Koerner, 40.
46 Dr. Martin Luther King, Jr., was assassinated in a plot that likely originated at the top levels of the FBI and was carried out in collusion with Memphis, Tennessee, police and U.S. military intelligence. See Jeremy Kuzmarov, "Did J. Edgar Hoover Order the Assassination of Martin Luther King Jr?," *CovertAction Magazine,* January 16, 2023.

CHAPTER 14

The Rockefeller Republic

Nixon Is Elected

In the fall of 1968, while I was a student at the College of William and Mary in Williamsburg, Virginia, presidential candidate Richard M. Nixon arrived to film a campaign promotion. He and his handlers, his Secret Service detail, and his filming crew bustled around the picturesque Old Campus, with its stately orange-brown brick colonial-era buildings. After Nixon did his cameos and left, I stood on the steps of the campus auditorium and made a short speech to a group of students, saying that if Nixon were elected, it would be a disaster for the U.S. I was right. It was. In their book *The Untold History of the United States*, Oliver Stone and Peter Kuznick gave Chapter Nine the title, "Nixon and Kissinger: The 'Madman' and the 'Psychopath.'" They were right too.

Vietnam: How Bad Can It Get?

During the 1968 campaign, Nixon ran against Democratic Party nominee Hubert Humphrey, after the party's frontrunner, Bobby Kennedy, had been assassinated in Los Angeles. Nixon was without a doubt the beneficiary of Kennedy's death. Of course, the cover story was another "lone gunman" lie put forward by the mainstream media and the Los Angeles police. There have been persistent allegations about the involvement of the CIA.[1]

In October 1968, South Vietnamese President Nguyen Van Thieu refused to join the opening of the Paris Peace Talks that were expected to bring the war in Vietnam to an end. Thieu's sudden refusal created a major crisis for the Lyndon Johnson administration. Nixon's camp alleged that the talks had been meant to throw the election to Humphrey, but obviously that now wouldn't happen.

President Johnson believed that Nixon had worked behind the scenes to get Thieu to sabotage the talks. By now, both Johnson and Humphrey had reached the conclusion that further military involvement in Vietnam was against the national interest, even though what was called the "Europhile lobby," headed by the British, wanted the U.S. to ensure continued Western dominance of Asia. These forces warned Nixon that Johnson was about to announce a peace agreement with Vietnam that would assure Humphrey's victory. Henry Kissinger then assured these forces that Nixon would pursue the Vietnam War for at least five more years. The Paris Peace Talks then broke up.[2]

Nixon deceived the voters by claiming he had a "secret plan" to end the war. After his narrow win over Humphrey, Nixon said his plan was to resume fighting, but with fewer American casualties. He declared that he refused to be the first American president to lose a war;[3] evidently, he didn't count the failure to prevail in Korea as a loss. Eventually, the U.S. and North Vietnam did meet in Paris, though it took five years to reach an agreement. Initially, the two sides argued about the shape of the conference table. The five ensuing years were marked by catastrophic violence.

Nixon called his plan "Vietnamization." Supposedly, the South Vietnam Army would now do the bulk of the fighting, using U.S. military aid. A lengthy draw-down of the U.S. force of 540,000 started, with over 58,000 Americans being killed by war's end.[4] It was now a proxy war. But Nixon and Kissinger decided that the only way to interdict supplies coming down from North Vietnam to the Viet Cong was to extend the U.S. bombing to Laos and Cambodia. The bombing took place secretly, without informing Congress. Then in 1970 Nixon announced that U.S. troops had invaded Cambodia. That nation was not fighting anyone, but was smashed to bits, preparing the ground for the Khmer Rouge insurgency and the destruction of Cambodian civil society. Communist forces in Laos also gained power. Today live cluster bombs are still killing civilians in Laos.

Nixon's Vietnamization program did not work. Neither North Vietnam nor its Viet Cong allies would give in. In order to uproot support for the Viet Cong among South Vietnamese civilians, the CIA launched Operation Phoenix. This was a four-year terror campaign, not disclosed to Congress and not reported in the press, involving intimidation, torture, and murder of civilians suspected of collaboration, as well as those identified as Viet Cong. The campaign was carried out methodically and ruthlessly, with villages being targeted and suspects identified and rounded up. In the end, the CIA and South Vietnamese military claimed to have "neutralized" over 81,000 people, including over 40,000 dead.[5]

Meanwhile, U.S. planes were wiping out forests, jungles, and agricultural land with herbicides like Agent Orange to deprive Viet Cong soldiers of cover. Twenty million gallons of Agent Orange were sprayed on Vietnam during the war. And the drugs continued to move out of CIA-controlled areas in Laos, Cambodia, and Burma—the famed "Golden Triangle"—to dealers in Vietnam who would supply American soldiers throughout the Far East with heroin. Some of these dealers were South Vietnamese military officers. Inevitably, the heroin also began moving into the U.S. mainland.

Notably, the Paris negotiations were only between the U.S. and North Vietnam, with the South Vietnamese mere spectators. U.S. troops remained in Vietnam until March 29, 1973, while 7,000 Americans stayed behind to help South Vietnam continue to fight. But on April 30, 1975, during the presidency of Nixon's successor Gerald Ford, Saigon fell to the North Vietnamese and was renamed Ho Chi Minh City. The greatest military catastrophe in U.S. history to date was finally over. Even so, there were plenty of critics who clung to the argument that the U.S. military could have won the war if they had not been held back by the politicians.

On Campus

The antiwar movement was growing, with groups like Students for a Democratic Society springing up. African-American revolts were also taking place in major cities. We now know that both the FBI and CIA were infiltrating these groups, on and off college campuses. Massive antiwar and civil rights demonstrations were taking place, and violence had started to flare up. The National Guard killed four students at Kent State University in 1970. Days later police killed two students and injured eleven at Jackson State University.[6]

The CIA was flooding campuses with LSD and other hallucinogenic drugs as a means of diverting student anger into drug-induced stupors. This happened at William and Mary, where I was studying. We knew that one of the history professors was CIA. Then one day a rather lame-brained young man from the town showed up with a large quantity of LSD. This quickly was passed around, and many of the more active students in anti-war activities began to ingest it. It happened to be laced with strychnine, a poisonous additive, so there were a lot of bad trips.

Williamsburg was also the location of Eastern State Hospital, a major Virginia mental institution that was now being phased out, as medical science had discovered that medication could relieve the government of the expense of keeping mentally and emotionally disabled individuals in institutions. Psychotropic drugs were now being prescribed that were so devastating that patients were turned into vegetables by the administration of "chemical lobotomies." Several of the students who'd had bad trips fell into the hands of doctors from Eastern State. Some were disabled for life, particularly those given the drug thorazine. Among the multitude of dangerous side effects of thorazine are *tardis dyskinesia,* paralysis of the cranial nerves, or even heart failure and sudden death.

There is no question in my mind that the medical abuse of psychotropic drugs is intended not just to save money, but also has political utility in population control. Some of these drugs, which are meant to calm people, actually cripple them mentally so that they are driven to suicidal or homicidal behavior.

Notably, older styles of psychiatric treatments involving analysis or behavioral therapy have largely disappeared, reducing many psychiatrists to pill-pushers.[7] Meanwhile, the profits of pharmaceutical companies are off the charts, including those convicted of pushing addictive opioids like Oxycontin. This profiteering even applies to vaccinations for school-age children, where pharmaceutical companies have been shielded from liability and unsafe combinations of ingredients have been linked to autism and other crippling disorders. An example is the use of mercury in childhood vaccines.

Back at William and Mary, I graduated with a Bachelor's degree with honors in June 1970. I had thought of graduate school, but with no money and college loans to pay, all I wanted was to go to work. I had made Phi Beta Kappa at a time the federal government waived the Federal Service Entrance Exam for top-ranked students. So in August my wife and I moved to Washington, DC, where I went to work for the U.S. Civil Service Commission at 1900 E Street, NW, a five minute walk from the White House.

I worked in or near downtown Washington for most of the next 37 years. At the Commission, they would come around with tickets to the White House grounds whenever an audience was needed for President Nixon to greet a foreign head of state. This included the time Queen Elizabeth came to visit. So I saw Nixon quite a few times over the next four years, until he resigned over Watergate on August 8, 1974, but never close up!

Nixon's Globalism

You can't say that President Richard Nixon and his National Security Adviser/Secretary of State Henry Kissinger didn't think big. In contrast to the U.S. government's doctrinaire anti-communism since the end of World War II, they tried to make peace with both the Soviet Union—*détente*—and Communist China—the "Opening to China."

With the Soviet Union, the U.S. began selling large quantities of wheat and signed the 1972 Berlin Agreement, leading to the diplomatic recognition of East Germany. Nixon also began arms control talks with the Soviets, the Strategic Arms Limitation Talks. An interim agreement, SALT I was signed in 1972, freezing ICBM deployment. It was the first arms control accord signed with the Soviets since the start of the Cold War.

It was the U.S. bankers that were starting to call for an accommodation with China. David Rockefeller's Chase bank was the first to open a branch in Hong Kong, still a British crown colony. With that foothold, Rockefeller now called for an opening of trade between the U.S. and mainland China. Later he opened the first U.S. bank in Moscow and the first in Beijing since the communist revolution.

Thus, it was the Rockefeller banking interests that provided the impetus for these huge changes in U.S. foreign policy. It was no accident that Henry Kissinger, the top U.S. diplomat, was a longstanding Rockefeller crony.

Nixon and Kissinger tried to navigate the murky waters of the Middle East by seeking a *stasis* between Israel and its Arab neighbors at the time of the 1973 Yom Kippur War, when Israel seized the Golan Heights and West Bank. In this the Nixon/Kissinger team failed, with Saudi Arabia and the other oil-producing states retaliating against the U.S. by launching an oil embargo. Then, once the oil taps were turned on again, U.S. oil companies maneuvered a hike in the price of oil of four times its previous cost. It was price-fixing on a global scale.[8] Here the Rockefeller oil interests profited handsomely.

Regime Change in Chile

Under President Lyndon Johnson, the Alliance for Progress in Latin America that Kennedy began had turned into an avenue for massive U.S. investments by large banks and corporations. Chase's David Rockefeller was like an American czar for Latin American finance, and Johnson turned a small U.S. police and military assistance program in Latin America into an opportunity for the sale of advanced weaponry to nations that were not at war with anyone, nor would be in the foreseeable future. For instance, Johnson approved the sale of fighter planes to Peru and of twenty-five Skyhawk jets to Argentina.[9]

The U.S. was also becoming deeply involved with Latin American militaries which it saw as natural allies against leftists or militants that might threaten U.S. corporate investment. In 1964, the Brazilian military overthrew the government of President João Goulart after he promised to nationalize the oil industry and implement a program of rent control. The U.S. Embassy asked the CIA to aid in the revolt, which left Brazil with a military government that lasted until 1985.

On September 11, 1973, President Salvador Allende of Chile was overthrown by a junta of military officers led by General Augusto Pinochet. A trained medical doctor, a Freemason, and a cultured political veteran, Allende belonged to the Socialist Party and favored what were seen in Chile as moderate socialist policies. As president, Allende pledged to nationalize major industries, expand educational opportunities, and improve the living standards of the Chilean working class. He had been elected president in 1970. But it was nationalization of industries that was anathema to the Chilean upper class, U.S. corporations, and the CIA.

Chile was a country with major U.S. investments, with the leading corporate actor being ITT. An ITT director, John McCone, once a Chevron oil executive,[10] had been head of the CIA under Presidents Kennedy and Johnson and the successor to Allen Dulles. McCone's wife was a major stockholder in Anaconda, a giant U.S. corporation with copper-mining operations in Chile.

In May, June, and July 1970, McCone had meetings with the new CIA Director, Richard Helms, where he suggested that Helms meet with ITT President Harold Geneen on the situation in Chile. Geneen offered to make a "substantial contribution" to the CIA's effort to block Allende from winning the upcoming presidential election. But Allende won the election, leaving the CIA feeling "it had its nose rubbed in the dirt" by the Chilean electorate.[11]

The coup d'etat in Chile resulted from what amounted to direct orders from President Nixon to get rid of a politician who not only threatened American financial interests but also gave a bad example to other Latin American countries that increasingly were subject to profit extraction by U.S. banks and corporations. Rather than submit to the humiliation of captivity by the U.S.-backed Chilean military, President Allende committed suicide the day of the insurrection in the presidential palace in Santiago. Pinochet's military junta dissolved the Chilean Congress, suspended the constitution, and began persecuting alleged dissidents. At least 3,095 civilians "disappeared" or were killed as Chile entered a dark age of suppression.

Rockefeller Banking Rules

Even though gold had been removed as currency by the Roosevelt Administration, the pegging of currencies to gold by the Bretton Woods agreements of 1944 had exerted a considerable amount of discipline on international trade and the movement of capital. It incentivized nations to balance their governmental budgets, with those that manipulated their currency to capture the gold market easily being found out and deterred. But the large

U.S. trade deficits during the Vietnam War required a constant shipment of gold abroad. All were reasons why the big American banks, particularly David Rockefeller's Chase Manhattan Bank and Walter Wriston's First National City Bank, lobbied President Nixon to abandon Bretton Woods and the gold peg. Nixon took this monumental step on August 15, 1971.

The world economy has never been the same. U.S. banks no longer had restrictions on their worldwide expansion. This was especially pursued in the face of U.S. inflation resulting from shocks caused by the quadrupling of the price of oil in the spot market. Higher prices meant more lending, particularly with interest rates staying low, at least until the Federal Reserve plunged the economy into recession through interest rate hikes at the end of the 1970s. Though popular memory pins the oil price shocks of the on the Arab oil producers, it was engineered entirely by the U.S. and British oil companies that were largely controlled by the big banks.[12] The Rockefeller interests had triumphed.

We now know that Kissinger made a deal with the Arab oil producers that the U.S.-instigated price hikes would be to their benefit, as long as the Arab nations deposited their oil earnings in the U.S. and British banks that controlled the entire charade. What came to be called the "Eurodollar" market could then be accessed from London and easily spent without going back through U.S. banks. Substantial amounts of oil money were also stashed in Treasury bonds. So oil was turned into dollars, allowing the dollar to retain its status, or even enhance it, as the world's reserve currency.[13]

With all oil purchases denominated in dollars, thus requiring countries to hold dollars in order to be able to purchase oil, this is what gave the dollar its crucial reserve status. The resultant need for the dollar enabled the U.S. to print money for which there would be enduring external demand. This facilitated the financing of the worldwide U.S. military garrison, eventually leading to 800 U.S. military bases and facilities abroad for further protection of its foreign prerogatives based on dollar hegemony. Not exactly the outcome de Gaulle sought when France, suspecting the value of the inflated currency, had refused to take dollars in payment. It is said that France's action is the event that caused Nixon to succumb to the pressure of abandoning the gold standard.

Indeed, the removal of the gold peg led to a vast worldwide inflation that is still continuing. With more devalued money being held by banks on deposit, their ability to make loans skyrocketed. Much of the new money went into speculation, including the nightmare creation of financial derivatives. This innovation made possible the frenzy for leveraged buyouts which began to emerge as a major destructive force. The growth of non-productive derivatives helped create today's radical separation of the *rentier* from the productive economy. It marked the birth of the world of all-powerful hedge funds such as Vanguard, BlackRock, and Goldman Sachs.

Government tax bases also expanded, which fostered the growth of the public workforce with regular inflationary pay raises. The inflation made possible today's trillion-dollar defense budget. It also pushed housing costs higher, lurching from one bubble to the next. If you want to quantify the rate of inflation, look at the 2022 spot price of gold at $1770 an ounce. That's more than a 5000 percent increase in the price of gold since 1971. It's not just three people who are to blame, of course, but feel free to keep Nixon, Kissinger, and David Rockefeller in mind when you ponder how we got into today's catastrophic mess.

Banking had long ceased being a national phenomenon. It was increasingly global, with the government taking on the job of protecting overseas investments, as it had done with Chile. Nor were the IMF, the World Bank, or even the Bank for International Settlements sufficient policy planning instruments to ensure the profits of the global financial community. A greater degree of organization was called for.

Thus, at the 1972 meeting of the Bilderberg Group, David Rockefeller proposed establishing his own instrument, the Trilateral Commission, intended to bring together policy makers from the U.S., Europe, and Japan. His motion was seconded by an academic from Columbia University named Zbigniew Brzezinski.[14] The purpose was to have a body outside the elected political system to protect open trade and capital flows. One of the early members was the Governor of Georgia, a peanut farmer and Naval Academy graduate named Jimmy Carter. Other early members were future Federal Reserve chairmen, Paul Volcker and Alan Greenspan, both Rockefeller protégés.

Indeed, the U.S. had become the "Rockefeller Republic."

Nixon's Domestic Programs

David Rockefeller was not impressed with federal government programs to relieve poverty. He characterized President Lyndon Johnson's Great Society benefits, including Medicare and food stamps, as "handouts."[15] Created during the Depression, Aid to Families with Dependent Children continued as the basic federal welfare program. But Nixon expanded Johnson's Great Society. Between 1965–1968, Johnson had doubled War on Poverty spending from $6 billion to $12 billion. By 1974, Nixon would double it again to $24.5 billion.

The Nixon administration also created OSHA and EPA, both badly needed as basic safety and environmental programs. Nixon also approved NASA's proposed space shuttle program, though ominously, the design was for a vehicle large enough to serve both scientific and military purposes. This ran contrary to the National Space and Aeronautics Act of 1958 that mandated a civilian space program for peaceful purposes.

President Nixon also proposed a Family Assistance Plan, which was a modified basic income program for families without means or with low income. It was defeated mainly by Southern Democrats who were alleged to be acting out of racial prejudice in not wanting to give blacks something for nothing. The Earned Income Tax Credit that is presently in existence came out of the effort as a compromise.

The idea of a basic income guarantee had been around for a long time. Thomas Paine had proposed it during Revolutionary War days. In the 1960s, so had Milton Friedman and Dr. Martin Luther King, Jr. With Johnson's Great Society, the nation had begun to take seriously the importance of helping people rise out of poverty. With the decline of family farming, and conversion to a monetary economy based on paid employment, along with the increasing dominance of large corporations, it became increasingly clear that modern capitalism had produced what was starting to be called a "trickle-down" economy, where the wealth of corporations supposedly would "lift all boats."

But many people were left out, through unemployment, under-employment, unpaid domestic service work like childcare or eldercare, etc., so the welfare state that Reagan and Clinton later derided was coming into being to help these groups especially. Furthermore, banks were still shunning opening branches in poorer or minority neighborhoods, a practice later known as "redlining," which kept these localities from creating a business base.

The argument against a basic income guarantee is that it demoralizes a population where people get "free money." But the problem of large numbers of people being effectively shut out of the modern financial economy has never come close to being solved, and with the advance of AI, will inevitably deepen. Even with a job or indeed two jobs, people working at or close to the "minimum wage" have been permanently consigned to poverty.

The 1972 Election

By the election of 1972, U.S. troops in Vietnam were down to 70,000, with Kissinger announcing that "peace is at hand."[16] This effectively undercut the campaign of the anti-war Democratic Party nominee, Senator George McGovern. McGovern also spoke in favor of a basic income guarantee, though the provision as part of Nixon's proposed Family Assistance Plan was dropped from consideration by Congress in October 1972, just before the election.

As he had done in 1968, Alabama Governor George Wallace mounted a third-party campaign for the presidency in 1972. Wallace's likely presence on the presidential ticket threatened to disrupt Nixon's "Southern Strategy," intended to take advantage of white resentment in the South against Democratic president Lyndon Johnson's civil rights and voting rights legislation. Johnson had been rightly convinced that these measures would assure permanent support of the Democratic Party by the nation's large and growing black electorate. The Republican Party's strategy would eventually assure that the "Solid South" became a Republican rather than a Democratic stronghold.

But Wallace was the fly in the ointment. In the 1968 presidential election, where Nixon defeated Democratic Party candidate Hubert Humphrey by an electoral college count of 301 to 191, Wallace had picked up 46 electoral votes from the states of Louisiana, Arkansas, Alabama, Mississippi, and Georgia. While not enough to change the outcome, the votes for Wallace demonstrated the threat he posed to the Southern Strategy.

Just as an assassination in 1968—that of Robert Kennedy—removed a Nixon competitor, so did the shooting of George Wallace on May 15, 1972, at a campaign rally in Laurel, Maryland. The alleged shooter was a drifter named Arthur Bremer. In a manner similar to future allegations with regard to the alleged assassin of Robert Kennedy, critics argued that there was never a meaningful investigation of the details of the shooting, that more shots were fired than Bremer's gun actually held, that Bremer had been seen previously with individuals that may have been associated with the CIA, and that he appeared to be in a hypnotic stupor the day of the incident. Like Kennedy's assassin, Bremer retained no recollection of the alleged shooting. Bremer was sentenced to prison and released on parole in 2007.

Watergate

Nixon defeated George McGovern with a popular vote of over sixty percent and a huge electoral landslide. But on June 17, 1972, five men had been apprehended at the Watergate complex in Washington, DC, and arrested on charges of burglary for a break-in at the Democratic National Committee headquarters. The men were working for the Nixon reelection campaign and were seeking to dig up dirt on Larry O'Brian, chairman of the DNC, and others. The attempt by Nixon and his aides to cover up their connection to the Watergate burglars led eventually to his resignation under threat of impeachment. Nixon didn't help his case by his extensive recordings of Oval Office conversations, later made public.

Nixon resigned on August 8, 1974. In addition to the five burglars, forty-three Nixon administration officials were convicted of crimes in connection with the scandal, including John Mitchell, Attorney General, and James Dean, White House Counsel.

Of the five burglars, all had some type of connection to the CIA, though none were considered active-duty CIA officers. E. Gordon Liddy, a long-time security operative, and E. Howard Hunt, a former CIA operative whose name is ubiquitous in annals of journalistic investigation of the Kennedy and King assassinations and other covert operations, directed the group.

The literature on Watergate is voluminous and will not be recapitulated here. I will mention one point of interest, however, that has never been resolved. This is the question of whether the CIA had targeted Nixon for removal, and whether the burglary at the Watergate itself was a set-up, deliberately planned to fail.[17] Had the forces that ran the country behind the scenes decided for their own purposes that Nixon had to go?

Gerald Ford

Republican Congressman and House Minority Leader Gerald Ford had been appointed Vice President of the U.S. in 1973 upon the resignation of then-Vice President Spiro Agnew, who was under investigation for corruption charges. Nixon said he had selected Agnew to be his vice president as "assassination insurance." But the insurance policy had been cancelled. Now, Ford stepped up to the presidency when Nixon resigned. Ford is the only person to have been president without ever being elected as either president or vice-president. He had also been a member of the Warren Commission and fully supported its false claim that Lee Harvey Oswald was the lone-nut assassin of JFK.

The dark forces lurking beneath the surface were reflected in Ford's appointments. One was his naming of Donald Rumsfeld as his chief of staff. Rumsfeld, a former Navy pilot, had become a congressman from Illinois at the age of thirty and was the lead backer in Congress for Ford's becoming House Minority Leader—Ford's stepping-stone to the White House.

Dick Cheney, a congressional aide, became Rumsfeld's deputy. A year later, when Rumsfeld was named Secretary of Defense, Cheney took the chief of staff position. Later, Cheney himself would be elected to Congress from Wyoming. Decades afterwards, Cheney, as George W. Bush's Vice President and Rumsfeld, as Bush's Secretary of Defense, would be primary architects of Bush's assaults on Afghanistan and Iraq, the two initial targets of the "War on Terror."

Ford issued a pardon to former President Nixon, and watched passively while the crowd of other Watergate participants went to prison. Ford also earned notoriety by turning a deaf ear to New York City's pleas for a federal financial bailout when it was going bankrupt in 1975. On October 21, 1975, the *New York Daily News* famously proclaimed: "Ford to City: Drop Dead." Ford later claimed that this headline cost him reelection as president in 1976.[18]

The Church Committee

The hearings and report of the *Senate Select Committee to Study Governmental Operations with Respect to Intelligence Activities*, otherwise known as the Church Committee, named after its chairman, Senator Frank Church of Idaho, was the most memorable event of Ford's presidency. After witnessing endless abuses, it was now time for Congress to take to heart the longstanding public disgust with the CIA and other U.S. intelligence agencies. The CIA was now to hit a speed bump.

The Church Committee was set up in 1975 to investigate abuses by the CIA, National Security Agency, FBI, and IRS. There was a House counterpart called the Pike Committee and a presidential commission headed by Ford's own appointed vice president, Nelson Rockefeller, to focus specifically on the CIA.

We saw previously how close Nelson Rockefeller was to the initial organization and operation of the CIA. The Rockefeller Commission now reported in June 1975 with a single-volume report viewed historically as a whitewash. It was edited by Ford's chief of staff Dick Cheney, with large portions of the original report deleted from the final version. Future president Ronald Reagan was a member of the Rockefeller Commission. Among other findings, the commission stated it found no evidence that the CIA participated in the assassination of President John F. Kennedy.

I have read the report and can only say that the Rockefeller Commission didn't look very hard. It was so glib in its assertions as to be an insult to the intelligent reader.

However, the findings of Church Committee were of an entirely different caliber. The Committee's report revealed Operation MKULTRA, which the CIA set up to drug and torture unwitting individuals for the purpose of mind control. This included the distribution of LSD on college campuses that I had witnessed at William and Mary. The FBI was named as running COINTELPRO that involved the surveillance of political and civil-rights organizations. "Family Jewels" was revealed as the program to assassinate foreign political and cultural leaders. The committee also exposed a number of CIA assassination techniques. Operation Mockingbird was named as a propaganda campaign that used U.S. and foreign journalists operating as CIA assets, with dozens of U.S. and overseas news outlets acting as cover for CIA activity. Another covert program was Project Shamrock, where the major telecommunications companies shared their traffic with the National Security Agency, starting in 1945. This was the first official confirmation of the existence of the NSA.

The Church Committee hearings revealed a vast array of covert surveillance, coercive, and punitive activities by agencies employing tens of thousands of people against ordinary citizens and public officials both of the U.S. and foreign nations. It showed a secret government at war, not just against foreign governments depicted as hostile, such as those of the Soviet Union and Communist China, but against our own people and those of supposedly friendly countries. In many regards the activities uncovered were illegal and/or unconstitutional.

These gangster-like abuses were traced back through every previous presidential administration to the days immediately following World War II. *Yet no one was ever held accountable for any of the abuses the committee found.* Much of its deliberations were never released to the public. And just because the committee examined certain areas in no way prevented their continuance to this day.

Operation Mockingbird in particular continues in full force or greater. We know for a fact that the mainstream media, all major publishing houses, and, today, all social media and many other internet outlets, are riddled with paid CIA operatives and other intelligence trolls.[19] So is foreign media, especially in Great Britain and Germany, where scarcely a word is published without intelligence agency scrutiny.[20]

The Church Committee's final report to the public appeared in April 1976 in six volumes. They also published seven volumes of hearing records. The committee had previously published an interim report on programs to assassinate foreign leaders, including Lumumba of the Democratic Republic of Congo, Trujillo of the Dominican Republic, Ngo Dinh Diem of South Vietnam, General René Schneider of Chile, and Castro of Cuba. President Ford tried but failed to get the Senate to keep the assassination report secret. He then issued an executive order banning CIA conduct of such assassinations. The committee also produced seven case studies on covert operations, but only the one on Chile was made public. The rest were withheld under pressure from the CIA.

It was later disclosed that the NSA had a secret Watch List containing millions of names, including a number of prominent U.S. citizens. Among them were Joanne Woodward, Thomas Watson, Walter Mondale, Art Buchwald, Arthur F. Burns, Gregory Peck, Otis G. Pike, Tom Wicker, Whitney Young, Howard Baker, Frank Church, David Dellinger, and Ralph Abernathy.

Senator Frank Church was shocked by what the committee learned about the immense operations and electronic monitoring capabilities of the NSA and the fact that its very existence had been unknown. He said:

> That capability at any time could be turned around on the American people, and no American would have any privacy left, such is the capability to monitor everything: telephone conversations, telegrams, it doesn't matter. There would be no place to hide.[21]
>
> I don't want to see this country ever go across the bridge... I know the capacity that is there to make tyranny total in America, and we must see to it that this agency and all agencies that possess this technology operate within the law and under proper supervision, so that we never cross over that abyss. That is the abyss from which there is no return.[22]

Today, the U.S. has fallen into that abyss. At the time of the Church Committee, there was no way to anticipate the vast increase in the potential power of covert agencies for extra-legal surveillance and mind control in the internet age. The potential for abuse was later to be revealed by Edward Snowden and others. It was clear then and is clear today that the U.S. government views every citizen of the U.S. as a potential enemy to be watched, restrained, controlled, and possibly apprehended, punished, or murdered as its policy considerations dictate. It is a certainty that the CIA has assassinated private U.S. citizens in their own country.

Thus, we have a government presided over by elected officials, but which is in thrall to the prevailing ideology cultivated and guarded in secret by what we now call the Deep State. The mass media are entirely under the control of Deep State covert propagandists. At the top of the pyramid are the oligarchs who control the economy, especially those who run the big financial institutions. Many of these have been named in this book.

The Church Committee had nothing to say about any possibility that the CIA had been dealing in drugs, whether in Southeast Asia, or elsewhere. But soon there would be indications that drug traffic coming out of Latin America into the U.S. had CIA involvement.

The findings of the Church Committee fully justify the title of an important book by researcher Douglas Valentine: *The CIA as Organized Crime: How Illegal Operations Corrupt America and the World.*[23] Valentine also covers the huge expansion of CIA drug dealing after the U.S. takeover of Afghanistan carried out in the aftermath of the 9/11 false flag attacks.

Philip Agee's CIA Diary

In 1975, as the Church Committee deliberated, ex-CIA agent Philip Agee (1935–2008) published a 630-page book exposing the inner workings of the CIA and listing dozens of names of individuals and organizations working undercover for the CIA in Latin America. The title of the book was *Inside the Company: CIA Diary.*[24] He later published a similar book on CIA activities in Western Europe.

The CIA pursued Agee across Europe and Central America. The State Department revoked his U.S. passport, a decision confirmed in a case Agee brought to the U.S. Supreme Court. He eventually found refuge in Cuba, where he lived until his death in 2008. He was a co-founder of the *CounterSpy* and *CovertAction* series of periodicals.

I was struck by the segments of *CIA Diary* where Agee was stationed for training at Camp Peary near Williamsburg, Virginia.[25] Agee was there in 1960, the year that my family moved to Williamsburg. Part of the training was to learn the complex system of coding to be used in secret dispatches. There was also physical training:

> We also have training at the gym in defense, disarming, maiming, and even killing with bare hands—just how and where to strike, as in karate and judo.[26]

As Agee describes it, the orientation of the CIA in "host" countries was that of a secret quasi-military force engaged in a fight to the death against not just the Soviet Union and Cuba, but communists, socialists, progressives, labor union leaders, leftists of any kind, land reformers, or anyone else who seemed to speak for "the people." On the side of the CIA were the host country's military, high government officials, the wealthy class, landowners, business owners, and, of course, U.S. bankers and businesspeople.

Any type of violent reprisal was allowed, so long as it could be done with what was termed "plausible denial," i.e., without being attributed to the U.S.—especially the manipulation of elections, distribution of propaganda, and bribes paid to officials. In Mexico, for instance, the CIA station was in the habit of giving money to high government officials so they could buy their girlfriends automobiles. But more ominously, there were assassinations. In fact, you are left with the impression that a CIA agent would just as soon kill you as look at you if it served his purposes.

Even half a century after publication, *CIA Diary* is a worthwhile read.

Working for the Government

But things weren't all bad, at least for me personally. Despite all that was going on in the world and the nation, I had a decent salary and got a lot of joy out of working for the federal government, as well as training in government operations. I started work with the U.S. Civil Service Commission in August 1970.

I was hired for my writing ability, and for my first assignment was asked to research and write a government regulation providing for training by employees to qualify for advancement as part of the government's upward mobility program. This was meant as a benefit primarily for women working as secretaries in Washington, DC.

I found the Commission to be a highly professional organization staffed by people with impeccable integrity and have always had a high regard for the quality of the federal civilian agency workforce.

I then spent two years teaching at a Washington, DC, private secondary school called the Field School. Returning to the Civil Service Commission, I worked on the administrative staff of the Bureau of Training and spent most of my time writing evaluation reports on central office and regional training centers. I then transferred to a job as a policy analyst on the Commissioner's staff at the Food and Drug Administration, just before Jimmy Carter assumed the presidency in January 1977 after a close election against incumbent Gerald Ford. I voted for Carter and had a lot of hopes for his presidency, after having spent six years in Washington witnessing the Nixon-era disasters. Nobody I knew even took Ford seriously.

Has a New Day Dawned?

By the election of 1976, there seemed to be new hope for America and the world. The Arab oil embargo was over. Gas prices had gone up but were manageable for most consumers, and homes were still affordable. Meanwhile, the Vietnam War was history, and the nation was at peace for the first time in ages. The CIA had been disciplined by Congress, so it seemed possible that covert violence would finally stop as well. Détente with Russia and the opening to China marked huge shifts in U.S. policy. Maybe the geopolitical direction was finally moving away from the danger of nuclear war.

Then there was Jimmy Carter himself. The U.S. political establishment had scoured the field in search of an outsider—a presidential candidate untainted by the horrors and abuses of the Johnson-Nixon-Kissinger era. While a member of David Rockefeller's Trilateral Commission, which made him even more of a safe establishment bet, Carter also had the benefit of being rather innocent-looking and soft-spoken.

Though he had credentials—a Naval Academy graduate and governor of Georgia—he was also a homespun peanut farmer with a downhome accent and a friendly smile. He spoke in favor of cutting the defense budget and reducing nuclear arms. He also advocated a foreign policy based on human rights, saying in his inaugural address, "Our commitment to human rights must be absolute."[27] He named Cyrus Vance, once a Kennedy administration Army Secretary and now a firm détente advocate, as his Secretary of State. And as his first official action as president, Carter issued a blanket pardon for all Vietnam War draft dodgers.[28]

Now let's fast forward four years to the 1980 White House Christmas Party, when I chatted briefly with President Carter on the South Lawn of the White House near the twenty-four-foot Christmas tree. I had spent the previous year working in the White House office for the president's Special Assistant for Consumer Affairs, Esther Peterson.

Carter had been defeated in the 1980 presidential election by Ronald Reagan, former governor of California. That balmy December night, Carter looked worn-out, a man under tremendous pressure. I am happy to say that as of this writing, at the age of 98, he is without a doubt the most successful ex-president in American history, still committed to his long campaign for human rights and dignity for all people—irrespective of the extent to which his human rights ideas were subsequently weaponized in support of U.S. geopolitical ambitions.

What Happened to Détente?

Carter tried, but the SALT II Treaty that he signed with the Soviets failed to put real limits on the arms race, was criticized in the Senate as going too far, and was never ratified. So the arms race continued, but with the U.S. remaining in the lead. While the U.S. protested the Soviet use of Cuban proxies in civil wars arising in Africa, the biggest conflict was over Afghanistan. In order to shore up a teetering communist regime, the Soviet Union sent 85,000 troops into Afghanistan in December 1979. This event clearly hearkened back to the "Great Game" between Britain and the Russian Empire of the 19th century over imperial control of Central Asia, which posited a potential Russian threat to India, a nation with already strong Russian sympathies.[29] Russia's action also brought to the forefront Carter's National Security Adviser Zbigniew Brzezinski, co-founder with David Rockefeller of the Trilateral Commission, who seemed increasingly to be acting as Carter's *alter ego* as the crises mounted.

Who Was Brzezinski?

Brzezinski was the Democratic Party's answer to Henry Kissinger. Like Kissinger, he was born in Europe, but in Poland, not Germany. Brzezinski's father was a Polish diplomat, who was stationed in Canada at the start of World War II, a time when Poland was being divided up by Germany and the Soviet Union. To return to Poland would have meant death, so the family stayed in Montreal, where "Zbig" began his studies at McGill University.

Later, with a doctorate from Harvard, Brzezinski was passed over for a faculty appointment in favor of the up-and-coming and already well-connected Kissinger, so he ended up at Columbia University in New York City. Brilliant and bitterly anti-communist, he became an adviser to President Lyndon Johnson but also grew close to the Rockefeller family. As stated above, with David Rockefeller, he helped found the Trilateral Commission in 1974 to coordinate the foreign policies of the U.S., Europe, and Japan.

After Carter became president, Brzezinski was so much in control that Secretary of State Cyrus Vance eventually resigned over what he viewed as Carter's recklessness in his deference to a man Vance called "evil." This "teamwork" was particularly evident in Carter and Brzezinski's dealings with the Iranian hostage crisis in 1979–1980. In 1979, the Iranians drove out the U.S.-supported Shah of Iran, declared an Islamic Republic, and took sixty-six American citizens hostage at the U.S. embassy. Carter and Brzezinski's plan for a military rescue in the desert failed disastrously. I'll return to the hostage crisis later in this chapter.

Geopolitically, Brzezinski had a longer-term obsession. He was determined to reverse the direction Nixon and Kissinger had taken in moving toward détente with the Soviet Union. He succeeded in embedding into the U.S. approach to the Soviets, almost as a reflex, that it had to be opposed at every turn, because communism was supposedly always inimical to "human rights." Naturally, this appealed to Carter's sensitivities.

Brzezinski had his biggest opportunity to face the Soviets directly through the latter's invasion of Afghanistan in 1979. Until the end of his life, Brzezinski and his apologists denied that prior to the Soviet incursion he had been running CIA operatives in that country in a deliberate attempt to goad the Soviets to send in troops. If that was what he was doing, he was risking all that Nixon and Kissinger had accomplished in normalizing relations with the

Soviets, but such scruples likely didn't matter to a fanatical warmonger who saw the dismemberment of the Soviet Union as his primary goal.

Brzezinski later spelled it all out in his 1997 book *The Grand Chessboard: American Primacy and Its Geostrategic Imperatives*. In that book, he likened the lives and destiny of billions of people to pawns in a board game. Nothing could be more revealing or better designed to thrill and motivate the CIA, the U.S. military, and the American politicians who lusted to straddle the globe than Brzezinski's philosophy. In 1979, an appallingly naïve President Carter took the bait and, with a major act of what we now call "virtue signaling," withdrew the U.S. from the Moscow Summer Olympics of 1980.

Meanwhile, Brzezinski worked with the CIA, Saudi Arabia, and Pakistan to arm a band of Afghan fighters known as the Mujahideen in order to engage the Soviets in a bitterly fought conflict that lasted until they withdrew from Afghanistan in 1989. By then, the Mujahideen were morphing into al Qaeda, the Islamic fundamentalist force that in various iterations terrorized the Middle East, usually with covert U.S. support, for decades. Along with perpetual conflict with Russia, sponsorship of Middle Eastern terrorism is also part of Brzezinski's terrible legacy.

Carter's Success with China, Panama, and the Middle East

Among Carter's successes was his announcement in 1978 that the U.S. and China would be extending full diplomatic recognition to each other, including the exchange of ambassadors. This was the solidification of America's "One China" policy that remains a legal reality, both in U.S. law and internationally, despite the U.S. provocations, *inter alia*, of shipping weapons to Taiwan. The U.S. withdrew from its 1954 mutual defense treaty with Taiwan and acknowledged Taiwan as part of China. At the time, conservative Republicans like Barry Goldwater and Ronald Reagan howled at the "betrayal" of Taiwan; now it appears the Democrats are courting the island's government as part of an anti-Chinese *cordon sanitaire*.

Carter also oversaw the Senate ratification of the Panama Canal Treaty that gave Panama full sovereignty over the Canal Zone. Negotiations had been underway since the Johnson administration. Though both Gerald Ford and Henry Kissinger supported the treaty, Reagan again denounced it. This was when Reagan was launching his presidential campaign with TV ads that began with the words, "I'm mad!" Apropos of the treaty, one senator pointed out, with regard to the origins of the Panama Canal, "We stole it fair and square."[30]

Carter was also able to bring a measure of peace to the Middle East, hosting talks at Camp David, Maryland, between Egyptian President Anwar Sadat and Israeli Prime Minister Menachem Begin. Egypt became the first Arab nation to offer Israel diplomatic recognition. Israel in turn withdrew its forces from the Sinai Peninsula, which they had held since the Yom Kippur War.

After Camp David, Carter flew to the Middle East in 1979 to meet separately with both leaders. Sadat himself was assassinated for his alleged betrayal by Egyptian soldiers at a military parade in 1981. Despite Carter's efforts, the issues involving Palestinian recognition and the Israeli occupation of the Golan Heights and West Bank remained unresolved. In his retirement, Carter published a book that characterized Israel as an "apartheid state."[31]

Financial Technology Accelerates

One of the less-noticed events of Carter's presidency was his effort to accommodate the banking industry in their push to modernize financial processing technology through the adoption of Electronic Funds Transfer (EFT). This began with the growing application of computers and the ability to move huge data files over T-1 trunk lines. For workers and consumers, this meant the faster availability of earnings through direct deposit into individual bank accounts. EFT was also viewed as contributing to the growth of overseas lending and the rapid international movement of financial balances.

Under the Carter administration, a new program was created within the Treasury Department to convert the government's entire payment and accounting system to EFT, an effort that took over two decades to complete. Among other things, the new system meant Treasury would no longer be required to send magnetic tapes with payment files all over the country in locked briefcases transported by airline couriers. Treasury could also upload its fund balances to the Federal Reserve nightly and earn interest payments on the transferred credits.

The Federal Reserve in turn could use the money as backing for more financial industry lending. The overall effect was to make it much easier for banks, businesses, and consumers to run up ever more debt. The biggest beneficiaries were the hedge funds. Later, I spent over twenty years at the U.S. Treasury Department, working in the Financial Management Service, the bureau that managed the government's EFT.

Less Developed Countries

Advancing technology also allowed what were now called Less Developed Countries (LDCs) to run up more debt, a situation that became dangerously acute during the mid-1970s. This was a direct result of U.S. government policy in facilitating bank lending abroad after World War I. There had always been the claim that the countries on the receiving end of U.S. bank loans would be able to pay them off while growing their economies thanks to American investment. But that formula didn't work, particularly with the American banks and corporations extracting so much profit from the debtor nations that the LDCs fell into a "debt trap."

The loans proffered by the IMF and World Bank, with their litany of neo-liberal conditionalities like privatization of utilities and cutting government employee salaries, were making the situation worse, not better. By 1977, the situation had become a crisis, with LDC budget deficits hitting $100 billion.[32] Yet in the eyes of Rockefeller-controlled banks like Morgan and Citibank, the LDC debt was a business opportunity, not a problem. Their solution was to lend even more. First the borrowing nations would pay off their old debt, then incur more to keep their economies running. But this didn't work for nations that did not have large oil revenues like those in OPEC. Countries without oil were in an ever-deepening hole. And nations that defaulted on their debt were in much bigger trouble than companies or individuals that could claim bankruptcy then start over again. Entire nations couldn't do that. Their entire populations paid the price.

Carter is Blindsided by the Federal Reserve

If the stock market is a valid measure of overall U.S. economic health, it is instructive to observe that the long-term growth starting at the end of World War II had peaked as the Vietnam War began to heat up in October 1965. Subsequently, there had been a steady, long-term *decline* in the Dow Jones Industrial Average that went on until March 1982, with a particularly steep drop during the four years of the Carter administration.

At the same time, inflation was a huge problem during the 1970s, both in the early years of the decade, with the oil industry's price hikes, and in 1979 when the crisis in Iran caused major oil supply disruptions and a new round of price increases. The combination of economic recession marked by falling stock valuations and inflation caused by oil market disruptions was called *stagflation*. Such is politically fatal, and so it proved during the last half of Jimmy Carter's term.

The Federal Reserve stepped in and implemented the disastrous policy of raising interest rates to historic highs. The Federal Reserve's prime lending rate on February 1977 was 6.25 percent. By November 1980, the month of the Carter-Reagan presidential election, the rate was at a historic high of 21.5 percent. The rate hike was devastating to small businesses which rely on reasonable interest rates for everyday operations and resulted in a record number of business failures. The hike accelerated the growing trend of shutting down and consolidating U.S. smokestack industries and was the start of the devastation of the so-called "Rust Belt" in the American Midwest. Though U.S. labor costs were higher than those in third-world countries which benefitted from factory relocation, government subsidies could have mitigated the disruption. But under the Federal Reserve, such subsidies had no place.

Meanwhile, the banks were quietly agitating for "deregulation." Accordingly, the Depository Institutions Deregulation and Monetary Control Act of 1980 repealed usury laws, authorized interest on checking accounts, and abolished interest rate ceilings.

The Federal Reserve interest rate hikes worked in that inflation did slow, even though the previous price increases, including family housing, merely stabilized. They did not come down. And the thousands of industrial firms that went out of business never came back. The long-term trend of the conversion of the U.S. from a manufacturing to a service economy now took off. The growth in aerospace and armaments industries from the Reagan military build-up that reversed Carter administration cuts eased the economic burden somewhat. This change favored those regions where military production was concentrated, which included, above all, California and the rest of the West Coast, then Texas and the rest of the South—the Sunbelt. For some reason, Reagan's home state made out like a bandit.

Delving deeper we can ask more specifically how, why, and by whom was all this devastation accomplished? William Greider addressed the mystery in his classic *Secrets of the Temple: How the Federal Reserve Runs the Country*.[33] Federal Reserve Chairman Paul Volcker (1927–2019) was instrumental in carrying out these epoch-making policies. Having been president of the Federal Reserve Bank of New York from 1975 to 1979 (which is owned by its member banks, chief among which are the Rockefeller banks), he was named by President Carter as Federal Reserve Chairman in 1979, then reappointed by President Reagan, serving two terms until 1987. Volcker had been a monetary tightener and inflation hawk since his undergraduate days at Princeton. Never mind that the inflation was largely caused by U.S. banks and oil companies.

Note that tight money and high interest rates were historically favored by the big banks. Armed with a master's degree from Harvard, Volcker spent his career shuttling among appointments with the Federal Reserve and the U.S. Treasury Department and assignments with David Rockefeller's Chase Manhattan Bank. He was instrumental in Nixon's decision to take the dollar off the gold peg and favored high interest rates as restoring monetary discipline in the absence of gold convertibility.

While high interest rates may be devastating to businesses and wage earners, they also attract foreign investors into U.S. domestic markets. Given the Fed's options of supporting the interests of big finance vs. those of ordinary people, the nod always goes to the former. So it was with what we now call the "Volcker Recession." And sure enough, starting in 1982, when the Dow Jones was at a low of 2494, it skyrocketed to 18348 by 1999, when the dot.com bubble finally burst. A substantial part of this phenomenal growth was itself the kind of asset inflation to which we are accustomed today.

The Volcker Recession returned the White House to the Republicans and Ronald Reagan. One of his campaign slogans was "Make America Great Again" (!). When Volcker took the pedal off the metal and pulled back interest rates, the new finance-led economy began to take off as the big banks resumed their consolidation and growth, wiping out the Savings and Loan industry along the way. But Reagan was able to declare prior to his 1984 reelection campaign, "It's springtime in America."

The Warmongers Coalesce

While all this was going on, the dark forces were starting to gather that would launch the U.S. on a new war footing during Reagan's first term. All these would eventually merge into what is today called the Neoconservative Movement.

Rumsfeld and Cheney

I have already mentioned the tandem of Donald Rumsfeld and Dick Cheney during the Ford administration. Now, after Ford lost the presidency to Carter, Rumsfeld was CEO for Searle Drugs but also served in numerous quasi-governmental roles, including as Reagan's go-between with Saddam Hussein, and became a member of the Council on Foreign Relations. Dick Cheney was elected as congressman from Wyoming in 1978, then served as Secretary of Defense under President George H.W. Bush during the first Gulf War. He later worked as CEO for Halliburton, a major military contractor. He was also a director of the Council on Foreign Relations. As CFR stalwarts, both were firmly in the Rockefeller camp.

Team B

After serving two terms as a congressman from Texas, George H.W. Bush, son of former U.S. Senator Prescott Bush, served as U.S. Ambassador to the UN, chairman of the Republican National Committee, and envoy to China before becoming President Ford's CIA director from 1976 to 1977. At the CIA, Bush created "Team B," which was a group of ultraconservative hardliners who set out to purge any analysts thought to be underplaying the threat to U.S. security from the Soviet Union. According to Stone and Kuznick:

> Harvard Russia historian Richard Pipes, a virulently anti-Soviet Polish immigrant, was put in charge of Team B. Pipes quickly recruited Paul Nitze and Paul Wolfowitz. According to Anne Cahn, who worked at the Arms Control and Disarmament Agency under President Carter, Team B members shared an "apoplectic animosity toward the Soviet Union."[34]

Team B ran into continued resistance from CIA analysts who were tracking the internal weakening of the Soviet Union during the 1970s and 1980s, when the agency failed to give proper notice of the events that would lead to the break-up of the Soviet Union in the early 1990s. Later, Robert Gates, CIA director under George H.W. Bush, took up the anti-Soviet drumbeat, overriding the pleas of his own analysts to take a more reality-based approach. Team B was the forerunner of the Committee on the Present Danger. One of its main tenets was that the Soviet Union was running out of oil, so it would be launching incursions into the Middle East to acquire the oil resources of countries like Iran and Iraq. This was a patent falsehood, as the Soviets had an endless supply of domestic oil, gas, and other energy sources.

Senator Henry M. "Scoop" Jackson

Jackson served a forty-two-year tenure in Congress as a representative (1941–1953) and U.S. senator (1953–1983) from Washington state. Jackson was a vehement supporter of the Vietnam War and a major congressional proponent of the military industrial complex. In fact, he was called the "Senator from Boeing." He was also a backer of military aid to Israel with strong support from groups like the Jewish Institute for National Security Affairs. But Jackson is most remembered today for providing a breeding ground for the Neoconservative movement. In addition to Richard Perle, neoconservatives Paul Wolfowitz, Bill Kristol, Charles Horner, and Douglas Feith were aides to Jackson, all of whom supported Ronald Reagan as president.

Committee on the Present Danger

The Committee on the Present Danger (CPD) was founded in 1976 at the end of the Ford administration and included conservative members of both political parties, including the "father of the hydrogen bomb," Edward Teller. At this time Teller was Director Emeritus of the federal government's Livermore Laboratory and Senior Research Fellow at the Hoover Institution. Another notable CPD member was Ronald Reagan, who was just finishing two terms as governor of California and running for the Republican nomination for president in 1976, where he finished a close second to Ford. Other members of the CPD were William Casey, Reagan's 1980 campaign manager, then Director of the CIA, Richard Allen, William Colby, General Maxwell Taylor, William Graham, NASA's acting administrator when Challenger blew up, and several dozen others. The main tenet of the CPD was the myth of the extent to which the U.S. lagged behind the Soviet Union in military might. The CPD's policy statement began:

> Our country is in a period of danger, and the danger is increasing. Unless decisive steps are taken to alert the nation, and to change the course of its policy, our economic and military capacity will become inadequate to assure peace with security.[35]

The CPD claimed that the Soviet Union was using arms control negotiations as a cover for its intent to fight and win an all-out nuclear war. These claims were known to be untrue, as shown by statistics coming out of the SALT negotiations, but the CPD propagandists persisted in their lies.

The 1980 Presidential Campaign

By early October 1980 it was starting to be clear that Carter was losing ground and that Reagan would be elected. Several factors were contributing to the malaise. One was the economic situation, with the recession deepening, inflation continuing, and the Federal Reserve raising interest rates into the stratosphere. At the Carter White House, we learned that Volcker had not even told Carter what he was going to do when interest rates began to soar out of sight in July. We on the Carter side of the fence were dismayed. Reagan seized the moment by blaming all of America's ills on the government, especially too much regulation and high taxes. The depth of Reagan's ignorance was remarkable, on an opposite pole to his political savvy.

Then there was the crisis with the Iranian hostages. After the April 24 fiasco in the desert, the failure of Carter and Brzezinski's rescue mission was followed by the resignation of Secretary of State Cyrus Vance. Negotiations now began between Iran and the U.S., with Algeria acting as a mediator. The U.S. promised to release about $11 billion in frozen Iranian assets, but the process for the transfer of funds dragged on for months.

The hostages were freed the day Reagan was inaugurated as president on January 20, 1981, after months of failed negotiations making Jimmy Carter and his negotiators look like chumps. It was never proven or even investigated, but we were sure that Reagan's operatives were in touch with the Iranians and were providing them with some kind of payoff to keep the hostages in custody until the right moment. With former intelligence operative William Casey as Reagan's campaign manager, it was a reasonable suspicion, despite the fact that such action on their part would have been illegal if not treasonous.

Finally, to top it all off, we learned that the Reagan campaign had stolen President Carter's briefing book as the presidential debates got underway. Of course, Reagan "won" the debates hands-down. Even Carter admitted it. Carter also failed to gain the support of followers of Senator Edward "Teddy" Kennedy, whom Carter defeated decisively in the Democratic primary.[36] And so the "Reagan Revolution" blasted off.

That said: during the Carter administration, the U.S. was not at war.

Flathead News

Meanwhile, in Montana, the government had come to terms with the fact that after the Flathead tribes signed the Hellgate Treaty in 1855, they did not receive fair compensation for the land they ceded.

In 1946 Congress had created the Indian Claims Commission. According to the Bureau of Indian Affairs,

> These claims represent attempts by Indian tribes to obtain redress for any failure of the Government to complete payments for lands ceded under treaty, for the acquisition of land at an unconscionably low price or for other failure to comply with a treaty or legislative action regarding Indian lands that grew out of the westward expansion of the United States.[37]

In 1965, the Indian Claims Commission ruled that the Confederated Salish and Kootenai Tribes had not been adequately compensated for the lands ceded in the Hellgate Treaty. The tribes had surrendered 12,005,000 acres to the government which were then worth $5,300,000. The total payment to the tribes, however, had only been $593,377.82, a tenth of that. After fees were taken out, the tribes received $4,016,293.29 in 1967. The amount of compensation was based on a determination of land value in 1855. No interest has ever been paid for the 112 years the tribes had been deprived of the use of that money.

In 1971, the U.S. Court of Claims handed down a ruling with respect to the 1910 Flathead Allotment Act under which "surplus" reservation land had been opened to white homesteading. This time, compensation to the tribes was based on 1912 land values totaling $7,410,000, of which only $1,783,549 had been paid by the homesteaders. The balance of $5,626,451 was paid by the government to the tribes a few years later.

Also in the 1970s, tribal elders were able to stop a major timber sale in the Mission Mountains, where the tribes were trying to create protected wilderness areas and a grizzly range. In 1975, the Two Eagle River School was founded on the reservation. This is a public high school for tribal members. Also in 1975, the reservation's Culture Committee was created and then divided into the Salish-Pend d'Oreille Culture Committee and the Kootenai Culture Committee.

Additionally in 1975, Congress passed the Indian Self-Determination and Education Act which recognized the right of Indian tribes to self-government as "domestic dependent nations," with tribes having the right to exercise inherent sovereign powers over their members and territory. This law is the basis for the Flathead Reservation police force and court system.

In 1978, the Salish Kootenai College was founded. Today the college grants associates, bachelors, and master's degrees, with specialties in native culture and natural resource management. In 1994 it was designated a land-grant college. Salish Kootenai College is a member of the American Indian Higher Education Consortium, a group of tribal and federally chartered institutions whose mission is to strengthen tribal nations. The college also has published an outstanding series of books on Flathead history and culture with distribution help from the University of Nebraska Press.

Reagan as President

Ronald Reagan was never a big hit as a Hollywood actor, but he would often say that his real interest was politics.[38] In the book I published in 2007 on the Space Shuttle Challenger disaster, I wrote the following:

> The election of Ronald Reagan as president in 1980 can be analyzed from many perspectives. From one point of view, it clearly represented a successful power play by what President Dwight D. Eisenhower had warned against in his 1961 farewell address—the

growing power of a permanent "military industrial complex." Reagan had already enjoyed the backing of this element when he ran for election and won as governor of California in 1966 and 1970.

Reagan seemed to thrive in the role of commander-in-chief. One of his heroes was British World War II Prime Minister Winston Churchill. Reagan's supporters loved his "great power" rhetoric. Yet he was not willing to repeat the disastrous experience of Vietnam by engaging in another major land war. Instead, wars were to be fought by proxy—this was the "Reagan doctrine"; i.e., the support of anti-communist military movements in third world countries. There was also an expansion of CIA covert action under Director William Casey. The prime example of Casey's influence was the creation of the Contra army by CIA operatives in Nicaragua starting in 1981, a by-product of which was the Iran-Contra scandal six years later.

As far as overt military action was concerned, the Reagan administration seemed to prefer easy targets, as in the invasion of the tiny Caribbean island of Grenada in 1983 or the bombing of Libya in 1986. One U.S. deployment backfired when the Marine barracks in Beirut were destroyed in 1983 by a suicide bomber, with the loss of 220 American lives.

But behind every target, the specter of the Soviet Union loomed. In its approach to the perceived Soviet threat, the U.S. remained a nation expecting total war.[39]

The détente with the Soviet Union achieved by Nixon and Kissinger was dead. Brzezinski had taken a big step toward its demise by pushing Carter to create the Mujahideen force against the Soviets in Afghanistan. This proxy war continued under Reagan and Bush I. Even though the Soviet Union withdrew from Afghanistan in 1989, the radical Islamist force the U.S. created remained, morphing into al Qaeda and later ISIS.

But Reagan's new crusade against what he called the Soviet Union's "evil empire" was the nail in the coffin of détente. Reagan had an army of Neocons to work his will (or was it vice-versa?) and a trillion-dollar military build-up to demonstrate his resolve. And even though the CPD and others were screaming about how far ahead the Soviet Union was in critical weaponry, there was also an underlying intention to shatter the Soviets by engaging them in an arms race they could no longer afford.

So the Reagan administration kicked into gear. One of the first things they did was start to sell weapons to *both sides* in the devastating war between Iran and Iraq that lasted seven years, celebrating the deaths of over a million people as a wise policy, keeping the combatants killing each other.

But Reagan himself was a cipher. He was good at making speeches and photo ops, but rarely discussed issues, held a press conference, or even stayed awake during long meetings. His grasp of issues was almost nil, and he was called by veteran presidential adviser Clark Clifford an "amiable dunce." His chief of staff said that in order to find a direction in making policy, the staff would listen to him make a speech then go off and try to come up with a program to fulfill it. What this meant in practice was that Reagan was totally a captive of the people around him, most of whom had their own agendas. The one who was most unscrupulous in using that power was CIA Director William Casey.

On March 30, 1981, Reagan was shot in an assassination attempt while coming out of a street-level conference room at the Washington Hilton Hotel. He arrived at the hospital near death but recovered after hours-long surgery done by the doctors at George Washington University Hospital. The shooter was a young man named John Hinckley, Jr., another of those familiar "lone-nut" drifters acting in a hypnotic stupor, acting on the implausible motive of seeking to attract the attention of then child actress Jody Foster. It turned out that John Hinckley, Jr., was known to the Bush family and that Hinckley's father was a Bush campaign contributor. Vice President George H.W. Bush, who would have become president on Reagan's death, was in charge of the investigation of the incident. But Hinckley was quietly confined to a mental hospital and life went on.[40]

Taxes and Finance

Reagan named Donald Regan, the CEO of Merrill Lynch, to be his Treasury Secretary, the first time that position was filled by the head of a Wall Street firm. Reagan also cut income taxes, with the largest decrease since the 1913 Income Tax Amendment. The top tax rate was cut from seventy percent to twenty-eight percent over several stages. The lowest rate was cut from fourteen to eleven percent. It is no exaggeration to say that due to his huge tax cuts, Reagan remains enshrined in the U.S. upper class pantheon of gods and heroes.

Almost up there with tax cuts in the latter's approbation was the firing of all the air traffic controllers who had been striking for higher wages, abolishing the Civil Aeronautics Board, the Community Services Administration and several other federal agencies loathed by conservatives, and other actions designed to punish the federal bureaucracy.

I had kept my civil service appointment in the White House consumer office and was now assigned to write speeches for Virginia Knauer, Reagan's newly-appointed Special Assistant for Consumer Affairs. I got in trouble when I objected to the idea that we were supposed to promote the claim that Reagan had "eliminated" inflation. I said that presiding over a recession didn't really qualify as "eliminating" inflation, but they were still irked.

What really ended the recession was cuts in the Federal Reserve's interest rates, though rates did not go back to what they had been in the days of John Kennedy. The banks were too greedy for that. Combined with the increases in military spending, the result of the tax cuts was the largest federal budget deficit in history. This was contrary to traditional Republican budget ideology and to Reagan's own claim to be cutting back on "big government." The higher military spending was obviously *growth* of "big government."

"Military socialism" was now fully entrenched in the U.S., with the revolving door between military service and the arms industry soon to be an everyday fact of life. Military socialism and the explosion of defense industries and think tanks transformed the once-semi-civilized town of Washington, DC, with its wide boulevards and peaceful neighborhoods, into the massively congested and frenetic megalopolis of today. The DC area now covers most of the land from the West Virginia panhandle to Delaware—a cancerous monument to the president who said, "Government is the problem, not the solution."

But the element of "Reaganomics" that explained everything else was the massive deregulation of the banking industry. This meant that the banks continued to grow through mergers, including gobbling up the Savings and Loan industry, and offering investment and credit services across state lines and international borders in a seamless array of loans, lines of credit, and money markets. This was made possible by the protection of depositor accounts by the Federal Deposit Insurance Corporation, protection that does not apply to mutual fund accounts managed by investment houses.

Apart from protected accounts, there was the massive growth in the 1980s of deferred tax plans such as 401ks that the Reagan administration used to replace the federal retirement trust fund and that served the same purpose for private businesses that wanted to get rid of their own employee retirement programs. Since then, Americans' retirement savings have been uninsured hostages of the financial markets. Cases of American workers losing their retirement savings when markets crash are legion. When the federal workforce was offered a choice of tax-deferred market-based annuities, I chose to stay with the old tried- and-true federal retirement trust fund.

Deregulation was also moving us into the era of junk bond scandals, with   starting to make the news in 1986, along with leveraged buyouts of corporations targeted for looting by investment firms. Bank lending was also growing for businesses that wanted to shut down their domestic manufacturing plants and move production overseas. The resulting unemployment was left to state governments to handle, though growth of the financial industry led to an increase in service economy jobs, especially accounting—not exactly a good match for former workers in manufacturing.

The growth of bank deposits gave the big banks a much larger lending capacity. As the crisis for Less Developed Countries deepened, the U.S. decided to increase loans through the IMF and the World Bank, thereby shifting the danger of default from the private banks to these quasi-public institutions, and via them, to the governments to whom they provided loans. The IMF and World Bank were able to make demands on governments that the banks could not, including privatizing public utilities, firing government employees, and cutting health and education expenditures.

The LDC community largely consisted of countries that had attained independence from European colonizers since World War II. But rather than these countries attaining economic independence and improved living conditions for their populations, they faced endemic poverty with growing wealth disparities between rich and poor. This led to chronic political instability, especially in Latin America, Africa, and the Middle East. The U.S. benefited from this instability by stepping up its weapons sales to these countries. All of them were "hosts" to CIA operatives devoted to helping governments fend off populist revolts. Increasingly, these countries would also host U.S. military facilities.

A full-blown Third World debt crisis emerged by the end of the Reagan administration and was bleeding over into the George H.W. Bush administration. Over the decade from 1980 to 1989, Latin American per capita GDP fell by eight percent, with riots now taking place over austerities ordered by the IMF. Nomi Prins sums it up by noting that "the banks had created a gigantic international debt bubble."[41] The debt that could not be paid was massive. The banks had expected that there would be rising commodity prices in products like oil and copper available for nations to service and repay their loans. But the banks then realized that they were overexposed so stopped lending.

This was yet another massive failure that can be blamed on the fractional reserve banking system. It was proving to be a monster that led to massive debt and had to continually be fed with new borrowers, i.e., new victims. The only alternatives were inflation or predatory warfare. This was one reason the George H.W. Bush administration tried to take down and absorb Iraq through the Gulf War which the U.S. instigated in 1991, wreaking immense damage on Iraq's civilian infrastructure. Next up on the bankers' agenda was Russia, which would soon be available for looting with the collapse of the Soviet Union in 1991.

All things considered, the "Rockefeller Republic" had triumphed.

ENDNOTES FOR CHAPTER 14

1 Koerner, 81. In his 2023 presidential campaign, Robert F. Kennedy, Jr. states that the convicted gunman, a Palestinian named Sirhan Sirhan, was *not* the person who shot his father.

2 Madhav Nalapat, "Shadow Men Work to Remove President Trump," *Sunday Guardian,* January 22, 2017. We'll hear more from Nalapat in Chapter Seventeen.

3 Ambrose and Brinkley, 244.

4 There were estimates that more than that number of U.S. troops committed suicide later in their lives due to Vietnam-induced PTSD. It was to be a norm. See Meghann Myers, "Four Times as Many Troops and Vets Have Died by Suicide as in Combat, Study Finds," *Military Times,* June 21, 2021. https://www.militarytimes.com/news/your-military/2021/06/21/four-times-as-many-troops-and-vets-have-died-by-suicide-as-in-combat-study-finds/. See also "2021 Name Additions and Status Changes on the Vietnam Veterans Memorial," *Vietnam Veterans Memorial Fund,* May 4, 2021.

5 Seymour Hersh, "Moving Targets," *The New Yorker,* November 20, 2013.

6 Samuel Momodu, "The Jackson State Killings, 1970," *BlackPast,* Sept. 9, 2017. https://www.blackpast.org/african-american-history/events-african-american-history/jackson-state-killings-1970/

7 This is not to deny that some people using psychotropic medications are enabled to live more normal lives.

8 See especially F. William Engdahl's discussion of "Saltsjoebaden: the Bilderberg Plot," 265ff. in *Gods of Money,* op.cit.

9 Prins, 265.

10 A Chevron oil tanker was named the *John A. McCone.*

11 Seymour Hersh, "The Price of Power: Kissinger, Nixon, and Chile," *The Atlantic,* December 1982.

12 For a superb analysis of the 1970s energy crisis and the role of the Rockefellers' oil and banking cartel, see Gibson, 107ff.

13 Engdahl, 269ff. Also see Kelly Mitchell, *Gold Wars: The Battle for the Global Economy* (Clarity Press, 2013).

14 Stone and Kuznick, 402.

15 Prins, 260.

16 Ambrose and Brinkley, 246.

17 Christina Pazzanese, "Is there anything to learn about Watergate? New history says yes," *Harvard Gazette,* June 14, 2022.

18 Prins, 299.

19 Attempts have been made by intelligence trolls to sabotage my own writings.

20 The literature on the subject is vast. See Ted Galen Carpenter, "How the National Security State Manipulates the Mass Media," *CATO Institute*, March 9, 2021. Also Caitlin A. Johnstone, "The CIA used to infiltrate the media. Now the CIA is the media," *MROnline*, April 28, 2021. Also, on Jonathan Cook's blog, "Is the CIA editing your newspaper?," December 19, 2016. An early work on the subject is Carl Bernstein's 1977 *Rolling Stone* article confirming that at least 400 American journalists had secretly carried out assignments for the CIA during the previous 25 years.

21 "The Intelligence Gathering Debate," *NBC News*, August 18, 1975.

22 James Bamford, "The Agency That Could be Big Brother," *New York Times*, December 25, 2005.

23 Douglas Valentine, *The CIA as Organized Crime: How Illegal Operations Corrupt America and the World* (Clarity Press, 2016).

24 Philip Agee, *CIA Diary: Inside the Company* (Harmondsworth, Middlesex, England: Penguin Books, 1975).

25 Ironically, the restoration of Williamsburg as the capital of colonial Virginia was financed by the Rockefellers under the leadership of John D. Rockefeller, Jr., and his wife Abbey Aldrich Rockefeller.

26 Agee, 49.

27 Ambrose and Brinkley, 282.

28 The electoral vote between Carter and incumbent Gerald Ford was close: 297 to 240. Ford himself had barely beaten challenger Ronald Reagan for the Republican nomination. Incumbent vice president Nelson Rockefeller had announced that he did not wish to run for reelection as Ford's running mate. Ford selected Senator Robert Dole instead.

29 Today, the U.S. is desperately trying to keep India "on-side" and not support Russia in the U.S. proxy war in Ukraine.

30 Ambrose and Brinkley, 291.

31 Jimmy Carter, *Peace, Not Apartheid* (Simon and Schuster, 2006).

32 Prins, 302.

33 William Greider, *Secrets of the Temple: How the Federal Reserve Runs the Country* (Simon & Schuster, 1989).

34 Stone and Kuznick, 396–97.

35 Richard C. Cook, *Challenger Revealed: An Insider's Account of How the Reagan Administration Caused the Greatest Tragedy of the Space Age* (Thunder's Mouth Press, 2007), 154.

36 Throughout the 1970s, Teddy Kennedy had been viewed as the leading Democratic senator for likely success in gaining the White House, even with George McGovern running in 1972. But the burden of his two brothers having been killed plus the overhang from the Chappaquiddick incident proved insuperable.

37 Bureau of Indian Affairs, "Indian Claims Commission Granted More Than $45 Million During 1969," April 6, 1970.

38 Patti Davis, *The Way I See It: An Autobiography* (Putnam Adult, 1992).

39 Cook, 128–29.

40 Russ Baker, "Bush Angle to Reagan Shooting Still Unresolved as Hinckley Walks," August 16, 2016. https://whowhatwhy.org/politics/government-integrity/bush-angle-reagan-shooting-still-unresolved-hinckley-walks/>

41 Prins, 355.

CHAPTER 15

"Springtime in America" Faces Early Frost

Goal of the CIA

Speaking at a 1981 White House meeting with President Ronald Reagan and his top advisers, CIA director William Casey said: "We will know our disinformation program is complete when everything the American people believes is false."[1]

The CIA, along with the rest of the national security state, including the military, had been working to further Casey's goal and has been ever since. In league with them? Government-supported "think tanks" like the Rand Corporation, as well as the mainstream media, the corporate PR machine, "non-profits" like the Council on Foreign Relations and the Atlantic Council, and, today, internet giants like Google, Wikipedia, etc. Politicians beholden to rich donors, particularly the giants of Wall Street, are fronting for all these. Lying to the American people is indeed a way of life and a lucrative profession.

Allied with the elites of other nations, their goal is to create a depleted world population of compliant zombies, drained of creative energy, incapable of independent thinking, riddled with fear, obedient to every conditioned impulse, running from every virus, hating every designated enemy, parroting the mass media, spending as advertising directs, numbed by drugs and medication, and ready to die at the whim of their masters.[2] Is this the New World Order?

Reagan Wrap-Up

National Endowment for Democracy

By the 1980s, the CIA had acquired a bad name due to the horrors it had been perpetrating for a quarter century. In the words of *Covert Action Magazine*'s Abby Martin:

> Brutal terrorism was unleashed on entire countries by U.S. backed and trained right-wing death squads that targeted poor people, peasants, farmers, students, all in the name of fighting communism."[3]

With the revelations of the Church Committee, a change of approach by the CIA was needed. *Too much bad press!* So "they" decided to use "softer" methods to enforce the ongoing illegal special operations abroad.

The chosen method was to create a "non-profit" congressionally funded organization, the National Endowment for Democracy (NED). Brought into being by legislation in 1983, NED became the most important of the "non-governmental" organizations (NGOs) that sprang up like weeds worldwide, whose mission was to impact, under the mantle of civil society discourse and activism, the political systems of dozens of governments targeted by the U.S.[4] The main job of the NED is to foment "color revolutions" in countries marked by the U.S. for "regime change."

Death Squads/Iran-Contra

President Reagan seemed to have a special place in his heart for right-wing military governments, especially in Central America, which were so oppressive that they have gone down in history as noted for their U.S.-trained "death squads."

We have seen how the Reagan team established a covert relationship with Iran during the hostage crisis. Then, in the fall of 1986, the revolutionary Sandinista government of Nicaragua, which had deposed the U.S.-sponsored Somoza dictatorship in 1979, came under attack by a rebel force known as the "Contras" that was receiving supplies and training from the CIA, even though U.S. funding of the Contras had been prohibited by the congressional Boland Amendment. When, in the fall of 1986, the Sandinistas shot down a CIA transport plane over Nicaragua carrying weapons, the surviving American crew member confessed that he worked for the CIA. A month later, an Arab magazine printed a story on covert U.S. arms sales to Iran to pay for Contra arms and munitions, naming the CIA as the likely architect of the Iran-Contra plot.[5]

Scandal now erupted in Washington, with Secretary of State George Schultz pleading ignorance. Overseeing the operation were Reagan's National Security Adviser, Admiral John Poindexter, and his deputy, Lieutenant Colonel Oliver North. "North immediately began shredding documents in his White House office, while the FBI called for a special prosecutor."[6]

The Tower Commission, under former Texas Senator John Tower, investigated. While it found that laws had been broken, it failed to link either Reagan or Vice-President George H.W. Bush to the operation.[7]

North and the rest of the crew escaped accountability, even after a Congressional select committee conducted hearings and charged the administration with "confusion and disarray at the highest level of government, evasive dishonesty, and inordinate secrecy, deception, and disdain for the law."[8] Reagan, the "Teflon president," wriggled off the hook.

The Contra terror in Nicaragua never got off the ground, though it did use its CIA training to cause the deaths of several thousand civilians. There were Contra-controlled airfields in Nicaragua, where the CIA landed shipments of weapons. The couriers then flew back to the U.S. with cocaine that originated in South America. The point-of-entry in the U.S. was the Mena Airport in Arkansas, shielded by Arkansas politicians, including, allegedly, Governor Bill Clinton.

There are reports that Clinton was observed in meetings at the Mena airport, where a top CIA official visiting from Washington told him he was under consideration to become president.[9] Thus the Iran-Contra affair may have had longer tentacles than was disclosed at the time. CIA drug trafficking from Latin America to the U.S., then into the hands of dope dealers in African American communities, was also part of the Reagan legacy. The trafficking was later admitted by the CIA's own inspector general.[10]

Even today, more than thirty-five years later, the U.S. is still attacking Nicaragua with economic sanctions.

The Challenger Disaster

On January 28, 1986, space shuttle Challenger blew up one minute after liftoff at the Kennedy Space Center in Florida, killing all seven astronauts, including Christa McAuliffe, the "Teacher-in-Space."

I had been working for NASA since the previous July as an analyst on the shuttle's program staff. After meetings with the solid rocket booster engineers at headquarters, I reported that safety of the O-ring joints on the booster rockets had been compromised and that catastrophic failure was possible. The engineers said they "held their breath" with each shuttle launch.

The flaws had been known for years, but NASA had failed to halt flights to make the needed repairs.

After the explosion, NASA began a cover-up. After failing to get anyone's attention, I took a folder of O-ring documents to *The New York Times*, which published them as the lead story on Sunday, February 9, 1986, along with my name and excerpts from one of my warning emails.

The story soon emerged that the night before the launch, engineers from the solid booster rocket contractor at Thiokol, Inc. in Utah, had protested vehemently that the launch must be stopped. The reason was that the O-ring joints were much more susceptible to failure when the weather was cold. The overnight lows at the Kennedy Space Center were expected to be well below freezing. The engineers' warnings were ignored.

During the investigation, I was questioned in public by the Presidential Commission convened to investigate the disaster and had several meetings with the Commission's staff and NASA's Office of the Inspector General. I reported information I had learned from a CNN executive that NASA had rushed the launch so Challenger would be aloft in time for President Reagan's State-of-the-Union address that night. CNN had been pressured by the White House not to report the story, and it didn't.

I never returned to NASA after my testimony. Later I wrote a book on the disaster, *Challenger Revealed: An Insider's Account of How the Reagan Administration Caused the Greatest Tragedy of the Space Age,* where I explained my view that NASA failed to stop flights because the space shuttle was being taken over by the Air Force and had to keep flying because the shuttle was being used as a testing platform for the President's Strategic Defense Initiative (SDI). Also known as "Star Wars," SDI was Reagan's scheme for putting weapons into space that supposedly would defend against a Soviet missile attack. Although Reagan touted SDI as a "defensive" system, it would have been able to launch space-based weapons for offensive purposes, including a nuclear first strike. I was able to develop this thesis by working with Dr. Robert Bowman, an Air Force lieutenant colonel and former head of the Air Force's advanced space programs.

With the halt in the space shuttle program after the disaster, Star Wars was put on hold. Space-based weaponry would not re-emerge as a serious military objective for another twenty years. I never returned to NASA after my testimony, so spent the next twenty-one years working for the U.S. Treasury Department's Financial Management Service. I had plenty of time to study the history of U.S. government finance. I learned of the disastrous consequences of fractional reserve banking, and also learned about President Lincoln's Greenbacks. Eventually I saw that only a modern Greenback system could cure the cancer that was destroying our country. My research contributed to the work of the American Monetary Institute that became Congressman Dennis Kucinich's NEED Act, covered in the Appendix to this book.

The INF Treaty

By the end of the Reagan administration, a history-making figure from the Soviet Union had appeared on the world stage: Soviet Premier Mikhail Gorbachev. Gorbachev had begun to move toward an accommodation with the U.S. to slow the arms race, along with introducing policies within the Soviet Union for individual freedom and for economic and social modernization and decentralization. He called his approaches *glasnost*—"openness"—and *perestroika*—"restructuring."

Gorbachev wanted to negotiate an arms *reduction* treaty with the U.S., as opposed to just another effort at arms *control*. Reagan was willing to meet him part-way. The result was the Intermediate Nuclear Force Treaty—INF. For the definitive history of the INF Treaty and an explanation of its importance, see Scott Ritter's *Disarmament in the Time of Perestroika: Arms Control and the End of the Soviet Union*.[11]

The treaty marked the first time that the U.S. and Soviet Union had agreed to reduce their nuclear arsenals, eliminate an entire category of nuclear weapons, and employ on-site inspections for verification. As a result, the two nations destroyed 2,692 short, medium, and intermediate-range missiles by the treaty's implementation deadline of June 1, 1991.[12]

The Trump administration unilaterally withdrew from the INF Treaty in 2019, claiming Russian violations.

The End of the "Reaganomics" Bubble

What Reagan touted as "springtime in America" in his 1984 campaign speeches was having problems.

Toward the end of his second term, the Reaganomics bubble was over. By the time George H.W. Bush was sworn in, the U.S. faced unprecedented government debt from tax cuts and military spending along with a growing trade deficit. The banking system was living off loan repayments from Less Developed Countries, but this source of revenue was drying up. When the stock market crashed in October 1987, newly appointed Federal Reserve Chairman Alan Greenspan took the step of pumping huge sums of credit into the banking system to forestall a major recession, a practice to be repeated in 2008.

This was more than a rescue. It enabled the banks to move full-speed-ahead in leveraged buyouts,[13] the creation of derivatives, and other risky trading that had everything to do with making money, but nothing to do with rebuilding U.S. manufacturing. This was the economic environment when George H.W. Bush became president.

For the U.S. financial system to work, the rest of the world had to be feeding money into the bankrupt U.S. government balance sheet that already had Reagan's doubling of the military budget to contend with. The government bailed itself out by selling Treasury bonds to foreign nations and international investors, borrowing that had increased dramatically since 1975. The federal government's debt doubled from 1975 to 1980, doubled again by 1986, and again by the end of Bush's single term in office in 1993.[14]

The Election of George H.W. Bush

George H.W. Bush was born into a prominent American banking family. His father, Prescott Bush, had made a fortune with Brown Brothers Harriman, a bank that was cited during World War II for collaboration with the Nazis.

Having served as one of the youngest Navy combat pilots during World War II, George H.W. Bush attended Yale University following his discharge, where he was elected to the Skull and Bones secret society. Yale was the nation's prime CIA recruiting ground, and researchers believe Bush joined before he moved to Texas to enter the oil business. There he ran the Zapata Petroleum Company. There is a body of opinion that believes Bush was involved in CIA activities against Fidel Castro. Some link him to the JFK assassination. Bush served two terms in Congress, then was CIA director under President Gerald Ford before becoming President Ronald Reagan's vice president for two terms. In 1988, Bush defeated Democratic Massachusetts Governor Michael Dukakis in a landslide and became president.

Bush and the Gulf War

Bush initiated America's relentless drive to overthrow the government of Iraq, with Desert Storm most indelibly marking his presidency. Desert Storm was not just a matter of giving America's bloated military something to do after Reagan's trillion-dollar build-up. The impetus for the war had started with the lies the CIA was telling about how much more powerful the Soviet military was compared to that of the U.S.,[15] though in point of fact, by 1989, the Soviet Union's East European satellites were electing democratic regimes, the Berlin Wall had come down, and the Warsaw Pact had dissolved.

The Soviet Union would collapse in 1991. This was the geopolitical environment when the Gulf War began.

The primary reason for war against Iraq was control of oil, though the interests of Israel cannot be discounted. The 1990 Iraqi invasion of Kuwait was merely the *casus belli* that allowed the U.S. to obtain a UN resolution under which thirty-five nations participated in the attack on Iraq. But Iraq had a *casus belli* of its own in Kuwait's unilateral provocation in undermining the OPEC oil price structure and in illegally pumping oil out of Iraqi oil fields by slant drilling. Both provocations had been instigated by the U.S. through the time-honored practice of getting an opponent to attack first. It was within this context that the U.S. invaded Iraq in 1991.

Five distinct threads of U.S. influence converged to enable the policy:

1. President George Bush was a leader of the Deep State and was supported by the emergence within the CIA of Robert Gates, who represented the extreme anti-Soviet faction.
2. The military was overseen by Secretary of Defense Dick Cheney and his deputy Paul Wolfowitz, two leading Neocon hawks.
3. Big banking and big oil, especially the Rockefeller interests, were heavily invested in their desire to seize the Iraqi oil fields.
4. The Soviets under Mikhail Gorbachev sided with the U.S. in the war, with Gorbachev believing U.S. Secretary of State Jim Baker's promise that NATO would not advance "one inch" eastward toward Russia with the Cold War ending.
5. The growing pro-Israel Neocon faction within the government and in league with JINSA, AIPAC, and other lobbying organizations were advocating stridently for the U.S. to remake the Middle East through controlling or overthrowing the governments of numerous Islamic states, especially Syria, Libya, and Iran. Saudi Arabia, the Gulf emirates, Egypt, and Turkey were already in the American camp.

The U.S. did not "win" this war. Saddam Hussein was not removed from power, and oil resources remained in Iraqi hands, though its infrastructure was severely damaged. The U.S. tried to weaken Iraq for another decade by conducting bombings, enforcing a no-fly zone, and applying economic sanctions, which had a severe effect on the health of the Iraqi people, causing the death of over 500,000 Iraqi children.[16] Finally, under Bush's son George W., the U.S. would use 9/11 as an excuse for total subjugation of Iraq in 2003. The result was total dismantling of Iraq's civil society that has never been redressed or repaired.

Recession and the Growth of Financial Predators

Bush's Gulf War was going on as the U.S. economy was sinking into recession. Speculation in real estate loomed large. Heavy bank dependence on the U.S. real estate market, both homeowners and commercial, would headline financial turmoil until today. Bank involvement in massive real estate lending pushed prices up with the interest on loans resulting in higher rents and mortgages.

In order to fill the gaps in their balance sheets, banks increasingly pressured the federal government to further reduce regulation, culminating in the Glass-Steagall repeal in 1999. The deregulation enabled massive trading in derivatives, such as commodity futures and currency and credit swaps. All these were "hedges," with nothing to do with actual economic productivity, and swelled the coffers of investment firms like Vanguard, BlackRock, and Goldman Sachs.

Investment banks also made money by helping corporations finance the movement of factories to Third World countries with vastly lower labor costs than in the U.S., thereby delivering a profound blow against American workers and the domestic manufacturing base. In 1992, just before leaving office, Bush initiated the North American Free Trade Agreement—NAFTA—that was implemented under Clinton. The result was massive loss of farming and factory jobs in both the U.S. and Mexico and a sharp decline in the Mexican standard of living. This led millions of impoverished Mexicans to cross the U.S. southern border illegally in search of jobs.

The Wolfowitz Doctrine

Paul Wolfowitz may have been the archetypical American Neocon of the late 20th and early 21st century. But in the interest of space, I will leave it to the reader to do further research. Wolfowitz today is a member of the Council on Foreign Relations, a former member of the steering committee of the Bilderberg Group—all "Deep State"—and a former president of the World Bank.

The Wolfowitz Doctrine is an unofficial name for the Defense Planning Guidance for the 1994–1999 fiscal years dated February 18, 1992, published by Wolfowitz as the Under Secretary of Defense for Policy, with his deputy Scooter Libby. Not intended for public release, it was leaked to *The New York Times* on March 7, 1992. The document brought the decades-old plan for total U.S. global military dominance up-to-date and has defined the direction of U.S. foreign policy to the present.

The document, viewed by many as outrageous, was hastily re-written under the supervision of Secretary of Defense Dick Cheney and Chairman of the Joint Chiefs of Staff Colin Powell before being officially released on April 16, 1992. Following are the provisions of the original Wolfowitz Doctrine[17] before the Cheney/Powell sanitized version came out:

> The U.S. must show the leadership necessary to establish and protect a new order that holds the promise of convincing potential competitors that they need not aspire to a greater role or pursue a more aggressive posture to protect their legitimate interests. In non-defense areas, we must account sufficiently for the interests of the advanced industrial nations to discourage them from challenging our leadership or seeking to overturn the established political and economic order. We must maintain the mechanism for deterring potential competitors from even aspiring to a larger regional or global role.
>
> Like the coalition that opposed Iraqi aggression, we should expect future coalitions to be *ad hoc* assemblies, often not lasting beyond the crisis being confronted, and in many cases carrying only general agreement over the objectives to be accomplished. Nevertheless, the sense that the world order is ultimately backed by the U.S. will be an important stabilizing factor.
>
> While the U.S. cannot become the world's policeman by assuming responsibility for righting every wrong, we will retain the preeminent responsibility for addressing selectively those wrongs which threaten not only our interests, but those of our allies or friends, or which could seriously unsettle international relations.
>
> We continue to recognize that collectively the conventional forces of the states formerly comprising the Soviet Union retain the most military potential in all of Eurasia; and we do not dismiss the risks to stability in Europe from a nationalist backlash in Russia or efforts to reincorporate into Russia the newly independent republics of Ukraine, Belarus, and possibly others....We must, however, be mindful that democratic change in Russia is not irreversible, and that despite its current travails, Russia will remain the strongest military power in Eurasia and the only power in the world with the capability of destroying the United States.
>
> In the Middle East and Southwest Asia, our overall objective is to remain the predominant outside power in the region and

preserve U.S. and Western access to the region's oil. We also seek to deter further aggression in the region, foster regional stability, protect U.S. nationals and property, and safeguard our access to international air and seaways. As demonstrated by Iraq's invasion of Kuwait, it remains fundamentally important to prevent a hegemon or alignment of powers from dominating the region. This pertains especially to the Arabian peninsula. Therefore, we must continue to play a role through enhanced deterrence and improved cooperative security.

Note the recognition that "Russia will remain...the only power in the world with the capability of destroying the United States." This is due to the Soviet nuclear arsenal and the reality of "Mutually Assured Destruction," which has loomed in the geopolitical background since the 1950s, yet today seems to be disregarded.

The Wolfowitz Doctrine was a legacy of the Reagan/Bush presidencies. Extreme as it appears, there is nothing essentially different from the days before the U.S. entered into World War II and the intent to establish total global military dominance. More than forty years had passed since then, but the basic policy was a continuum, and it was now fully public. The difference was that the "Wolf's" teeth were bared and its growl had become deafening.

Note that the main motivation of the original Council on Foreign Relations policy proposals was to protect U.S. overseas investments. The Wolfowitz Doctrine had the same intent—to protect U.S. global financial hegemony and control of oil. These were also the central tasks of the CIA.

It could be argued that the Wolfowitz Doctrine is patently in defiance of the UN Charter that outlaws aggressive war. Patrick Buchanan called it a "formula for endless American intervention in quarrels and war when no vital interest of the United States is remotely engaged."[18] I would call the Wolfowitz Doctrine the product of dangerous madmen. It has served to entrench endless fear and suspicion, a constant siege mentality, and an acceptance of perpetual war within the American polity. The financiers and the military industrial complex may love it, but by the 1990s its ramifications became manifest in the growing threat of U.S. government bankruptcy due to the massive budget deficits of the Reagan-Bush presidencies.

The Election of Bill Clinton

In 1992, Bill Clinton, Governor of Arkansas, defied George H.W. Bush's quest for reelection. If the CIA had indeed promised Clinton at the Mena Airport that he would someday be president, they kept their word.

Clinton may have been elected due to the presence of Texas businessman Ross Perot, who split the Republican vote, allowing Clinton to win with only forty-three percent of the popular vote.

Obviously, this interpretation is speculative, as Clinton was able to take advantage of Bush's weakness, which was his total identification with the worlds of war and intelligence through the Gulf War and his prior role as CIA director. With the Soviet Union on the verge of collapse, signaling the end of the Cold War, domestic issues moved to the forefront. The new emphasis was, as Clinton's advisers put it, "It's the economy, stupid!" Bush was a relic of a bygone era. Clinton and his pro-business "Third Way" looked to the future, not the past. This would mean the further unleashing of big finance in its drive to reshape the world, with David Rockefeller its symbol and driving force. The Clinton years saw no major wars, eight years of growing prosperity, and for the last three, a balanced federal budget. But behind the façade, an image the Republicans relentlessly attacked by investigations and impeachment, trouble was brewing.

Clinton's Presidency

It's convenient to divide a president's term in office along the lines of domestic vs. foreign affairs. With President Bill Clinton, he said he took almost no interest in the latter. Three months into his presidency he said, "Foreign policy is not what I came here to do."[19] Perhaps in order to claim plausible deniability, he later punctuated his passivity when he said of the 1999 NATO bombing campaign against Serbia, following the CIA project of the 1990s to break up the nation of Serbian-led Yugoslavia, that he did not participate in any of the bombing decisions.

Clinton was able to cut the national crime rate by a Draconian program of imprisonment of minor offenders and reduced welfare benefits for blacks, forcing many young mothers into low-paying service jobs. The implementation of NAFTA continued the outsourcing of factory jobs, and speculative investment continued to grow, especially with repeal of Glass-Steagall in 1999. The Dot.com bubble was among the onerous results. Clinton failed at one of his premier priorities, the creation of a universal health care system for the U.S.

We don't know the extent to which Clinton was an obedient bystander in foreign policy decisions or how much was done for him with his connivance—by his National Security Advisors William Lake and Sandy Berger, his Secretaries of State Warren Christopher and Madeleine Albright, a string of CIA directors, ending with George Tenet, and his Defense Secretaries and military commanders. But we have come to learn that despite the relative quiet on the foreign affairs front, and the anxious questioning among liberal-left observers about whether there really would be a "peace dividend" with the Soviets now beaten, a growing number of low-grade confrontations and interventions portended the future outbreak of today's era of endless war.

Clinton's role in these events has been documented by Jeremy Kuzmarov in a new book. As James Bradley, author of *Flags of Our Fathers: The China Mirage*, writes in a review of the Kuzmarov book:

> Who first set us on this disastrous road of endless war and imperial overreach? Who won over liberals by saying that military interventions were for humanitarian purposes? Who first raised false fears of WMD in Iraq and set in motion the U.S. invasion? Who first carried out the odious practices of extraordinary rendition and drone attacks in the War on Terror? Who first violated the U.S. pledge [not to extend] the orders of NATO and triggered the new Cold War with Russia? Jeremy Kuzmarov brilliantly answers these questions in his stunning new book, *WARMONGER: How Clinton's Malign Foreign Policy Launched the U.S. Trajectory from Bush II to Biden.*[20]

A Neocon Volcano

These were critically important events, indicating an underground river of deep and murky magma—a Neocon volcano—that would erupt after Clinton left office and his successor, George W. Bush, stepped onto the stage. This eruption, of course, was the "War on Terror" that followed 9/11. The river of magma had several sources—the more public part included the Neocon press led by Jewish pro-Israeli stalwarts like William Kristol, editor of the *Weekly Standard*, and Robert Kagan, a war hawk and *Washington Post* columnist. William Kristol was the son of Irving Kristol, the "godfather" of the Neoconservative movement. Another key figure was journalist Norman Podhoretz. Both came out of the New York City communist/socialist cabal often characterized as "Trotskyite," but heavily infiltrated by the FBI and CIA.

This bevy of leftist radicals have become reliable promoters of U.S. political and military hegemony. "Trotskyite" in this context refers to Leon Trotsky's goal of worldwide revolution as opposed to Stalin's policy of "socialism in one country" that blended over time with Russian nationalism.

Other contributions to the Neocon magma flowed from the political ambitions of conservative pro-Israel politicians like Speaker of the House Newt Gingrich, Senator Joseph Lieberman, and Christian evangelicals like Jerry Falwell. The most deeply hidden current involved a Deep State component that came into being in order to plan Continuity of Government (CoG) operations in case of nuclear war that involved suspension of the Constitution, abrogation of civil rights, the rounding up of dissidents, etc. The CoG was described in an article on "The Hidden Government Group" by author/activist Peter Dale Scott:

> Going one step further, Andrew Cockburn quoted a Pentagon source to support a claim that a COG planning group under Clinton was now for the first time staffed "almost exclusively with Republican hawks." In the words of his source, "You could say this was a secret government-in-waiting. The Clinton administration was extraordinarily inattentive, [they had] no idea what was going on."
>
> The Pentagon official's description of COG planners as a "secret government-in-waiting" under Clinton (which still included both Cheney and Rumsfeld) is very close to the standard definition of a cabal, as a group of persons secretly united to bring about a change or overthrow of government.[21]

The Prelude to 9/11

During the Clinton administration, a succession of false flag events made "terrorism" and "terrorists" a constant presence in the mainstream media and thereby in the public mind. In order to create the myth of "al Qaeda" as a living terrorist organization or movement and not simply a band of guerrillas created by the CIA as a militia assembled to fight the Soviets in their occupation of Afghanistan, the media elevated Osama bin Laden, who was cited as their leader, to the status of a universal bogeyman. It began by media circulation of such horrifying pictures of terrorist preparation as the BBC's image of hooded men training by swinging across monkey bars, marching in rows, and firing rifles while prone.[22]

But then matters escalated:

A huge truck bomb went off in the basement of the World Trade Center on February 26, 1993. Tapes cited in court indicated that this plot was organized and facilitated by the FBI, though the tapes themselves were never made public.[23]

The U.S. embassies in Nairobi, Kenya, and Dar es Salaam, Tanzania, were bombed on August 7, 1998.

Near the end of 1999, the Clinton administration, working with the government of Jordan, claimed to have detected and thwarted terrorist plots to detonate bombs at New Year millennium celebrations around the world.

On October 12, 2000, the *USS Cole* was bombed in the harbor of the Yemeni port of Aden. The attack killed seventeen Navy sailors. A week after George W. Bush assumed office, the CIA claimed that Osama bin Laden had done it.

It's always good to have a specific person, if demonization is the intent; witness the current treatment of Putin, now being blamed for everything. The public were told by the mainstream media that the CIA had confirmed that Osama bin Laden, leader of al Qaeda in Afghanistan, was behind all these plots. Various short biographies of the evil doer were provided. Osama bin Laden was from a prominent Saudi family and had fought against the Russians in Afghanistan. It was acknowledged that he was a CIA asset at the time. What caused him to "turn against" the U.S.—*inter alia*, the U.S. troop presence in Saudi Arabia—was never satisfactorily explained. Thus, the usual claim was trotted out in every instance that "they hate us for our freedoms."

President Clinton launched cruise missiles into Afghanistan and Sudan in order to kill Osama bin Laden or his lieutenants, but the strikes failed to produce results. Clinton later said he called off another missile strike in Afghanistan out of unwillingness to kill civilians. Of course, these failures contributed to complaints that Clinton was "soft" on terrorism. No matter that under Clinton, the U.S.-instigated sanctions on Iraq produced an admitted half a million casualties among children under five, and that the U.S. and UK had been regularly bombing Iraq in support of their illegally initiated "no fly" zone....

Senator Bob Dole, Clinton's opponent in the 1996 presidential election, also charged that Clinton had been dragging his feet on NATO expansion. Against the advice of George Kennan, deviser of Soviet containment, Clinton oversaw the 1998 admission into NATO of Poland, Hungary, and the Czech Republic, an action Russia saw as a betrayal of promises given to Gorbachev years earlier.

Kennan wrote in 1998:

I think it is the beginning of a new cold war. I think the Russians will gradually react quite adversely, and it will affect their policies. I think it is a tragic mistake. There was no reason for this whatsoever. No one was threatening anybody else."[24]

Similarly frank sentiments were expressed by Clinton's secretary of state, Madeleine Albright, who wrote in her memoirs that:

> [Russian President Boris] Yeltsin and his countrymen were strongly opposed to enlargement, seeing it as a strategy for exploiting their vulnerability and moving Europe's dividing line to the east, leaving them isolated.

Deputy Secretary of State Strobe Talbott attested to the fact that:

> Many Russians see NATO as a vestige of the cold war, inherently directed against their country. They point out that they have disbanded the Warsaw Pact, *their* military alliance, and ask why the West should not do the same.[25]

But the Neocons would not back off. In the July-August 1996 issue of *Foreign Affairs*, the house organ of the Council on Foreign Relations, William Kristol and Robert Kagan attacked Clinton's cautious approach to engaging foreign adversaries, calling for a "heroic" foreign policy based on "elevated patriotism" that promotes "the virtues of militarism" as opposed to the "cowardice and dishonor" in not undertaking to "destroy many of the world's monsters."[26] There have been few better expressions of unbridled Neocon celebration of aggression. Goebbels would have been proud. So, likely, was David Rockefeller, always the chief CFR patron.

In 1996, Senator John McCain chimed in by saying that, "My biggest criticism is that this Administration lacks a conceptual framework to shape the world going into the next century" and fails to "explain what threatens that vision."[27] Criticism also came from Henry Kissinger and Jeanne Kirkpatrick, Reagan's UN Ambassador, who charged that Clinton "had to resort to Band-Aid diplomacy in the absence of a grand design."[28]

But what was the "grand design" these worthies were promoting? Clinton had said that his vision of the U.S. was "like a big corporation competing in the global marketplace."[29] A cornerstone was the elimination of tariffs and the creation of a global trading regime under the auspices of the World Trade Organization, with the banks and investment houses of Western nations presiding over cheap labor and resource extraction from the Third World.

But this was not enough for Clinton's critics, especially of the Neocon stripe. Their idea of the "grand design" went much further, which was why they went to so much trouble in the 1990s to stoke the fear of terrorism.

For "terrorism" was to become the new *casus belli* for the 21st century. Essentially, war was being declared against any nation the U.S. didn't happen to like or whose resources it coveted.

Combined with the FBI's actions against supposed domestic terrorists like those who allegedly bombed the Murrah Federal Building in Oklahoma City on April 19, 1995, the American people, by the end of Clinton's term, had been thoroughly brainwashed to hate and fear terrorists of every stripe. The Deep State had moved well along toward CIA Director William Casey's aforementioned goal: "We will know our disinformation program is complete when everything the American people believes is false."

The Neocons were at the forefront of the attack on Clinton over the White House intern scandal. Clinton's 1999 impeachment trial had been preceded by five years of distracting publicity growing out of investigations by independent counsel Kenneth Starr, who had been examining the suicide death of White House counsel Vince Foster and the Clinton Whitewater real estate allegations. Clinton was thus neutralized by becoming the most investigated president in history to date, while the stage for 9/11 was being set. Taking the lead in attacking Clinton over Monica Lewinsky was William Kristol's *Weekly Standard*.

The Collapse of the Soviet Union

World history at the end of the 20th century cannot be understood without examining the collapse of the Soviet Union and the travails of Russia and other Soviet component states during the 1990s. This includes the partnership between U.S. financiers and Russian oligarchs in hijacking the Russian economy and running it into the ground.

The rise to power of Soviet Premier Mikhail Gorbachev resulted in epoch-making events, including the signing with President Reagan of the breakthrough INF Treaty in 1985, the Soviet withdrawal from Afghanistan in 1988, and the dissolution of the Warsaw Pact in 1989. That year the Berlin Wall came down, and in 1990 Gorbachev agreed to the reunification of Germany, with the U.S. affirmation that NATO would not expand. President George H.W. Bush announced, "The Cold War is over."

Gorbachev tried to enlist the help of the U.S. in creating a post-Cold War Europe of peace and stability, traveling in May 1990 to Washington, DC for a summit meeting with President Bush. Now-public transcripts show that Gorbachev repeatedly addressed the looming economic crisis in the Soviet Union and the need for support of his planned reforms. He even asked Bush for loan guarantees, a request the U.S. rejected. A 2010 report for the

National Security Archive by Svetlana Savranskaya and Thomas Blanton lays out Gorbachev's vision:

> The documents show that Gorbachev came to Washington determined to make one final push for his idea of a European security structure, or the "common European home." He envisioned a gradual transformation of NATO and the Warsaw Pact into political organizations and their subsequent dissolution as the Conference on Security and Cooperation in Europe (CSCE) would become institutionalized and subsume NATO security functions. For Gorbachev, this was the answer to Soviet Union's pressing issues—modernization and integration into Europe.[30]

Bush and Secretary of State James Baker failed to respond to Gorbachev's concept, but Baker did give Gorbachev verbal assurance that NATO would not expand "one inch" beyond the German border into Eastern Europe. Gorbachev agreed to German reunification, but the real U.S. answer came in 1991, when the Soviet Union collapsed and U.S. economists whose efforts were skewed by CIA subversion watched the Russian economy tumble off a cliff.

Even as Gorbachev and Bush were meeting, not only were the nations of Eastern Europe overturning the rule of their communist parties and embracing multiparty elections, the constituent republics of the Soviet Union were moving toward independence. This began with the Baltic states of Estonia, Latvia, and Lithuania. Gorbachev aroused the anger of conservative forces within Russia by letting them depart without a fight.

In August 1991, figures in the Red Army and Communist Party arrested Gorbachev in a coup. In Moscow, Boris Yeltsin, now President of the Russian Republic—Russia being the largest of the Soviet states—called for a general strike and opposition to the *putschists*. The Moscow military garrison refused to attack the defiant Yeltsin, and in three days the coup collapsed.

Gorbachev then resigned as president of the Soviet Union. Yeltsin and his supporters abolished the Communist Party, and the leaders of Russia, Belarus, and Ukraine declared that "the USSR as a subject of international law and geopolitical reality is ceasing its existence."[31] The constituent Soviet states were now free to organize separate governments. The U.S. recognized and established relations with the former Soviet republics on Christmas Day 1991.

Twenty-five million Russian-speaking nationals now found themselves separated overnight from Russia itself. The Russian economy moved quickly into crisis. Yeltsin issued invitations to U.S. economists to remake the Russian system into a Western-style free enterprise model. U.S. Undersecretary of the Treasury Lawrence Summers and other U.S. financiers and academics worked to impose a regime of "shock therapy"—austerity, deregulation, and a vast program of privatization of state-owned enterprises. Stone and Kuznick write:

> In what Russians called the "great grab," the nation's factories and resources were sold off for a pittance to private investors, including former Communist officials, who became multimillionaires overnight.[32]

Thus came into existence the famous Russian oligarch class. The sale of state assets to the oligarchs was overseen by the U.S. economists working within the Yeltsin government. There are claims that the purchases were made with gold stolen from Soviet vaults. Large amounts of the profits realized by the oligarchs in buying and selling assets ended up in their purchase of large tracts of London real estate and monies laundered through offshore tax havens. Among the money laundering locations were the British Virgin Islands, Guernsey Island, Monaco, Cyprus, and the Cayman Islands.

Boris Yeltsin was elected president of the Russian Federation in 1992. Faced with crisis, he abolished the constitution and ruled by decree. Meanwhile:

> Russia's economy collapsed. Hyperinflation wiped out people's life savings. Tens of millions of workers lost their jobs. Life expectancy plummeted from sixty-six to fifty-seven years. By 1998, more than eighty percent of Russian farms had gone bankrupt. Russian GDP had been cut almost in half.... Russia was rapidly becoming a third-world nation.[33]

Many of those who lost their jobs committed suicide. Similar conditions prevailed in the other former Soviet republics, with the number of people living in poverty jumping from 14 million in 1989 to 147 million in 1998. Russian novelist Alexander Solzhenitsyn wrote in 2000:

> As a result of the Yeltsin era, all the fundamental sectors of our state, economic, cultural, and moral life have been destroyed or looted. We live literally amid ruins, but we pretend to have a normal life...great reforms...being carried out in our country...were

false reforms because they left more than half of our people in poverty.... Will we continue looting and destroying Russia until nothing is left?... God forbid these reforms should continue.[34]

Yeltsin's popularity collapsed, though he was reelected president in 1996, aided by the CIA and U.S. money. Today it is easy to see that the U.S. government deliberately allowed the Russian economy to enter free-fall. Meanwhile, a number of companies from the U.S., UK, and Europe entered into partnership with firms that now were owned by the Russian oligarchs. These included BP, Shell, Exxon, and BASF. Another prominent company was Halliburton, where Dick Cheney was CEO in the 1990s.

A sub-plot relating to Russian resources now came into play. Going back to the start of the Age of Petroleum in the 19th century, Russia has been a potential source of almost infinite resource wealth, particularly hydrocarbons. Germany learned early on that Russian resources could potentially fuel its own industrial expansion. Hitler was well-aware of the potential from the time he began to formulate his ideas related to *drang nach osten* – the "drive to the east."

Meanwhile, Britain and the U.S. saw the necessity of gaining control of oil in the Middle East. This had been the primary motivation for the CIA overthrow of the government of Iran in 1953, repeated in President George H.W. Bush's Gulf War and later with President George W. Bush's post-9/11 "War on Terror" attacks on Iraq and Afghanistan.

Beyond the Middle East lay Central Asia, with vast reserves of oil and gas in the Caspian Basin and the central Asian republics now spun loose from Soviet control. The collapse of the Soviet Union put the resources of this huge reservoir of the earth's hydrocarbon supply up for grabs.

As the Russian economy continued to languish, Russia went into default on its foreign debt, the ultimate humiliation. In 1999, Yeltsin resigned, turning over the reins of government to acting-president Vladimir Putin, a relatively unknown administrator, hand-picked by Yeltsin as his successor. Yeltsin told his Western contacts not to worry: Putin was a reliable quantity.

Russia's recovery and eventual reentry into great power status would begin with Putin. Putin took several decisive steps:

1. He banished those oligarchs who refused to join in rebuilding the Russian economy.
2. The oligarchs who remained were required to pay their fair share of taxes and work within the government's regulatory framework.
3. He allowed Western-owned businesses to remain.

4. He re-nationalized critical utilities and energy infrastructure, such as Gazprom, accepting foreign investors as minority partners in state-owned enterprises.

5. He worked through the nationalized Russian Central Bank. This was perhaps the most important and far-reaching step, for if a government has no control over its monetary system, it has no control over anything. Of course, the U.S. government does not control its own monetary system. It's the bankers that create the money who control the nation. The U.S. lost its sovereignty to big finance long ago. Putin resisted.

The Death of John Fitzgerald Kennedy, Jr.[35]

Some wondered whether the National Transportation Safety Board was being truthful when it claimed that the crash of John F. Kennedy, Jr.'s Piper Saratoga light plane on July 16, 1999, off the coast of Martha's Vineyard, killing him along with his wife, and his wife's sister, was caused by pilot error due to spatial disorientation, and contemplated various possibilities: JFK Jr. assassinated through an explosion, either an on-board bomb, or being shot out of the air by a missile or shoulder-held rocket.

JFK, Jr. had intended to enter politics, specifically to run for New York governor in 2002 and would certainly have succeeded. He might then have gone on to become president, as there have been a number of New York governors, even short-timers, who became president, including Theodore Roosevelt and Grover Cleveland. JFK, Jr. did not believe the government's lie that his father was killed by a lone sniper, suspecting that his father was assassinated in a plot involving the CIA.

Working for the U.S. Treasury

After I testified before the Presidential Commission on the Space Shuttle Challenger accident, I transferred to the U.S. Treasury Department, where I spent the next twenty-one years until retiring in 2007. I worked as a policy analyst for Treasury's Financial Management Service (FMS) that was responsible for processing most of the federal government's financial transactions, including payments to employees, contractors, Social Security recipients, vendors, retirees, and a multitude of other recipients. We managed the government's checkbook to the tune of several trillion dollars a year.

While such a mission might evoke yawns, you'd be surprised at how much trouble accountants can get themselves into, exacerbated by the breakneck pace by which the government was trying to automate all its financial systems in order to maximize the use of Electronic Funds Transfer (EFT).

One of our tasks was to assure that every night, after the close of business, the government's entire cash balance was transferred to the Federal Reserve, which would pay the government interest on the money, then use that money as reserves against which the banking system would lend. We were leaders in this practice which soon would be imitated by every corporation in the country.

The balances were called "repos," or "repossession agreements." By the end of the decade, the financial system had an unfathomable amount of money available to lend which engendered in turn an explosion of hedge funds and other types of derivatives trading. All due to the legalized practice of fractional reserve banking, now on steroids with EFT. The world economy—at least that of the U.S. and other Western nations—was being transformed into a gigantic gambling casino. In 2008–2009 it collapsed and likely will do so again.

My most interesting project concerned the "Cobell" project. Recall that the Dawes Act, when passed in 1887, broke up the Indian reservations into individual allotments to be held "in trust" by the federal government. Many of these allotments, typically 160 acres, were never occupied by the Indians, who were expected to use them to become farmers. But the government would often sell the Indians' land or lease it with mineral rights to white ranchers and miners at fire sale prices. Often the land was in arid, inhospitable areas, unsuitable for farming, so not attractive to the Indians as subsistence farms. Payment for this land went to the federal government, as the Indians whose reservations were being subdivided had no way to handle the funds.

The idea behind making the Indians subsistence farmers was to destroy "inferior" native cultures. Related to this was sending Indian children to white-run schools where they would be transformed into "good Americans."

By the 1920s and 1930s, the government realized that this policy had failed. The remaining reservation land gradually began to be returned to the tribes as tribal governments were formed under the auspices of the Bureau of Indian Affairs. The principle was established that the tribes held limited sovereignty, were not under the jurisdiction of state and local laws, and were entitled to manage their own affairs.

Meanwhile, the "Individual Indian Allotments" still held by the original recipients were passed down from one generation to another, with allotments often being divided among eligible heirs. So the allotments became "fractionated." In some cases, a single parcel might have over 100 joint owners. This fractionation made it impossible in most cases for the land to be used, sold, or administered. By the late 20th century an estimated 300,000 Indians were the owners of some amount of property, though it continued to be held "in trust" by the federal government.

The Indians finally gained redress for this longstanding historic disaster in 2009 with a $3.4 billion settlement from the federal government, following a thirteen-year journey through the federal court system in response to a class action suit filed by a heroic Indian woman named Elouise Cobell. She was a member of the Blackfeet Tribe of Montana and had been founder of the nation's first tribally owned commercial bank.

I became involved in the case after the Secretary of the Treasury and Secretary of the Interior had been cited for contempt of court by the U.S. District Court for the District of Columbia. Treasury had a fiduciary responsibility for the funds that should have been paid to the Indians for more than a century for leases on their land. Unfortunately, Treasury had no records of where this money might have gone. At meetings I attended with the Bureau of Indian Affairs in Albuquerque, New Mexico, we were told that there had once been records but that most of them had been lost in floods, eaten by rodents, or couldn't be found.

The Treasury and Interior Departments now gave up on ever being able to reconstruct what the Indians were actually owed for the use of their land. It became obvious that whatever compensation the Indians would receive would be an arbitrary figure based on whatever Mrs. Cobell and the other plaintiffs would agree to.

The government's settlement offer was negotiated by the Obama administration, with Congress approving the $3.4 billion figure. Of that amount, $1.4 billion would be divided among 263,500 claimants, amounting to $5,313 each. The remaining $2 billion was allocated to tribal governments to buy back fractionated allotments from entitled heirs. This land would be added to communal reservation land already owned by the tribes.

Mrs. Cobell passed away in 2011 as an honored elder of the Blackfeet nation. The University of Montana has an Elouise Cobell Land and Culture Institute. An excellent article on her life story may be found at "Elouise Cobell ('Yellow Bird Woman') 1945–2011" by Emma Rothberg.[36]

My trips to Albuquerque were always interesting. I met Native Americans who earned their living by working with Indian tribes on creating tribal constitutions and legal structures. I was also able to rent a car and drive through the beautiful but austere Southwestern landscape.

One time I drove west from Albuquerque into the chaparral to the ancient native pueblo of Acoma, said to be the oldest continuously occupied town in the U.S. On the way to the pueblo was a tribal casino on the main highway that had been built as a result of the 1988 Indian Gaming Regulatory Act.

Acoma Pueblo sat on a mesa that could only be reached by a bus that departed from a visitor center. The pueblo consisted of a grid of one- or two-story adobe houses and a church. Automobiles were not allowed.

A few miles from the current pueblo was a much higher mesa that was the former home of the tribe until it was attacked by the Spanish in 1598 during the Pueblo Revolt. After what was called the Acoma Massacre, the natives rebuilt their pueblo at today's site which survives to the present day. Acoma is another example of the resilience of Native American life.

Project for the New American Century

The Neocons began to make their move toward seizing control of U.S. foreign policy by forming PNAC—the Project for the New American Century (PNAC) in 1997. It was a lobbying group, a proponent of total American militarism and control, a virtual brand-name, a spearhead for U.S. seizure of the world's petroleum assets, a sponge for right-wing donations, the Wolfowitz Doctrine on steroids, and much more. First and foremost, PNAC wanted a renewed war against Iraq.

A large number of foreign policy "hawks" signed-on as the principal sponsors of PNAC, but the ones most in charge were the two intellectuals mentioned earlier: William Kristol, editor of the *Weekly Standard*, and Robert Kagan, former State Department official, leading Neocon theorist, and, today, *Washington Post* columnist and husband of Victoria Nuland. Nuland later was a principal U.S. government architect of the 2014 Maidan coup in Ukraine and leader of the current U.S. proxy war in Ukraine against Russia. She was also, allegedly, part of the planning team[37] that blew up the Nord Stream pipelines between Russia and Germany in 2022.

The initial PNAC report speaks of a universal and endless *Pax American* that "must have a secure foundation on unquestioned U.S. military preeminence." PNAC then prophesied that "the process of transformation is likely to be a long one, absent some catastrophic and catalyzing event—like a new Pearl Harbor."

PNAC also called for an expanded and much more sophisticated use of space weaponry and of biological weapons. In its 2000 report on *Rebuilding America's Defenses,* PNAC wrote that, "Advanced forms of biological warfare that can 'target' specific genotypes may transform biological warfare from the realm of terror to a politically useful tool."

It's worthwhile to do some serious reading about what exactly PNAC was advocating, especially as a guide to subsequent events, not only 9/11, but also the wars against Serbia, Afghanistan, Iraq, and Libya, the Patriot Act, creation of the Department of Homeland Security, the militarization of America's police forces, Obama's drone assassination program, the activities of the military's Joint Special Operations Command (JSOC), U.S. incursions into Syria under Trump, endless CIA/NED operations and color revolutions, growth of the CIA's drug business to include Afghanistan as well as the Golden Triangle and Latin America, creation of the Five Eyes network, implementation of total NSA surveillance systems, the ascendancy of the World Economic Forum, more out-of-control derivatives trading, and fostering of the COVID pandemic, up to and including Biden's proxy war against Russia now taking place in Ukraine, as well as U.S. threats against China.

All of this is "PNAC," even though the organization itself ceased operations in 2006. By then, PNAC's takeover of the U.S. government was so complete that an external organization was no longer needed.

The U.S. Supreme Court Names George W. Bush President

Before the Neocons could make their next move, they needed a president who would front for them. Not even Spielberg could have come up with a better actor than George W. Bush. A Yale Skull and Bones man like his father, he knew how to keep secrets. As former owner of the Texas Rangers baseball team, he knew how to entertain crowds. As managing general partner of the team, he enjoyed sitting in the stands and mingling with the fans. Just a "regular guy."

Bush II first strode the halls of power as two-term governor of Texas. For his first term, he won a close election against incumbent Democrat Ann Richards. He prevailed when his campaign spread rumors that she was a lesbian. Then he won his quest for reelection by a landslide by declaring June 10, 2000, to be "Jesus Day" in Texas. Seven weeks later, he was named the Republican presidential nominee. In his acceptance speech he said of the Clinton administration, "They've had their chance. They have not led. We will." That's inspiration!

But Bush still needed to win the 2000 presidential election against an attractive, brilliant, well-spoken, and squeaky-clean appearing competitor, Vice-President Al Gore. So this is what the Republicans did: When the election results came in, with Gore leading in the popular vote, but Florida teetering, they were able to send in a mob to stop a valid recount, and get the Republican Florida secretary of state to allow Bush, with aid from the U.S. Supreme Court, to cheat his way into an electoral victory. George W. Bush's brother Jeb Bush happened to be governor in Florida. You can read about it in greater detail in Stone and Kuznick's book.[38]

Florida's removal of 50,000 registered blacks from the voting rolls by what the state claimed was a purging of "convicted felons" had a decisive impact on the vote. Many people found their names removed who had never been convicted of a crime. A subsequent lawsuit by the NAACP and an investigation by the U.S. Civil Rights Commission found that at least 12,000 of the voters removed should have been allowed to vote, a number that easily would have swayed the election to Al Gore.[39]

So Bush and Cheney became president and vice-president respectively. Donald Rumsfeld was named Secretary of Defense. And that's all it took.

ENDNOTES FOR CHAPTER 15

1 From White House staffer Barbara Honegger: "I am the source for this quote, which was indeed said by CIA Director William Casey at an early February 1981 meeting of the newly elected President Reagan with his new cabinet secretaries to report to him on what they had learned about their agencies in the first couple of weeks of the administration." November 25, 2014. https://kundaliniandcelltowers.com/Did%20CIA%20Director%20William%20Casey%20really%20say%20We%20will%20know%20our%20disinformation%20program%20is%20complete%20when%20everything%20the%20American%20public%20believes%20is%20false-Quora.pdf>

2 Shades of *1984*.

3 Abby Martin, "Empire Files: The School that Trains Dictators and Death Squads," *The Real News Network,* December 5, 2015. https://therealnews.com/amartin1204ef13soa

4 Another non-profit committed to such subversion is George Soros's Open Society Foundation that often works closely with the NED.

5 Peter Dale Scott, "The Hidden Government Group," *WikiSpooks,* Nov. 15, 2015. https://wikispooks.com/wiki/Document:The_Hidden_Government_Group>

6 Ambrose and Brinkley, 339.

7 Tower died in a 1981 plane crash. The next day Senator John Heinz also died in a plane crash in Pennsylvania. Like Tower, Heinz had chaired a study group appointed by Reagan to examine Iran-Contra.

8 Ambrose and Brinkley, 340.

9 Jeremy Kuzmarov, "There is Absolutely No Reason in the World to Believe That Bill Clinton Is a CIA Asset—Except for All the Evidence," *CovertAction Magazine,* January 3, 2022.

10 Ryan Devereaux, "Managing a Nightmare: How the CIA Watched Over the Destruction of Gary Webb," *The Intercept,* September 25, 2014. https://theintercept.com/2014/09/25/managing-nightmare-cia-media-destruction-gary-webb/

11 Scott Ritter, *Disarmament in the Time of Perestroika: Arms Control and the End of the Soviet Union* (Clarity Press, 2022).

12 Arms Control Association, "The Intermediate-Range Nuclear Forces (INF) Treaty at a Glance," August 2019.

13 Leveraged buyouts are standard practice with equity firms. They purchase a company, strip it of assets, saddle it with the debt that was used to make the original purchase, then sell it and walk away. This type of legalized larceny is common. The Toys "R" Us chain is among the many companies destroyed in this way. East Coast fishing fleets have also been bought up and stripped by equity firms. Among the biggest predators is Goldman Sachs, which furnished many executives to the Clinton presidency. Such "vulture capitalism" is organized crime with Congressional complicity.

14 By 1996, with Clinton now in office, the U.S. foreign debt was over a trillion dollars, with the biggest creditor being Japan. By the end of 2021, under Biden, the debt stood at $7.4 trillion, with the biggest creditor now being China. The only thing preventing collapse is the petrodollar and its use by other nations as a reserve and trading currency, which translates into the unending necessity for military control of oil-producing nations in the Middle East.

15 These lies were manifested by the appointment of Robert Gates as CIA director. See Charley Reese, "Don't Confirm Gates for the CIA—Abolish the Agency Altogether," *Orlando Sentinel*, October 3, 1991.

16 U.S. Secretary of State Madeleine Albright told CBS News anchor Leslie Stahl that the death of 500,000 Iraqi children due to malnutrition and disease caused by U.S. sanctions "was worth it."

17 The official version may be found on *Wikipedia*. https://en.wikipedia.org/wiki/Wolfowitz_Doctrine

18 Stone and Kuznick, 483.

19 Ambrose and Brinkley, 399.

20 Clarity Press, 2023.

21 *Wikispooks*, Peter Dale Scott.

22 BBC, March 2, 2011. https://www.bbc.com/russian/multimedia/2011/05/110502_v_al_qaeda_training>

23 James Bovard, "30 Years Ago the FBI might have had its biggest bomb: The World Trade Center Attack," *New York Post*, February 26, 2023.

24 Ted Galen Carpenter, "Ignored Warnings: How NATO Expansion Led to the Current Ukraine Tragedy," The Cato Institute, February 24, 2022. https://www.cato.org/commentary/ignored-warnings-how-nato-expansion-led-current-ukraine-tragedy#

25 Ibid.

26 Ambrose and Brinkley, 425.

27 Ibid., 425.

28 Ibid., 403.

29 Ibid., 402.

30 Svetlana Savranskaya and Thomas Blanton, "The Washington/Camp David Summit 1990: From the Secret Soviet, American and German Files," The National Security Archive, June 13, 2010. https://nsarchive2.gwu.edu/NSAEBB/NSAEBB320/index.htm

31 Ambrose and Brinkley, 373.

32 Stone and Kuzinck, 485.

33 Ibid., 485.

34 Stephen F. Cohen, *Failed Crusade: America and the Tragedy of Post-Communist Russia,* (W.W. Norton, New York, 2001), 4–5.

35 This section draws on John Koerner, *Exploding the Truth: The JFK, Jr., Assassination* (Chronos Books, 2018).

36 Emma Rothberg, "Elouise Cobell ('Yellow Bird Woman') (1945–2011)," National Women's History Museum. https://www.womenshistory.org/education-resources/biographies/elouise-cobell-yellow-bird-woman>

37 As reported by Seymour Hersh.

38 Stone and Kuznick, 492–94. Also see reports on how famed neocon John Bolton was a key player in throwing the 2000 election to Bush. See John Nichols' *CBS News* column from April 13, 2005, "John Bolton vs. Democracy": "'I'm with the Bush-Cheney team, and I'm here to stop the count.' Those were the words John Bolton yelled as he burst into a Tallahassee library on Saturday, December 9, 2000, where local election workers were recounting ballots cast in Florida's disputed presidential race between George W. Bush and Al Gore. Bolton was one of the pack of lawyers for the Republican presidential ticket who repeatedly sought to shut down recounts of the ballots from Florida counties before those counts revealed that Gore had actually won the state's electoral votes and the presidency." A veteran of the Reagan and Bush I administrations, Bolton became a director of the Project for the New American Century. Under Bush II, he was named Under Secretary of State for Arms Control and International Security Affairs (2001–2005) and United States Ambassador to the United Nations (2005–2006). Under President Donald Trump, Bolton served for a year as national security advisor before being fired, though Bolton claimed he had resigned. He remains one of the biggest cheerleaders for the U.S. proxy war against Russia in Ukraine.

39 A Supreme Court worthy of the name would have demanded a re-run of the election in Florida.

CHAPTER 16

The Bush/Cheney Catastrophe

Who Is Running Things?

At the turn of the millennium on January 1, 2000, the "Rockefeller Republic" seemed to be history. Nelson Rockefeller had been dead for over twenty years. His younger brother David was eighty-four years old and would die in 2017. True, there were still rich bankers with their fingers in every pie, like Jamie Dimon, today head of JP Morgan Chase. But there is also the Rockefeller Foundation, run by family members and one of the most powerful behind-the-scenes institutions in the world. Besides its public persona of funding for science, health, and educational research, it has ownership stakes in companies like General Mills, Kellogg, Nestle, Bristol-Myers Squibb, Procter and Gamble, and Roche and Hoechst. In fact, between the Rockefeller Foundation and JP Morgan Chase, the Rockefeller empire controls half the U.S. pharmaceutical industry and much else besides.

But now a new class of oligarchs was appearing on the scene, enabled by the fabulous amounts of money unleashed on the world by financial deregulation, the growth of Electronic Funds Transfer, soaring consumer and government debt, and massive corruption. These were people who had made their billions not in banking, but in computers like Bill Gates; in e-commerce like Jeff Bezos; in social media like Jeff Zuckerberg; in stock trading like Warren Buffett and Michael Bloomberg; in currency speculation like George Soros; in industry like the Koch brothers; in venture capital like Marc Andreessen; in surveillance capitalism like Peter Thiel; in hedge funds, like Larry Fink; in entertainment, like Ted Turner and Oprah Winfrey; in real estate like Donald Trump; and in high-tech like Elon Musk. Plus scores of multimillionaires from the top of the pharmaceutical and armaments industries and wired-in politicians like the Bushes, Nancy Pelosi, Clinton, Obama, and Biden families. All wanted a piece of the action. How their "new money" related to the "old money" of the Rockefeller Republic is still being sorted out.

The question is whether there is a core group that intermingles all these power centers and that runs things and makes the big decisions.

But first, let's recap.

Alexander Hamilton and Thomas Jefferson were empire-builders, though the American imperial project could not have been carried forward without the victory of the North in the Civil War. But it was Cecil Rhodes and Nathaniel Rothschild who formulated and financed the plan to "recover America for the British Empire." This was accomplished through machinations occurring over two world wars. The key turning point had been the assassination of President William McKinley and his replacement by anglophile Theodore Roosevelt.

Eventually the U.S. took over the driver's seat, though London kept a role as a Eurodollar/money laundering center, while the British "Tavistock" media kept up its *1984*-style cheerleading for endless war by the Anglo-American military imperium. On Britain's role in shady financial dealings, *The New York Times* wrote:

> Taken together with its partly controlled territories overseas, Britain is instrumental in the worldwide concealment of cash and assets. It is, as a member of the ruling Conservative Party said last week, "the money laundering capital of the world." And the City of London, its gilded financial center, is at the system's core.[1]

So Britain has not disappeared, but back in the U.S., the drive for global military dominance was bolstered by lobbying from the Rockefeller-controlled Council on Foreign Relations at the start of World War II. The CFR was run by the New York financial elite that included a powerful element of Jewish/Zionist influence going back to the Rothschilds/Belmonts, and the Schiffs/Warburgs. The Eastern establishment banking families revolved around the Morgan/Rockefeller interests and new players like Goldman Sachs, another Jewish investment house. As noted in 2010 by no less a source than *The Atlantic*:

> Is it legitimate to think of Goldman as a Jewish firm? Messrs. Goldman and Sachs, who founded the firm in the nineteenth century, were Jewish, as have been most of its partners since then, almost all of its leaders, and its current CEO (Lloyd Blankfein).[2]

The Atlantic goes on to explain. "It was founded because Jews were excluded from other firms." Indeed, such a concentration reflects the history of Jewish banking since the Middle Ages, compounded by both Christian and Islamic prohibitions of the practice of usury. The world's banking system subsequently ballooned beyond domination by any specific players, though the Rothschilds still have seats atop the globalist banking world.

By the late-20th century, Fed Chairman "Maestro" Alan Greenspan's policies enabled U.S. bankers and investors to vacuum up wealth from other nations through control of the earth's money and resources. The CFR's concept of military power was meant to advance and protect this process. *All U.S. wars from the end of World War II to the Ukraine conflict today are enabled in this way and have global domination as their purpose.* London tags along, hiding behind the American shield.

Enter the debt trap. A set of international organizations was created to facilitate dollar hegemony—the IMF, the World Bank, and later the WTO, WHO, GATT, NAFTA, etc. But debt is the hidden time bomb. Even low central bank interest rates are ruinous for borrowers when compounded, and can lead, as they did in Greece, to the sell-off of public assets in repayment of debt contracted, not by the state, but by private entities the state is obliged to protect. Compound interest must always be fed by new sacrifices of treasure and blood in order to generate the escalating amount of money required solely to pay off debt. The result is that through becoming so bogged down in chasing after profits through lending and interest, *humanity has failed to create a financial system that fairly harnesses the incredible power of modern industry and technology* and provides its benefits to everyone.

There's more: The U.S. government fought the second world war with the objective of world conquest, though Roosevelt at times envisioned a United Nations that reflected multiple power centers. But Roosevelt died before the end of the war, so that execution of the post-war program was placed in the hands of the national security state set up under the neophyte Truman in thrall to the military, the CIA, the NSA, and, domestically, the FBI run by J. Edgar Hoover. Today we call this body of policy makers, analysts, and career soldiers the "Deep State."

The Deep State is allied with figures in Congress beholden to plutocrats and donors connected with the military industrial complex and with the American oligarchy.

The Deep State has run all the wars of the post-World War II period.

The Deep State is seamlessly joined with organized crime, which enforces the less savory side of the agenda, including the CIA's worldwide drug trafficking. The CIA may have destroyed more lives with drugs than all the American wars. Ethnic drug gangs around the world, now increasingly from Mexico, are part of this politically protected network.

During the post-World War II period, the ideology of world conquest received further promotion, justification, and definition from intellectuals like Henry Kissinger, Leo Strauss, Bernard Lewis, Zbigniew Brzezinski, Samuel Huntington, Francis Fukuyama, and many others. This is bolstered by the "think tank" world: the Rand Corporation, the Heritage Foundation, the Atlantic Council, and always, the CFR. The personnel merge with "NGOs" like the NED and Open Society and Ivy League universities. Yet another layer: The Neocon movement that rose out of the New York City communist/socialist *demimonde* was turned by the CIA and FBI into a powerful cadre of national security bureaucrats and lobbyists. Their initial focus was to gain control of the Middle East. The Neocon movement was closely allied with Israel and funded by the U.S. Jewish/Zionist lobby.

The Neocons are a "wild card" with respect to the original CFR plans. While now the Neocons are leading the charge against Russia via the Ukraine proxy war, it appears the Neocon/Zionist lobby is also involved in pushing for the "woke" movement to infiltrate the political policy spectrum and demoralize the general American population. The BLM, Antifa, and World Economic Forum are part of this, as are false flag biological threats like COVID. Entertainment venues like Netflix and social media platforms like Facebook, Twitter, Google, etc. deluge the population with propaganda, some of it wildly beyond common sense, such as the "Russiagate" narrative, sung for years in the mainstream media even after the Mueller and Durham[3] investigations disproved it. The Musk/Taibbi "Twitter Files" project has exposed some of the free speech suppression machinations; another of these now is *Wikipedia*, with full-time intelligence operatives massaging every article to assure compliance with the Deep State's narrative.

The power of the Deep State is projected and enhanced by its control of the mass media, as well as of the academic, medical (with Big Pharma among the most powerful), and educational captive communities.[4] *The New York Times, Washington Post, The Atlantic, The New Republic,* the broadcast networks, *CNN*, and many smaller outlets like the now-defunct *Weekly Standard* are all Deep State/Neocon/Zionist propaganda organs. With the firing of Tucker Carlson, even *Fox News* has caved in. *The New York Times* in particular has been a CIA mouthpiece for decades. Abroad, Britain's *The Economist*, run by the Rothschilds, is among the most influential outlets. The entire mass media of U.S. vassals like Britain and the EU are Deep State-controlled.

Behind everything else are the big tax-exempt foundations that provide ideological direction to the entire system. The Big Three have traditionally been the Rockefeller, Ford and Carnegie foundations. In recent years they have been joined, or surpassed, by influential newcomers, the two most prominent being the Bill and Melinda Gates Foundation and the Open Society Foundations. The foundations all possess endowments with massive portfolios in the financial markets where they dominate if not control Big Pharma, Big Oil, higher education, etc.

But can the oligarch controlled fractional reserve banking system swollen by money printing and compound interest indefinitely continue the exponentially increasing transfer of wealth to the richest members of society? Since the rich resist taxation and every tangible product has already been turned into a bubble, including the family homes of ordinary people, the only remaining source of funds is public debt. We have entered a period of catastrophic debt crises resulting in worldwide inflation that threatens to destroy everyone's wealth, rich and poor alike.

The financialized world system is imploding before our eyes. As Michael Hudson has so ably pointed out,[5] the parasites are destroying the host. The only nations that seem to be escaping are those the West has forced out of the system—especially Russia, China, and Iran—who are now working towards a multipolar "Fair World Order." This is why the West seeks to destroy them. The key to their capacity to resist the West's financial onslaught has been their *government-owned central banks* that control credit-creation and put it towards the benefit of society as a whole, rather than the global financial oligarchy.

Bush and Cheney

One of the challenges in writing this book has been to give adequate expression to the utter catastrophe visited upon our country and the world by the administration of President George W. Bush and Vice President Dick Cheney. The two have been joined at the hip by the general understanding that, contrary to the typical back seat role given to most vice presidents, it was Cheney who ran the show.[6] We are still living under the shadow of what these two infamous crime lords wrought.

According to Bush's first Secretary of the Treasury, Paul O'Neill, the number one topic from the start in Bush's National Security Council meetings was Iraq—the need to attack Iraq, the need to get rid of Saddam Hussein, how to administer Iraq after the conquest, and how to divide up Iraq's oil resources. Specific contractors were named to do the post-war work, and a map was produced of designated areas within Iraq for new oil exploration. Iraq possessed the world's second largest deposits of oil reserves in the world after Saudi Arabia.[7]

When President Saddam Hussein announced that Iraq was considering getting off the petrodollar as the sole means for trading in oil, this was anathema to the U.S. which feared a domino effect throughout the Middle East. It could influence Saudi Arabia to do the same. De-dollarization was also high on Iran's agenda.

Retention of the U.S. dollar as the world's reserve and oil trading currency was the chief driving force enabling U.S. foreign policy. It was the dominance of the dollar that generated the wealth the U.S. needed to police the world with its military machine. Forcing states to hold dollars in reserve in order to buy oil is key to preservation of the dollar's hegemonic status.

In his 1997 book, *The Grand Chessboard: American Primacy and Its Geostrategic Imperatives,"* Zbigniew Brzezinski wrote that military bases in Central Asia would be essential for "America primacy," due to the large oil reserves around the Caspian Sea. Afghanistan was the gateway to former Soviet regions deep within Central Asia.

With Britain's post World War II decline, the imperial project of Central Asian control was taken up by the U.S. The collapse of the Soviet Union in 1991 provided the opportunity. U.S. conquest of Afghanistan now seemed possible.

Preceding and possibly giving rise to the idea that would soon be promulgated by the Project for the New American Century, Brzezinski wrote in 1997 that the American public "supported America's engagement in World War II largely because of the shock effect of the Japanese attack on Pearl Harbor." Brzezinski too had suggested that Americans would support military operations in Central Asia "in the circumstance of a truly massive and widely perceived direct external threat."[8] Brzezinski also pointed to Ukraine as key to control of the Eurasian "heartland."

Now all of Moloch's prophets were pointing to a "New Pearl Harbor."[9]

9/11

On September 11, 2001, less than eight months after George W. Bush and Dick Cheney were inaugurated, the Twin Towers and WTC-7 of the World Trade Center in New York City were destroyed, along with a portion of the Pentagon Building outside Washington, DC.

The Bush administration swiftly claimed, without evidence, that the Twin Towers and the Pentagon were flown into by commercial jetliners piloted by Middle Eastern hijackers, and that a fourth hijacked jetliner crashed in a field near Shanksville, Pennsylvania. While it was claimed that 2,977 people died in the attacks—and that seems to be true, as those people have indeed vanished—the exact circumstances of how this took place remains in question.

At none of the four crash sites was there the kind of evidence of a commercial jetliner going down that normally would be seen—human remains, luggage, smoldering airplane parts, "black boxes," etc. While the Twin Towers disintegrated (along with Building 7, which had not been hit), the Shanksville crash site should have yielded some clues. But all the sites were secured by security personnel against impartial expert inspection. Independent journalists were barred from entry. Journalist Hunter S. Thompson may have been about to report these findings before he died by "suicide" on February 20, 2005.[10]

Due to the implausibility of so much of the official accounts—that a hijacker's passport turned up, etc., that cell phone calls were made from the hijacked planes, calls that the current technology would not have allowed—many questions arose, and many alternative narratives, such as that if planes did hit the Twin Towers, they could have been drones. At least one eyewitness reported seeing a military cargo plane striking one of the towers, not a passenger plane. At the Pentagon, comments by eyewitnesses and the building damage indicated a strike by a missile, not a jetliner.

Many first responders reported hearing successive blasts of explosives going off and feeling the vibration. The buildings collapsed at close to free fall, possible only if they had been "pulled." Many independent researchers state that the Twin Towers and WTC-7 were destroyed by implanted explosives, citing evidence of thermite amid the rubble. Another researcher built on that explanation of the destruction of the WTC buildings but argued:

> ...they used hydrogen thermobaric bombs...they planted hydrogen tanks in the elevator shafts, released the hydrogen, waited until it reached explosive air mixture and then set them off in sections down the building. While there may well have been some thermite or thermate used for critical beams, neither of those two burn fast enough to cause the explosions we see. If it had been massive thermite or thermate, we would have seen the sun-bright level burning when they set them off. Hydrogen burns with a pale blue flame and in broad daylight would be invisible. It would have been impossible to wire the twin towers with explosives to produce what we see, and we would have seen the explosion flashes. Building 7 was a conventional, controlled demolition. [The] Pentagon had explosives inside and was hit by a drone.[11]

Irrespective of how the destruction was accomplished, many researchers believe that 9/11 was an "inside job," a false flag carried out by elements within the federal government. The governments of Israel and Saudi Arabia have been named as possible co-conspirators.

A substantial 9/11 truth movement sprang up, which remains active to this day. The best consolidation of evidence concerning the 9/11 attacks is now being presented by a new academic organization run by 9/11 scholars, the International Center for 9/11 Justice,[12] reporting its developments on Twitter. Its website features three pillars of the 9/11 truth movement: the peer-reviewed *Journal of 9/11 Studies*;[13] the final report of the seven-year 9/11 Consensus Panel project, *9/11 Unmasked: An International Review Panel Investigation*;[14] and the 2012 *Report of the 9/11 Toronto Hearings*.[15] The latter anthologizes the writings of twenty-three independent researchers well-versed on 9/11 with a broad spectrum of expertise.[16]

We are still waiting for a definitive official account of the 9/11 conspiracy. The report of the government's 9/11 Commission was a joke, repudiated even by Commission members. A real report would be a massive undertaking, especially considering allegations that the attack on the Twin Towers was cover for a huge gold theft from the Fed storage vaults that had taken place earlier, that fortunes were made through stock market manipulations, that the Pentagon strike killed people investigating a multi-trillion dollar theft of government funds, and that the destruction of WTC-7 caused the loss of active documentation of financial crimes being investigated by government agencies.

Would our government lie to us to this extent? Consider the many incidents cited in this book and elsewhere that involve dishonesty and abuse of trust, from Custer's attack on the Sioux to get himself elected president, to the sinking of the Maine in Havana harbor that began the Spanish American War, to the German attack on the Lusitania that may have been allowed in order to get us into World War I, to President Franklin Roosevelt's actions in enticing the Japanese to attack Pearl Harbor, to President Lyndon Johnson's false claims with respect to the Gulf of Tonkin incident, to Nixon and Kissinger's scuttling of Vietnam peace talks so the war would go on for five more years, to PNAC's calling for a New Pearl Harbor, etc.

Would our government actually engage in activities that would knowingly harm American citizens in order to provide a *casus belli*?[17] What about Operation Northwoods? This proposal by the Joint Chiefs of Staff and CIA to carry out fake attacks on U.S. citizens and Cuban refugees, then blame those attacks on the Castro government, is not a "conspiracy theory." Operation Northwoods has been fully disclosed in U.S. government documents released to the public. It didn't happen because President Kennedy disapproved it. But it would have been of a scale similar to 9/11 and could have brought on a nuclear war with the Soviet Union.

So could our own government have lied to us? You tell me.

The night of the 9/11 attacks President George W. Bush told the nation:

> Today, our fellow citizens, our way of life, our very freedom came under attack in a series of deliberate and deadly terrorist acts.... Today our nation saw evil, the very worst of human nature.... The search is under way for those who are behind these evil acts. I've directed the full resources of our intelligence and law-enforcement communities to find those responsible, and to bring them to justice. We will make no distinction between the terrorists who committed these acts, and those who harbor them.[18]

Later on the night of 9/11, Bush wrote in his daily diary, "The Pearl Harbor of the 21st century took place today." Wonder where he got that idea? Meanwhile, the media assured us, Cheney was safe in an underground bunker in Washington and was directing the U.S. response.[19]

Only three days after 9/11, "Congress approved $40 billion to avenge the victims of Tuesday's terrorist attacks." Within *three days* the U.S. government's "War on Terror" was underway.[20]

And less than two months after 9/11, the massive Patriot Act was passed which few if any members of Congress could possibly have read. One of the most gung-ho was Senator Joe Biden. From *The Intercept:*

> Biden was a passionate promoter of the Patriot Act and repeatedly claimed that it was based on his proposals from the 1990s, including the Antiterrorism and Effective Death Penalty Act of 1996. He has bragged that he passed earlier surveillance legislation even though "civil libertarians were opposed to it." And Biden chided his colleagues for not supporting even further-reaching measures that he wanted at the time. "There were those who decided that the threat to Americans was apparently not serious enough to give the president all the changes in law he requested," Biden said.[21]

My Reaction to 9/11

When 9/11 happened, I had begun a program where the Treasury Department provided the equipment and phone link for me to work at home two days a week. Work-at-home had started being approved for selected senior staff members.

I was at our home in King George, Virginia, when the 9/11 attacks took place. I followed the news all day on *CNN*. When I returned to work in downtown Washington the next day, my colleagues told me they had heard the explosion at the Pentagon.

As claims began to be released of the hijackers and their plotting, I was increasingly skeptical. That Friday, when I heard President George W. Bush's remarks from the Memorial Service at the Washington National Cathedral, I saw that the tragedy was being used as a pretext for the immediate launching of a war—against somebody. Bush's speech at the Memorial Service was a reprise of what he said the night of the 9/11 attacks.

It didn't take long to see that 9/11 was a fraud, perhaps the biggest in history. My attitude to the Bush administration's story quickly became, "How stupid do you think we are?"

Ramping Up "Security"

Soon we would have a gigantic new Department of Homeland Security to reorganize the civilian side of government and its response to 9/11. Three former Treasury bureaus would be folded into DHS—the Customs Bureau, BATF, and the Secret Service. The massive Patriot Act had already been written and was ready to go. For some reason, immunity for Big Pharma against lawsuits alleging damage to children from vaccines would also be part of the Patriot Act. And the mass media was fully on board.

Soon metal detectors would appear at the entryways to every federal office building in downtown Washington. We would empty our pockets of keys and other items before walking through. Briefcases would be scanned. Visitors also had to be carefully screened, and emergency escape plans would be devised. New internal security programs would be implemented with a plethora of newly minted security specialists. Surveillance cameras were installed everywhere. We employees were being seen as potentially dangerous individuals. The message was: you never know who might be sneaking into the building with a bomb. The message became expansive: You never know where or when terrorists might attack. Maybe even Peoria.... "Rogue" elements armed with suitcase-sized nukes, we were told, were not out of the question.

At my own office at Treasury's Liberty Center across the street from the Tidal Basin and Jefferson Memorial, we conducted drills where the entire building would be evacuated. Of course, it had to be done quickly. The lights and elevators were turned off to simulate blackout conditions, so we had to take care not to tumble down the crowded stairways. Employees with canes or walkers would get an escort. Care was taken not to conduct the drills on rainy days.

Next, we would cross a side street and assemble on a soccer field while unmarked vans pulled up along the curb. As we watched, persons designated as "essential" would get into the vans and be driven away to a "secret site." The rest of us would wait around, crack a few jokes, then be told to return to work.

Inside the building, interior conference rooms were designated as sites to "shelter-in-place." We were each given a pack with a gas mask, a water bottle, a nutrition bar, and a flashlight in case we had to lay low for more than a few hours or stay overnight.

At the time, I was managing a contract that involved several contractor employees. I had to tell the lead contractor that I was sorry but budget constraints meant that her staff would not be getting an emergency pack. These were only for the career staff. The contractors would be on their own, locked out of the "shelter-in-place" rooms. We just laughed.

At one point I spoke to the Director of Security for the Financial Management Service. This was a person who had no security experience but had taken some training courses the government had conjured up on short notice. I suggested that we buy a fleet of motorboats, moor them at the nearby Washington Harbor marina, and keep them ready to zoom down the Potomac River to where I was living in King George if we found ourselves under attack. I said I would volunteer as the fleet captain and put people up for the night. She did not laugh.

I was also assigned to a Treasury Department committee to examine the government's financial systems for "vulnerabilities." Gven the growth of Electronic Funds Transfer, there were obviously a million ways that electronic systems could be hacked and money siphoned off. So we ended up declaring the entire U.S. financial system, public and private, as "critical infrastructure."

Everything was considered a potential terrorist target. And indeed, an actual crisis in protecting electronic systems emerged and is still going on, with the security problem never as yet solved and perhaps beyond solution. Vast amounts of money were paid to contractors and consultants to figure things out. Today financial cybersecurity is a massive industry. So is financial cybercrime.

Afghanistan

The first nation to taste Bush/Cheney "revenge" was Afghanistan, ruled by the Taliban. The U.S. and Britain invaded Afghanistan on October 7, 2001, twenty-six days after 9/11. Bear in mind that *nothing* on this scale can happen in twenty-six days without months or even years of pre-planning; it was just waiting for an appropriate pretext.

British Prime Minister Tony Blair was now among the most vociferous supporters of the "War on Terror." The British poodle was barking loud! The U.S., Britain, and several other "coalition" nations were joined in the attack by various Afghan factions defeated earlier by the Taliban in the civil wars that went on after the Soviet Union withdrew in 1989. The Taliban were a fundamentalist Islamic movement that originated among refugees from the civil wars who had gathered across the Pakistani border. The Taliban government called itself the "Emirate of Afghanistan." *They'd had nothing to do with 9/11.*

No matter. The U.S. demonized the Taliban, accusing them of carrying out atrocities against civilians, etc., but so was every faction engaged in combat from the 1970s until today, including the American occupiers and the national government they set up to rule as proxies. But part of the Afghan landscape was al Qaeda, left over from the U.S.-supported Mujahideen that fought as a guerrilla force against the Soviets. It was al Qaeda that the U.S. accused of carrying out 9/11, as well as the terrorist incidents of the 1990s. Their leader was claimed to be—who else?—Osama bin Laden.

No one has ever demonstrated any connection between al Qaeda and the Taliban. The former consisted of what seemed to be small bands of heavily armed guerrillas, recruited from outside the country. The Taliban were people native to the country whose ancestors had lived there for centuries. The Taliban offered to turn over bin Laden to a neutral third country—*if* the U.S. would provide evidence of their guilt. Needless to say, the Saudi passports that miraculously survived the disintegration of the Twin Towers that the U.S. put forward to the American public as incriminating evidence would not suffice. As David Ray Griffin wrote in 2011:

> The public's belief that there were al-Qaeda terrorists on the planes was bolstered by the claim that some of their passports had been found at crash sites. But were these reports believable? For example, the FBI claimed that, while searching the streets after the destruction of the World Trade Center, they discovered the passport of Satam al-Suqami, one of the (alleged) hijackers on American Flight 11, which had (reportedly) crashed into the North Tower.[22] For this to be true, the passport would have had to survive not only the fire ignited by the plane's jet fuel, but also the disintegration of the North Tower, which evidently pulverized

almost everything in the building into fine particles of dust. But this claim was too absurd to pass the giggle test: "[T]he idea that [this] passport had escaped from that inferno unsinged," remarked a British commentator, "would [test] the credulity of the staunchest supporter of the FBI's crackdown on terrorism."[23]

By 2004, the claim had been modified to say that "a passer-by picked it up and gave it to a NYPD detective shortly before the World Trade Center towers collapsed." So, rather than needing to survive the destruction of the North Tower, the passport merely needed to escape from al-Suqami's pocket or luggage, then from the plane's cabin, and then from the North Tower without being destroyed or even singed by the giant fireball that erupted when this building was struck. (In *Flat Earth News*, Nick Davies reported the opinion of some senior British sources that "the discovery of a terrorist's passport in the rubble of the Twin Towers in September 2001 had been 'a throwdown,' i.e. it was placed there by somebody official.")[24]

It took the U.S., British, and allied Afghan forces only two months to defeat the Taliban and set the stage for the UN Security Council to start setting up the Afghan Interim Administration under U.S. pal Hamid Karzai. Most of the al-Qaeda and Taliban fighters escaped to Pakistani border regions or to rural or mountainous areas in the southern provinces of Afghanistan.

From December 6–17, 2001, CIA and U.S./UK/German military forces engaged al Qaeda in the Battle of Tora Bora in a border region of fortified caves, supposedly to capture Osama bin Laden, an attempt which failed. Osama bin Laden may not have even been in the vicinity.

In 2002, Taliban leader Mullah Omar launched an insurgency against the Afghan provisional government and the U.S.-led International Security Assistance Force. The war went on for almost twenty years, with massive civilian casualties, an epidemic of targeted drone assassinations directed personally by President Barack Obama, a huge growth in CIA-managed heroin production originating in Afghan poppy fields, failure at any substantive efforts at "nation building," and the expenditure by the U.S. of $2.26 trillion, according to Brown University's Costs of War Project.[25]

In 2021, after the U.S. withdrew under orders from President Joe Biden, the Taliban succeeded in less than a month in re-establishing their control of Afghanistan. The 300,000-man Afghan army evaporated within days, with President Ashraf Ghani fleeing to Tajikistan, reportedly with a suitcase full of American cash.

The CIA reaped a bonanza with the cultivation of poppies and the subsequent manufacture of heroin in Afghanistan. This heroin exploded across

Asia into Russia as a tool for demoralizing Russian society. But this was only part of what Douglas Valentine says is the CIA's $300 billion a year that it brings in from its worldwide illicit drug trade. This money is greater than the government budgets of half the world's nations.[26]

It was an ignominious end to the centerpiece of the Bush/Cheney "War on Terror" and the longest-running conflict in U.S. history.[27] In the eyes of the world, it was a defeat of the U.S. on a scale of its defeat by Vietnam.

CIA Torture

With the attack on Afghanistan underway, the CIA began to establish an extensive network of "rendition" sites, i.e., places where they would hold and torture victims snatched from the streets of other countries, in disregard of their sovereignty. The types of torture to be used were approved in bizarre detail by Secretary of Defense Donald Rumsfeld. A committee meeting at the White House that included Cheyney, Rumsfeld, National Security Advisor Condoleezza Rice, Secretary of State Colin Powell, CIA Director George Tenet, and Attorney General John Ashcroft were said to be deciding *personally* on specific types of torture for specific captives.[28] It was like scenes from a horror movie.

Though President George W. Bush declared, per Donald Rumsfeld, that torture would not violate the Geneva Convention because captives from the "War on Terror" were "enemy combatants," not prisoners of war, what were arguably U.S. war crimes, which since Operation Phoenix had been the prerogative of the CIA, were now on the daily agenda of the country's top political leaders. General Barry McCaffrey said, "We tortured people unmercifully. We probably murdered dozens of them...both the armed forces and the CIA."

The torture program was disclosed publicly in 2004 with leaks of cruel and ghastly photos from the Abu Ghraib prison in Iraq. That Americans could perform such horrible acts on other human beings was image-shattering. The U.S. base at Guantanamo in Cuba became the main detention center for the torture program. President Barack Obama later promised to shut it down, but it is still in operation.

Unfortunately for its credibility, the U.S. has *never* been able to find anyone who could unequivocally be linked to the actual 9/11 attacks. Only two people have ever been subjected to a courtroom trial for alleged crimes related specifically to 9/11. One was arrested in Minneapolis and one in Germany. They were both convicted of having links to al Qaeda but not to any known 9/11-related activity. Meanwhile, hundreds of thousands or more of foreign nationals have been killed in battle or extrajudicially by

U.S. drones and by CIA/military death squads in Iraq and Afghanistan, or indiscriminately through large-scale aerial bombing attacks.

We are now awaiting the trial at Guantanamo of the alleged 9/11 "mastermind," Khalid Sheikh Mohammed. On February 11, 2008, the U.S. Defense Department charged Khalid Sheikh Mohammed and four alleged accomplices with planning the 9/11 attacks. Now, more than fourteen years and multiple government shifts in direction later, we are waiting for the five men to go on trial before a military court at Guantanamo in a specially constructed courtroom, where the press will be separated from the court proceedings by a thick plastic barrier. The barrier will allow the judge to press a button to censor what the approved-in-advance journalists are allowed to hear.

The government claims that Khalid Sheikh Mohammed has "confessed." It has also been established that *he was waterboarded 183 times*,[29] so that a question at trial is likely to be the admissibility as evidence of a confession extracted by repeated torture sessions over a decade ago. Naturally, some observers doubt such a trial will ever be completed. Maybe this is why the Biden administration is reported to be strenuously plea bargaining with the defendants.[30]

The Iraq War

Run-Up to Congressional Approval

As the attack on Afghanistan got moving, preparations to sell the planned war against Iraq began to gel, including a September 19–20, 2001, recommendation for war by Neocon stalwart Richard Perle's Defense Policy Board that reported to Secretary of Defense Donald Rumsfeld. The party line was the supposed possession by Saddam Hussein of weapons of mass destruction, particularly the nuclear kind.

The chief propagandist, making appearances on *Meet the Press* and other venues, was Vice President Dick Cheney. Later, Cheney made threats that the U.S. itself might use nuclear weapons to deter Saddam Hussein, even as Rumsfeld was asking the CIA to have *its* analysts certify Saddam's possession of WMD, which none ever did.

Well in advance of the war, however, the CIA's operations directorate was already at work recruiting Iraqi army officers with promises of Swiss bank accounts and high positions in a post-war administration if they would lie to U.S. media that Saddam Hussein was secretly training anti-American terrorists. This was reported in *The New York Times* in November 2001.[31]

It was clear that Iraq posed no military threat to the U.S., despite reference to a potential "mushroom cloud" by Secretary of State Condoleezza Rice. In order to secure congressional approval of a war resolution, the Bush/Cheney administration timed the vote just before the 2002 midterm elections. That way, Democrats opposing the war could be accused of being "soft on terrorism."

Nevertheless, the vote was far from unanimous. On October 2, 2002, the Senate passed the resolution by a vote of 77–23. The House voted 296-133 in favor. The resolution alleged that Iraq and al Qaeda were connected, but no evidence ever emerged to prove it. Voting for the Iraq war resolution were Democratic Senators Joe Biden, Hillary Clinton, and John Kerry, among others. Among those voting against were Bernie Sanders and Paul Wellstone.

Nine days after the resolution was passed, progressive Democratic Senator Paul Wellstone, running for reelection, died in a plane crash along with his wife, daughter, three campaign staffers, and the pilot and co-pilot. Suspicions of assassination have never gone away. Journalist Jackson Thoreau wrote in a December 30, 2003, report:

> Shortly before he died in a mysterious airplane crash eleven days prior to the 2002 elections, Minnesota Sen. Paul Wellstone met with Vice President Dick Cheney....
>
> At a meeting full of war veterans in Willmar, Minn., days before his death, Wellstone told attendees that Cheney told him, "If you vote against the war in Iraq, the Bush administration will do whatever is necessary to get you. There will be severe ramifications for you and the state of Minnesota"....[32]

It has also been alleged that the government launched what are now suspected as false flag anthrax attacks originating at Fort Detrick, Maryland, against senators who were not foursquare behind the Patriot Act in Fall 2001.[33]

The Invasion of Iraq

With Congress on board, the pressure for an invasion of Iraq ratcheted up, with lobbying groups like the American Israel Public Affairs Committee (AIPAC) chiming in to laud Bush's appointment of leading Neocon warhawk Paul Wolfowitz to once again serve as Deputy Secretary of Defense, the post he had held during the first Iraq War when he promulgated the Wolfowitz Doctrine. Now, President George W. Bush told Congress in his January 2003 State of the Union that Saddam Hussein had been buying weapons-grade uranium from Africa. It was later exposed as an utter lie.

On February 5, 2003, Secretary of State Colin Powell addressed the UN General Assembly for seventy-five minutes, even waving a vial of anthrax-like powder, as he claimed that Iraq was preparing biological and chemical weapons and that Saddam Hussein "remains determined to acquire nuclear weapons." More exposed lies.

When Powell testified before the Senate Foreign Relations Committee the next day, ranking Democratic Party member Joe Biden enthused, "I'd like to move the nomination of Secretary of State Powell for president of the United States."[34] Biden had become the Senate's leading proponent of the war.[35]

Meanwhile, Bush was proposing possible false flags to provoke an Iraqi attack, including "painting a U.S. surveillance plane in UN colors to draw Iraqi fire, producing a defector to publicly disclose Iraq's WMD, and assassinating Saddam."[36]

On March 19, 2003, the U.S. invaded, supported in its "shock and awe" campaign by troops from Britain, Australia, and Poland. The Iraqi forces were overwhelmed, the government collapsed, Saddam was captured and later executed, and the U.S. took over the country, ruling through its Coalition Provisional Authority. The Ba'ath Party was made illegal, and the entire Iraqi army was dissolved—providing combatants for the lengthy insurgency against U.S. occupation that followed. The war didn't stabilize until withdrawal of most U.S. forces by President Barack Obama in 2011. But by then the damage was done.

Iraq was devastated by the war with up to a million civilians killed. Total military casualties on both sides were well over 200,000. U.S. killed and wounded were 37,000.[37] The war also gave birth to ISIS—the Islamic State of Iraq and Syria—that launched a guerrilla insurgency of Islamic fundamentalists that would transform into mercenaries paid by the CIA to fight against the elected Syrian government and now is said to be supplying fighters to support Ukraine in the U.S. proxy war against Russia.

When all was said and done, Iraq became a colony of the U.S., with the largest embassy in the world built within Baghdad's Green Zone. There are still combat troops in Iraq today, though their mission has been changed to "combat ISIS."

Ken Adelman, President Reagan's director of the Arms Control and Disarmament Agency, considered a leading Neocon, had told the *Washington Post* on February 13, 2002, "I believe demolishing Hussein's military power and liberating Iraq would be a cakewalk." His words were repeated by many in the Bush/Cheney administration. He was wrong.

Douglas Valentine in his book *The CIA as Organized Crime* describes the reign of terror it took for the U.S. to exert control over Iraq. Part of the strategy was to get U.S.-controlled Iraqi police and paramilitary squads to do the dirty work, a strategy identical to Operation Phoenix in Vietnam:

> After the CIA death squads eliminated the senior leadership of the Iraqi government in 2003, they targeted "mid-level" Ba'ath Party members—a large portion of Iraq's middle class...a needless rampage.... All of it covered up. Not one victim featured on TV. All you'll ever see is ISIS beheading people....

But that was just the beginning. Valentine also writes of *New York Times* reporter Judith Miller, whose reporting was "a major pretext for the war on Iraq":

> The *Times* never explains that every unit in the [Iraqi Army's] Special Commandos has a CIA adviser handing out hit lists to its counterpart American "Special Police Transition Team." Up to forty-five U.S. Special Forces soldiers work with each Iraqi unit. These teams are in round-the-clock communication with their CIA bosses via the Special Police Command Center. There is no record of the Special Police or Special Commandos ever conducting operations without U.S. supervision, even as they massacred tens of thousands of people.[38]

As many had warned, a beneficiary of the war, apart from the U.S. firms sucking billions of dollars out of the Iraqi economy, was Iran, now the strongest nation in the region, directly competing for power and influence with Israel and Saudi Arabia. Iran has also made inroads within Iraq itself through its Shi'ite allies. Due to Shi'ite influence, "Iraq's parliament has passed a law that makes it a crime to normalize ties with Israel, and violations of the law can be punishable with a death sentence or life imprisonment."[39]

In fact, an assault on Iran itself has always been a part of the "War on Terror." General Wesley Clark confirmed the plan for a wider war when he saw a document in the Pentagon in November 2001 that listed six nations targeted for takedown: Iraq, Libya, Somalia, Sudan, Syria, Iran. The slogan was being bandied about that, "We are on our way to Baghdad, but real men are going to Tehran."

The cost of the Iraq War to the U.S. eventually reached at least $3 trillion in addition to the $2.6 trillion for Afghanistan mentioned previously.[40]

The U.S. also used huge amounts of depleted uranium munitions that have polluted the Iraqi landscape to this day. More than 300,000 depleted uranium rounds were fired during the war.[41]

"Color Revolutions" and Renewed Conflict with Russia

Under Clinton and Bush/Cheney, the U.S. launched a series of covert conflicts to take advantage of Russia's economic and political collapse of the 1990s. The long-range goal was dismemberment of Russia, first by turning the former Soviet republics on the Russian periphery into weak pro-American remnants, then dividing Russia itself into component sub-states. The larger goal is to commandeer Eurasian resources.

The fly in the ointment was that Russia was still in possession of nuclear weapons. Despite Russia's travails, it never relinquished command of its nuclear arsenal. So the U.S. had to tread carefully by chipping away at Russian sovereignty and territory without inciting nuclear war.

U.S. actions began with NATO/U.S. air strikes to assure the breakup of Yugoslavia, leading to the so-called Yugoslav Bulldozer Revolution of 2000, and covert aid to Islamic jihadists in the two Chechen wars aiming at secession that Russia crushed decisively. There followed the "Color Revolutions" in nearby states: Georgia's Rose Revolution (2003), Ukraine's Orange Revolution (2004), and Kyrgyzstan's Tulip Revolution (2005).

We can't say that Russian President Vladimir Putin didn't give the U.S. a chance. After 9/11, Putin contacted Bush and offered to help track down the alleged terrorists, including a pledge to open Russian air space to U.S. military flights.

In 2001, Bush took the U.S. out of the 1972 ABM Treaty as part of a move to restart Reagan's Star Wars space weaponry project. This action was a milestone. Going back decades, limitation of nuclear or even conventional weapons had always been more important to the Soviet Union than to the U.S. The proximity to Russia of U.S. weapons staged in Europe was an immediate and visible threat to Russian security. Reduction in those weapons would automatically reduce tensions.

In March 2004, Bush took steps to further expand NATO by the admission of Bulgaria, Romania, Slovakia, and Slovenia, plus the three former Soviet republics of Lithuania, Latvia, and Estonia. In 2008, Croatia and Albania also joined. Georgia and Ukraine were next on the list. Russia itself had once expressed interest in joining NATO, but that had been rebuffed. But with Georgia and Ukraine now on the table, Russia was outraged.

The U.S. moved to threaten Russia with a nuclear first strike through the 2002 Nuclear Posture Review that asserted the right of the U.S. to use nuclear weapons (a) if WMDs were used against it, (b) to penetrate hardened or underground targets not reachable with conventional weapons, or (c) if the U.S. was faced with "surprising military developments."[42]

Even more provocative was a March-April 2006 article in *Foreign Affairs*, the mouthpiece of the Council on Foreign Relations, by scholars Keir Lieber and Darryl Press, asserting that the decline of Russia's nuclear arsenal and the weakness of China's would make it impossible for either nation to retaliate against an American first strike.[43]

Putin finally had enough. In 2005, he said in a speech that the breakup of the Soviet Union, with twenty-five million Russians now living beyond the borders of the Russian Federation, was "the greatest geopolitical disaster of the twentieth century." At the 2007 Munich Security Conference, Putin shocked U.S. attendees John McCain and John Kerry among others, by speaking out against the unipolar world view that the U.S. was trying to enforce.

In February 2009, just days after Obama's inauguration, U.S. Ambassador to Russia William Burns wrote a secret memo, later outed in *Wikileaks*, entitled "Nyet Means Nyet." The memo underlined a Russian warning to U.S. decision makers that if the U.S. continued to pursue inclusion of Ukraine in NATO, Russia would have no choice but to take military action and that Ukraine would lose, at a minimum, Crimea and Donbass.[44] Bush/Cheney had left the Obama administration with the worst relations with Russia since the height of the Cold War and a ticking time bomb in Ukraine.

After Vladimir Putin's ascent to the Russian presidency in 2000, Russia decided to modernize and expand its nuclear weapons arsenal, along with developing high-powered conventional weapons and delivery systems, in order to be equipped to fight what appeared to be a coming full-scale war against it by the U.S. Thus, by the early 2000s, Russia began to resume the status of a nuclear superpower that was lost when the Soviet Union collapsed.

Economic warfare also became part of the equation. Apart from the overt wars of the Bush/Cheney years—Afghanistan and Iraq—the covert ones were just as important. By now, economic sanctions had become one of the U.S.'s most practiced tools, being levied by successive U.S. administrations against Iran, Cuba, Nicaragua, and Venezuela. And now, they were levied against Russia.

The U.S. Leads the World Over a Financial Cliff

It was under President George W. Bush and Vice-President Dick Cheney, along with Federal Reserve Chairman Alan Greenspan, that the U.S. led the world over the cliff of global financial collapse through the abuses, bubbles, and fraudulent escapades of 2001–2009. The fiasco culminated in the Great Recession and gigantic bailouts perpetrated by President Barack Obama. By rescuing the banks "too big to fail" that are still perched at the top of the Western financial system, he enabled the continuance of the earlier noxious banking practices and a financial regime that is imploding through unstoppable inflation and debt overhang.

In response, countries like Russia, China, and Iran are working together to build their own sustainable financial system to replace the one overseen by the U.S. that appears to be headed toward oblivion.

Big Bankers and Financial Disaster

Nomi Prins explains in *All the President's Bankers* that the linkages between government and the big bankers are stronger in the U.S. than in any other country. The leverage the bankers exert over government encompasses the military and intelligence establishments, which both view as central to their mission of total control by U.S. banking of the economies of the world. Plus, as I saw at Treasury, the U.S. financial system is *critical infrastructure, to be protected at all costs*.

This legacy is due in no small part to the roles played by the Rockefellers in their influence over government either through holding high official positions as did Nelson Rockefeller, from funding of influential organizations like the Council on Foreign Relations, to presiding over institutions with a quasi-official policy role like David Rockefeller's Trilateral Commission. This was what I called the "Rockefeller Republic."

We have seen bankers' influence at play throughout this book, but by the end of the Clinton administration at the turn of the millennium in 2000, this influence was off the charts. After decades of consolidation and mergers, the "Big Six" were now Rockefeller flagship JPMorgan Chase, Citigroup, the investment banks of Morgan Stanley and Goldman Sachs, plus two relative newcomers: Bank of America and Wells Fargo. Considered "too big to fail," not only did they dominate worldwide investing, but they were safe havens for money coming from abroad. This made them major participants, with London, in the realm of international money laundering.

There was a continuing need for this inflow of foreign assets to feed the always-growing pyramids of debt on which the banks were perched. The elimination of the distinction between investment and commercial banks with the repeal of Glass-Steagall and the proliferation of derivatives, supposedly to reduce the risks of shaky lending, only made matters worse.

According to Nomi Prins, during the sixteen years of the Bush/Obama presidencies, Bush and Obama were "followers and reactors to bankers' whims." The power of the bankers "dwarfed central banks and presidents."[45] According to Warren Buffet, derivatives in particular were "weapons of financial mass destruction."

By 2012, the Big Six held $9.5 trillion in assets, an amount equivalent to sixty-five percent of U.S. GDP. But the system was becoming increasingly insolvent, because the massive amounts of money the banks lent was into a world economy which, after the bursting of the Clinton dot.com bubble, had ceased to grow. Because the Federal Reserve had gradually allowed the bankers to reduce the amounts of money held in their reserves, the system increasingly floated in mid-air. The only backing was "the faith and credit of the United States."

The banks rolled over their loans as long as they could, earning commissions and fees with each rollover, but when borrowers went broke, nothing more could be done. It took over *$19 trillion* in Federal Reserve and U.S. government subsidies for the system just to survive the catastrophic failure of the crash of 2008–2009.[46] This time bomb of debt remains embedded in the world's financial system.

Bush/Cheney, away fighting their "War on Terror," didn't seem to have noticed. They left the financial details to a cadre of managers from the world of Wall Street finance led by Larry Summers—at first, chairman of Goldman Sachs, then Secretary of the Treasury.[47] Running interference were Federal Reserve Chairmen Alan Greenspan and Ben Bernanke and president of the New York Federal Reserve Timothy Geithner, later Obama's Treasury Secretary.

How Did the 2008–2009 Crash Happen?

Weaknesses had begun to appear by the end of 2000, with Clinton leaving office and Bush waiting to step in. JPMorgan Chase was continuing its decades-long crusade to buy up any other banking enterprise available, but now it began laying off thousands of employees—not a good sign. Its stock suddenly plummeted from $207 to $157 a share. A decade later, it stood at $40.[48]

Bad news kept coming. "Corporate bankruptcies struck new records in 2001 and again in 2002."[49] With Clinton's legacy dot.com bubble now bursting, the biggest firm to collapse was telecom giant WorldCom, whose stock went from $64.50 to $0.09 a share. Many other telecom companies went under, with the total stock market loss hitting $2 trillion.

When half a million jobs disappearing, Fed Chairman Alan Greenspan cut interest rates to a record low of 1.5 percent. With heavy implications for the future, one of the dot.coms that survived and eventually flourished was Amazon. Another was Google.

Still, no one at the top paid much attention. The day after Bush invaded Iraq on March 20, 2003, Congress approved Bush's $2.2 trillion federal budget, including $726 billion in tax cuts. Bush was now as much a hero in the upper-class pantheon as Reagan was in his day. This was a script the Republicans had down pat. Later, Donald Trump would also read from it. But at the time, under Bush, it was payday again for the country club set—even though SEC Chairman Harvey Goldschmid was warning, "If anything goes wrong, it's going to be an awful big mess."[50]

Nomi Prins writes: "Now Wall Street began minting toxic securities lined with subprime loans and wrapped up in derivatives."[51] The biggest market for subprime loans was for home mortgages. These included millions of "liars' loans" based on information provided by mortgage applicants known by the bankers to have overstated or falsified their income, expenses, credit history, etc.

Why would so many individual home buyers participate in this essentially criminal practice? When I was writing articles on the subprime mess, I interviewed a mortgage broker from Washington State who told me she was *ordered* by the banks providing the funding to deliberately falsify mortgage applications to ensure acceptance. We agreed that this could not have been done without the knowledge of federal regulators. "Tranches" of these toxic securities would then be bundled and sold in European markets, where the purchasers had no way of knowing it was a scam.

Surely the regulators must have known that this would give rise to a housing bubble. Commercial real estate joined the party created by Greenspan's interest rate cuts, as the ease of obtaining liars' loans continued to push up housing prices.

Greenspan encouraged the banks to focus their mortgage lending on Adjustable Rate Mortgages at a time when he must have known that Federal Reserve increases in interest rates would then force many buyers of subprime loans into foreclosure. When that happened, the original mortgage note had been sold, resold, and bundled so many times, the original paper documents were lost. This often made foreclosures unenforceable, so homeowners felt comfortable simply dropping the keys to their house at the bank and walking away.

Still, denial was the norm. Timothy Geithner of the New York Fed said there was "no such thing" as a housing bubble, while by the time Ben Bernanke replaced Greenspan as Fed Chairman in 2006, the damage had been done. There were more than 1.2 million home foreclosures in 2006, up more than 42 percent from 200. Now in office as Secretary of the Treasury, Larry Summers said the U.S. economy was "very healthy" and "robust."[52] He kept up the façade until investment giant Bear Stearns collapsed in June 2007, with investors losing $18 billion almost overnight. With the Citigroup mortgage securities business also going under, CEO Charles Prince resigned with a $99 million severance package. Even with bad news, the bankers were taking care of their own.

On October 9, 2007, the Dow closed at 14,164, an all-time high, but within eighteen months it would lose over half its value. When on December 21, Bear Stearns posted another loss of $1.9 billion, the unprecedented took place. JPMorgan Chase, now operating under the "King" of Wall Street, Jamie Dimon, got a loan of $29 billion to buy Bear Stearns from—the Fed! The Fed was now an investment bank, "printing money" out of thin air enabling a major bank to purchase a competitor! Next to go bankrupt was another venerable institution, Lehman Brothers. No one wanted to buy it, so it just wrote off its investors' losses and disappeared down the memory hole.

But insurance behemoth AIG was another story—talk about too big to fail. AIG had already exhausted its bank credit following collapse of its investment portfolio. So the New York Fed authorized a loan of $85 billion in return for a 79.9 percent equity interest. On October 8 it provided an additional $37.8 billion for a total in subsidies of $182 billion. And who got the money? Goldman Sachs got $12.9 billion, Merrill Lynch $6.8 billion, Bank of America $5.2 billion, Citigroup $2.3 billion, Société Générale and Deutsche Bank $12 billion each, Barclay's of England $8.5 billion and UBS $5 billion.

We now see the Federal Reserve, on its own, "printing money" enabling the purchase of entire companies, with money created totally out of thin air. The Bernanke-era public term for this practice would become "quantitative easing," not that anyone ever understood what that meant—though Bernanke gave a clue when he referred to "helicopter" money in a speech he made in 2002, declaring that "a money-financed tax cut is essentially equivalent to Milton Friedman's famous 'helicopter drop' of money." The nickname "'Helicopter' Bernanke" stuck as the unprecedented period of bank bailouts loomed.[53]

But what all this meant was the ultimate "bubble-ization" of the U.S. economy, meaning that the powers-that-be who run the financial system can create any amount of money they want and do with it whatever they choose—war, peace, larceny, you name it. It's all done through the Federal Reserve Bank of New York, where open market operations are conducted. The Federal Reserve Bank of New York is *owned* by the Wall Street bankers, not the U.S. government.

J.P. Morgan's Money Trust had now arrived at its ultimate destination, as on September 29, 2008, the Dow Jones Industrial Average dropped 778 points. The crash had hit the stock market. After a visit from Secretary of the Treasury Henry Paulson, where, as reported by *The Guardian*,[54] he jokingly dropped to one knee in supplication, Nancy Pelosi and the congressional Democrats approved a huge bank bailout package of $700 billion by the name of TARP—the Troubled Asset Relief Program.

TARP was part of the Emerging Economic Stability Act of 2008 signed by President George W. Bush on October 3. This worked out to be only three percent of the eventual bank bailout and subsidy program.[55] With TARP in place, the Dow shot up 936 points, the largest one-day increase in history. But soon the markets went south again and the Great Recession of 2008 was on.

The recession spread worldwide. With the economic decline, the banks hoarded their bailout money, tightened their lending criteria, and decided to ride it out. Consumers under distress stopped borrowing. Housing prices plummeted. Those able to borrow could take advantage of the December 16, 2008, Federal Reserve rate cut to 0%. This had never happened in U.S. history. Savers were devastated, including many elderly on fixed incomes who yet had money in the bank. Their interest earnings ceased entirely.

In 2009 there were five million home foreclosures in the U.S. Actual unemployment reached seventeen percent. Overseas investors lost billions in the now worthless mortgage tranches purchased from U.S. banks. But not a single banker went to jail. *JPMorgan Chase CEO Jamie Dimon was handed a $9 million dollar bonus for 2009 for keeping his bank solvent.*

Overseas, unemployment reached fifty percent in some places as "country after country, from Greece to Spain to Ireland, struggled under immense debt and crippled economies." Some southern European countries have yet to recover. In the U.S., the Occupy Wall Street movement sprang up in 2011, with protests around the country. The movement was suppressed by police action with over 8,000 people arrested in New York's Zuccotti Park.

The Fed's multi-year bailout program showed the seemingly endless possibilities of "money printing" and eventually also facilitated the trillions of dollars paid to individual taxpayers and small businesses during the COVID pandemic. But the end result eventually was gargantuan debt and galloping inflation.

George W. Bush left office with the lowest popularity ratings of any president in history and deservedly so. By the end of 2008, his approval had dropped to 22 percent. Cheney's stood at 13 percent. Never had a president and vice-president been viewed with such loathing.

Obama Arrives

The Bush/Cheney catastrophe had the effect of discrediting the now-aging generation of Neocons. Some joined "think tanks" and academic posts, some retired, some seemed to go into hiding, but some found their way into Obama's incoming Democratic administration.

Who could have thought that a man of mixed race, which in the U.S. made him black—one with scant political experience—would be elected as the next president of the U.S.? But seemingly all it took for an unknown out of Chicago named Barack Obama to gain victory over Senator John McCain in 2008 was to utter two words: "Hope" and "Change." But like 9/11, these words were also frauds. Of course, McCain had nothing to offer but a discredited status quo and a big ego. As far as anyone could see at the time, especially after the John McCain/Sarah Palin clown show in the 2008 elections, where even Bush and the banks abandoned the Republican Party, that party had been turned into roadkill.

Under Obama, the Democrats became the bankers' party, the war party, and the party of the Deep State and has remained so ever since, despite the Donald Trump speed bump. Obama's largest contributor in the 2008 election was Goldman Sachs, among the world's largest financial predators—though probably surpassed today by BlackRock and Vanguard.

ENDNOTES FOR CHAPTER 16

1 Nicholas Shaxson, "The City of London is Hiding the World's Stolen Money," *The New York Times*, October 11, 2021. Also see Martin Sandbu, "How London and the U.S. became safe havens for dirty money," *Financial Times*, March 3, 2022.

2 Michael Kinsley, "How to Think About: Jewish Bankers," *The Atlantic*, January 29, 2010. https://www.theatlantic.com/national/archive/2010/01/how-to-think-about-jewish-bankers/346827/

3 Kimberley Strassel, "Inside the Clinton dossier and the con behind the Russiagate scandal," *New York Post*, November 5, 2021. https://nypost.com/2021/11/05/inside-the-clinton-dossier-and-the-con-behind-the-russiagate-scandal/

4 See Joan Roelofs, *The Trillion Dollar Silencer: Why Is There So Little Anti-War Protest in the United States?* (Clarity Press, 2022).

5 Michael Hudson, *Killing the Host: How Financial Parasites and Debt Destroy the Global Economy* (ISLET, 2015). https://michael-hudson.com/2015/09/killing-the-host-the-book/

6 See David Griffin's impressive work, *Bush and Cheney: How They Ruined America and the World*, Olive Branch Press, 2016. Also see Oliver Stone's and Peter Kuznick's *The Untold History of the United States*. Chapter Thirteen: "The Bush-Cheney Debacle: 'The Gates of Hell Are Open in Iraq'" offers a fifty-page chronicle of the Bush/Cheney era that is a "must-read."

7 *Before Scott McClellan, there was Paul O'Neill (1 of 2)*, erkd1, video, 6:25, June 24, 2008. http://www.youtube.com/watch?v=gpaq3Vr95oU

8 Ibid.

9 President Bush's future national security adviser, Jake Sullivan, repeated this trope in 2020, anticipating the U.S. proxy war against Russia in Ukraine. They just can't seem to get Pearl Harbor out of their heads.

10 Jeremy Kuzmarov, "Did Legendary 'Gonzo' Journalist Hunter Thompson Frighten Those in the Deep State So Much He Had to be Taken Out?", *CovertAction Magazine*, February 25, 2023.

11 Email to author. Also see video at: https://rumble.com/vlw7f3-911-lies-disproved-in-5-minutes.html [link broken]

12 International Center for 9/11 Justice. https://ic911.org/

13 *The Journal of 9/11 Studies*, International Center for 9/11 Justice. https://ic911.org/journal/>

14 David Ray Griffin and Elizabeth Woodworth, *9/11 Unmasked: An International Review Panel Investigation* (Olive Branch Press, 2018).

15 James R. Gourley, *The 9/11 Toronto Report: International Hearings on the Events of September 11, 2001* (independently published, 2012).

16 Twenty-three people with varying professional backgrounds came together to apply disciplined analysis to the verifiable evidence about the 9/11 attacks. This panel included people from the fields of physics, chemistry, structural engineering, aeronautical engineering, piloting, airplane crash investigation, medicine, journalism, psychology, and religion. Consensus 9/11 also has seven honorary members, including: Ferdinand Imposimato, the Honorary President of the Italian Supreme Court; the late biologist Lynn Margulis; and the late Hon. Michael Meacher, the longest-sitting member of the British House of Commons.

17 See also Rick Anderon, *Home Front: The Government's War on Soldiers* (Clarity Press, 2004).

18 "All You Need to Know About 9/11 to Prove It Was An Inside Job," *Global Research*, September 23, 2011, on *WikiSpooks.com*.

19 Ibid.

20 Ibid.

21 "2001: September 11 and the Patriot Act: Joe Biden did not just proudly vote for the Patriot Act, he took credit for many of its provisions," *The Intercept*, April 27, 2021.

22 "Ashcroft Says More Attacks May Be Planned," *CNN*, September 18, 2001. http://edition.cnn.com/2001/U.S./09/17/inv.investigation.terrorism/index.html; "Terrorist Hunt: Suspects ID'd; Rescue Efforts Go On; White House Originally Targeted" *ABC News*, September 12, 2001. http://911research.wtc7.net/cache/disinfo/

deceptions/abc_hunt.html.

23 Statement by Susan Ginsburg, senior counsel to the 9/11 Commission, at the 9/11 Commission Hearing, January 26, 2004. http://www.9-11commission.gov/archive/hearing7/9-11Commission_Hearing_2004-01-26.htm. The Commission's account reflected a CBS report that the passport had been found "minutes after" the attack, which was stated by the Associated Press, January 27, 2003.

24 Nick Davies, *Flat Earth News: An Award-Winning Reporter Exposes Falsehood, Distortion, and Propaganda in the Global Media* (Random House UK, 2009), 248.

25 Christopher Helman and Hank Tucker, "The War in Afghanistan Cost America $300 Million Per Day for 20 Years, With More to Come," *Forbes,* August 16, 2021.

26 Valentine, 214.

27 Today Bush is an instructor in "Authentic Leadership" for MasterClass.com.

28 Stone and Kuznick, 510–11.

29 Ibid., 510.

30 "Guantánamo prosecutors are exploring plea deals in 9/11 case after years of setbacks," *National Public Radio,* March 21, 2022.

31 Valentine, 123.

32 Jackson Thoreau, "Cheney Threat to Wellstone Over Iraq Vote Before Crash" [email], December 30, 2003. https://mn.politics.narkive.com/yvnp6kJw/cheney-s-threat-to-wellstone#post1

33 See Graeme MacQueen, *The 2001 Anthrax Deception: The Case for a Domestic Conspiracy* (Clarity Press, 2019). Also Francis A. Boyle, *Biowarfare and Terrorism* (Clarity Press, 2005).

34 Karen DeYoung, *Soldier: The Life of Colin Powell* (Alfred A. Knopf, New York, 2006), 450–51.

35 Scott Ritter writes: "Joe Biden is one of the main reasons why the United States went to war in Iraq. He likes to pretend that he didn't play such an important role, but I'll say it now and I'll say it forever: Joe Biden's a liar.... Joe Biden's a man who allowed thousands of Americans to be sacrificed for his pride, for his arrogance, for his narcissism, because he knew that one day he was going to be President of the United States." Simes Diitri, "Scott Ritter: Biden Led the Charge to Invade Iraq," *Sputnik News,* March 19, 2023.

36 Stone and Kutznick, 521.

37 Despite Tony Blair's cheerleading, only 179 British soldiers died. The poodle had lots of bark but little bite. This should not detract, however, from our appreciation of the profound influence of the British on U.S. foreign policy or of agencies like MI6 on covert action. According to researcher Matthew Ehret, this influence may be decisive. See Matthew Ehret, "The British Imperial Hand Behind Russiagate and Global Governance Exposed Again," *Matt Ehret's Insights,* June 13, 2023. https://substack.com/inbox/post/127846046. Ehret also documents the influence of the British ideology deriving from Cecil Rhodes and the British Roundtable on U.S. foreign policy during the Clinton, Obama, and Biden administrations. This dovetails with the anti-nationalist bent of the World Economic Forum and its interest in eugenics, pandemics, and green energy as preparing the ground for its Great Reset. See Matthew Ehret, "The Rhodes Scholars Guiding Biden's Presidency," *Unlimited Hangout,* March 25, 2022. https://unlimitedhangout.com/2022/03/investigative-

reports/the-rhodes-scholars-guiding-bidens-presidency/>

38 Valentine, 150–51.

39 Ibid., 113. As though to confirm the complete failure of the U.S. intention to absorb Iraq into its Middle East security structure, in March 2023 Iraq signed a mutual security agreement with Iran against the U.S. and its terrorist proxies in the region. Iraqi Prime Minister Mohammed Shia' al-Sudani said that the two countries were now "united as one." *IANS World News,* March 20, 2023.

40 Ray McGovern made his statement on *The Duran.* The cost figures for Afghanistan and Iraq are conservative estimates. *U.S. Intelligence community & conflict with Russia - Ray McGovern, Alexander Mercouris & Glenn Diesen,* The Duran, video, 2:14:05, December 6, 2022. https://youtu.be/qXw6kYLYRrs?si=GUB-lDZdA3nlbtAU

41 Cynthia Chung, "Ukrainian Nationalism as a Cold War Weapon," *The Saker,* January 25, 2023. Depleted uranium is made from nuclear waste that leaves a hazardous dust residue.

42 Stone and Kuznick, 538.

43 Ibid., 539.

44 *Scott Ritter 17 November Russia Ukraine war update about missile attack on Poland,* Straight Talk With Douglas MacGregor, video, 19:38, November 17, 2022. https://www.youtube.com/watch?v=ncC80NbQeeM

45 Prins, 394.

46 Ibid., 411: "At the height of the bailout period, $19.3 trillion of subsidies were made available to keep (mostly) U.S. bankers going, as well as government-sponsored enterprises like Fannie Mae and Freddie Mac."

47 Summers was an architect of the pillage of Russia after the Soviet Union collapsed. In his tenure as president of the World Bank, he also suggested that "the World Bank be encouraging MORE migration of the dirty industries to the LDCs." See "Summers Memo," *Wikipedia.* https://en.wikipedia.org/wiki/Summers_memo

48 Prins, 396.

49 Ibid., 399.

50 Ibid., 402.

51 Ibid., 402.

52 Ibid., 405.

53 Jeff Thomas, "Ben Bernanke-Revisiting the Helicopter Speech," *Doug Casey's International Man,* accessed January 12, 2023.

54 Suzanne Goldenberg, "A desperate plea – then race for a deal before 'sucker goes down,'" *The Guardian,* September 27, 2008. https://www.theguardian.com/business/2008/sep/27/wallstreet.useconomy1

55 Prins, 411.

CHAPTER 17

U.S. Governance Hits the Wall

Who Is Obama?

The Bush-Cheney "War on Terror" was a disaster. The world economy was in meltdown. *Nothing* in Barack Obama's background showed the likelihood of competence or experience in resolving the epochal mess. The only rational conclusion is that Obama was judged to meet the needs of whatever shadowy forces picked him for the job.

Obama resembled Bill Clinton in being designated for the presidency by being compliant, quick on his feet, and photogenic. It has been argued that Obama's mother, Ann Dunham, and his grandfather, Stanley Armor Dunham, were likely CIA assets.[1] His mixed race capitalized on the demographic shift away from a predominantly white electorate, especially in key states like California.

After graduating from Columbia University in 1983, Obama worked for Business International Corporation, a CIA front. For four more years, he was a "community organizer" on Chicago's South Side, which decked him with progressive-sounding credentials.

Evidently identified by "somebody" as a rising star, Obama then attended Harvard Law School, becoming president of the *Harvard Law Review*. Returning to Chicago, he worked as a "civil rights attorney," while teaching constitutional law at the University of Chicago.

In 1995, Obama published the autobiographical *Dreams From my Father*, later viewed as ghost-written with exaggerated or falsified information.[2] Following the book's publication, he was elected to the Illinois state senate. In 2003, he announced his opposition to the Bush/Cheney invasion of Iraq, again bolstering his "progressive" creds.

Obama's rise was "meteoric." He was elected U.S. senator in 2004 in a campaign where his initial Republican opponent, Jack Ryan, saw his compromising divorce records opened by a Los Angeles court after parties alleged to be Obama's supporters disclosed the "leaked" contents to the press. Almost immediately after being elected to the U.S. Senate, Obama decided to run for president of the United States.

In 2008, Obama defeated Hillary Clinton in the Democratic Party primary and was elected president in a landslide against a comically hapless John McCain/Sarah Palin ticket. For more on Obama's astonishing rise to power, see Wayne Madsen's *The Manufacturing of a President: The CIA's Insertion of Barack H. Obama Into the White House*.[3]

As half African but not an indigenous African American, Obama was nonetheless defined as Black, a manifestation of "identity politics," where the Democrats believe their route to electoral success leads through appealing to a mixture of racial minorities, the Jewish bloc, feminists, highly educated white liberals, government employees, and gays. Working class issues matter little anymore; radical change, not at all. The Democrats have largely eliminated worker and consumer concerns, along with anti-war sentiment, from a party presided over by a Democratic National Committee hardwired into Wall Street donors. Bernie Sanders and his supporters would learn this the hard way in 2016 and 2020.[4]

While a candidate, Obama attended a secret White House meeting with George W. Bush and Secretary of the Treasury Henry Paulson where he promised, once elected, to bail out the banks from the ongoing financial collapse.[5] Bush was okay with Obama's plan for the biggest government bailout in history. So was Goldman Sachs, which donated a fortune to Obama's war chest.

As president, now with Joe Biden his vice president and Hillary Clinton his secretary of state, Obama retained several members of Bush's national security team, most notably Secretary of Defense Robert Gates, who was Rumsfeld's successor in prosecuting the "War on Terror." Under Obama, troop "surges" took place in Afghanistan and Iraq, though most U.S. military personnel were chased out of Iraq by insurgents by 2014.[6]

Obama would become especially known for a major initiative to kill our "enemies" in Afghanistan and Pakistan by drones operated remotely by CIA personnel safe in faraway Tampa—while also frequently killing civilians and giving renewed use to the Rumsfeld-coined obfuscation, "collateral damage." Obama personally approved of the designated targets. By 2015, civilian deaths in Afghanistan stood at 11,000 a year.

Regime Change in the Middle East

Obama's wars took place in the context of what the media celebrated as a "pro-democracy" movement termed the "Arab Spring"—during which the

CIA, its NGOs, and its hired terrorists attempted regime change, with varied success, against Tunisia, Libya, Egypt, Yemen, Syria, and Bahrain.

Siding with Susan Rice, Hillary Clinton, and Samantha Powers against his own secretary of defense, Robert Gates, and national security adviser, Thomas E. Donilon, Obama oversaw NATO's 2011 bombing of Libya, one of Africa's most prosperous countries, along with the overthrow and assassination of Libyan President Muammar Gaddafi, leading to the total destruction of that nation's infrastructure and way of life.[7]

Notably, Libya's central bank at the time was state owned. Furthermore,

> Qaddafi promoted pan-African unity, a United States of Africa he hoped to lead against Western powers wanting balkanized easily-controlled states. Libya was central to Africa's independence, including its freedom from predatory central banks and international lending agencies, acting as loan sharks of last resort. He also funded Africa's only communications satellite, thereby saving users hundreds of millions of dollars for low-cost incoming and outgoing calls. In addition, he allocated two-thirds of the $42 billion needed to launch a public African Central Bank (headquartered in Nigeria), an African Monetary Fund based in Cameroon, and an African Investment Bank headquartered in Libya.[8]

Gaddafi's chief offense was his proposed gold dinar:

> In 2009, Colonel Gaddafi, then President of the African Union, suggested to the states of the African continent to switch to a new currency independent of the American dollar: the gold dinar. The objective of this new currency was to divert oil revenues toward state-controlled funds rather than American banks. In other words, to stop using the dollar for oil transactions. Countries such as Nigeria, Tunisia, Egypt and Angola were ready to change their currencies. Unfortunately, in March 2011, the NATO-led coalition began a military intervention in Libya in the name of freedom....[9]

"Freedom" in the U.S. foreign policy lexicon has always meant freedom for U.S. banks and corporations to set up shop abroad with the aim of extracting maximum wealth from victim nations. Invariably this means control of those nations' ruling elites, with the CIA as the enforcer. Obama was 100 percent in accord with this paradigm.

In 2015, Obama declined to launch a full-scale assault on Syria and opted for proxies, infiltrating U.S. advisers and weapons to constantly morphing groups of jihadi fighters and ISIS. All were trying to destroy the democratically elected government of Bashar al-Assad. President al-Assad had refused to step down under pressure from the Bush administration in the face of false flag chemical attacks involving "White Helmet" provocateurs.

In a major geopolitical move, Russia sent forces into Syria, which so far have prevented it from suffering the same fate as Afghanistan, Iraq, and Libya.[10] Russia's actions in Syria, and its naval base on Syria's Mediterranean coast, fed the ire of the U.S. in the run-up to the U.S. proxy war against Russia in Ukraine that had its genesis under Obama.

Persecuting Whistleblowers

Nobel Peace Prize-winner Obama stepped up the persecution of U.S. government whistleblowers, like the Army's Chelsea Manning, who disclosed U.S. torture and abuses in Iraq and Afghanistan. Prosecuted under the World War I Espionage Act, Manning was finally released from detention in 2020. Obama was also central to the pursuit of Julian Assange, publisher of *Wikileaks,* which had published the Chelsea Manning leaks. Assange was similarly pursued for charges under the Espionage Act, enduring a long period of refuge in the Ecuadoran Embassy in London before transfer to the UK's Belmarsh prison. A decision on whether Assange will be extradited by Britain to the U.S. to stand trial is still pending.[11] Chelsea Manning returned to prison rather than testify against Assange. *The New York Times*—one among several mainstream papers that similarly published the leaks—was not charged and did not protest Assange's charges in defense of freedom of speech.

In 2013, CIA contractor Edward Snowden fled the U.S. as thousands of documents he had obtained, demonstrating NSA/CIA spying on American citizens, were released publicly by journalist Glen Greenwald. Formerly, the government required a warrant to gain access to individuals' electronic media. Now, the massive warrantless surveillance by intelligence agencies revealed the extent of its disregard for Americans' privacy since 9/11 and the Patriot Act.

The U.S. Department of Justice stated its intention of also prosecuting Snowden under the 1917 Espionage Act, but Snowden ended up in limbo in the Moscow International Airport after the U.S. revoked his passport. Six months later, Russia granted him asylum. Later he obtained Russian citizenship.[12]

Full-Spectrum Dominance

When discussing the Obama presidency, reference should also be made to the military doctrine of "Full-Spectrum Dominance" that the Pentagon has promulgated in various formats, as well as their ongoing designation of Russia and China as our two main "adversaries."

First officially formulated in April 12, 2001, the U.S. Department of Defense in its 2010 *Dictionary of Military and Associated Terms* defines Full-Spectrum Dominance as:

> The cumulative effect of dominance in the air, land, maritime, and space domains and information environment, which includes cyberspace, that permits the conduct of joint operations without effective opposition or prohibitive interference.

The U.S. military defines as its purview the entire globe, every country, outer space, and all people, claiming the right to rule over it all with premeditated violence. Obama was a prominent enabler. We have every right to ask to what extent Full-Spectrum Dominance fulfills Obama's "Hope" and "Change" campaign slogans.

Planned Takedown of Russia

President Barack Obama was the instrument of the Deep State. He obeyed instructions with the financial bailouts of his first term. He obeyed in continuing and extending the failing "War on Terror." He obeyed in forwarding the anti-Russia project by means of the 2014 Maidan coup, leading to today's U.S. proxy war against Russia in Ukraine.

While plans against Russia moved forward, the growth of China's economic and military power was attracting increasing attention. The U.S. had shifted a huge amount of its manufacturing capability to China during the Clinton and Bush II administrations as the cost of labor—inescapable due to the overblown price of housing—made goods produced in the U.S. uncompetitive. U.S. deindustrialization resulted in a large negative balance of trade between the U.S. and China/Japan/Korea, with those nations using their earnings to purchase U.S. Treasury bonds needed in order to buy petroleum products only available in dollars on world markets. This in turn enabled the U.S. to spend its bond revenues on furthering its overseas military footprint.

China also began to compete in space by putting men into orbit and undertaking lunar exploration. It also had begun planning an enormous Eurasian infrastructure project, the Belt and Road Initiative. This would allow China eventually to move its Eurasian trade overland without danger of being interdicted by the U.S. Navy.

An eponymous article in *Foreign Affairs* foreshadowed Obama's "Pivot to Asia." Later, a serious attempt at challenging China began during the Trump administration. But for now, with Russia the immediate enemy, the Chinese were placed on the back burner.

We have already recounted the story of Nixon's détente, the end of the Cold War, and the 1991 Soviet collapse, leading eventually to the accession of Vladimir Putin as president of a resurgent Russia. By 2014, more than halfway into Obama's presidency, Ukraine, formerly a key component of the Soviet Union, had emerged as the crossroads of conflict. Obama had already pushed the region toward armed confrontation by setting up American bases in Poland and Romania that could launch cruise missiles with nuclear warheads against Russian cities.

But Ukraine was the prize, the launch pad and battering ram for foiling Russia's resurgence by regime change against Putin and gaining control of Russia's vast resources.

Russia's Comeback

In 2012, two years after pro-Russian Viktor Yanukovych was elected president of Ukraine, Vladimir Putin was elected to his third term as president of Russia. He had been elected previously in 2000 and 2004. The Russian constitution prohibited a third consecutive term, so Prime Minister Dmitri Medvedev, a protégé of Putin, had been elected to the presidency in 2008, with Putin shifting to prime minister. The question became whether Putin would retire gracefully into the shadows.

Putin decided otherwise and was elected to his third presidential term in 2012. This was after Medvedev, with Putin in tandem, had maneuvered Russia through the worldwide financial crisis of 2008–2009 and the brief war against Georgia over that country's armed excursion into Russian territory in South Ossetia.

The U.S. was apoplectic when Putin became president for the third time, even more so when Putin ran for and won his fourth term in 2018. By then the Russian constitution had been changed to a six-year presidential tenure. Relations between the U.S. and Russia were already on shaky ground as Russia continued to carry out Putin's economic revival and military rearmament.

Even though Russia was no longer communist, Putin's power left it open to charges of being "authoritarian." Russia was different from China, a one-party state. But the two were now lumped together as "threats" or "challenges" to democracy—even though "democracy" in the U.S. means control of electoral politics by the financial/corporate elite, especially after the 2010 Citizens United decision, when the Supreme Court declared that corporations have the same constitutional right of "free speech" as individuals. Going forward, corporations could lawfully be political campaign donors.

In rebuilding Russia after the 1990s economic collapse, Putin retained both multiparty elections and a market economy. But he reasserted state control in key respects, taking back ownership of critical industrial firms like Gazprom, though private investors, including from the U.S., held minority ownership stakes. The government, not stockholders, also controlled the armaments industry.

Russia's consumer economy was largely deregulated, as long as businesses paid taxes. The result was a wave of prosperity built on Russia's powerful base of natural resource wealth. It was prosperity, with rising wages and tax cuts, that kept Putin's polls in the seventy percent approval range, along with crackdowns on terrorist violence, on the oligarchs and crime bosses.

The water was muddied by suspicions within the Russian security apparatus that the unrest in Chechnya and elsewhere was being instigated by the CIA, as were attempts to generate color revolutions in former Soviet states adjacent to Russia. Putin was determined to resist. He resolved the unrest in Chechnya by addressing local grievances and rebuilding Grozny,[13] so successfully that Chechen forces under Ramzan Kadyrov are now fighting on the Russian side in Ukraine, with the Chechens feeling they had formerly been U.S. dupes.

Putin's most important reforms came in the financial sphere. He relied on a strong central bank that was state-owned and answered to the central government, not the financiers. China has done the same.

It's government control of central banking that separates Russia and China from the West, where banking oligarchies are engaged in unrestricted extraction of wealth from society at large by fractional reserve banking and usury based on compound interest. The Western banking parasites have been increasingly sapping the strength of their nation-state hosts even as challenges from Russia and China have grown. Transformation of the U.S. into a finance-dominated service economy relying on dollar hegemony abroad enforced by military might has been the catastrophic result. The U.S. has been walking zombie-like into this finance-created trap for decades. The shift was facilitated by the conversion of the Democrats into a pro-business party in the 1990s by Bill Clinton, *et al.*

Meanwhile, Putin moved toward a balanced government budget. Government control of gas and oil and revenues from hydrocarbon products supplied to Europe kept the economy growing. Putin also pulled the government's cash reserves out of the private banks in a similar fashion to the U.S. sub-treasury system of the 19th century. Again, it was in the sphere of government control of banking and finance that Russia represented the biggest threat to U.S. hegemony.[14]

Finally, Putin understood something few modern leaders have realized: that if a nation is to enjoy individual freedom and prosperity, a strong but popular leader whose primary concern is the wellbeing of the country is required. This is because there are so many centrifugal forces pulling away from the center, each with its own demands, that a firm hand is needed to balance these forces and effectively "mind the store." In the U.S., one who understood this principle was George Washington. Another was Abraham Lincoln. A third was Franklin Roosevelt. It is no accident that these were considered our greatest presidents. All were able to secure their authority by the benefits they brought to the lives of ordinary citizens.

Russia has its own gallery of heroes. The foremost is Peter the Great. Another is Catherine the Great. Even with Stalin having ruled by terror under a system that eventually failed, the Communist Party remains the second largest party in Russia. Putin has been compared to Peter the Great, never to Stalin. Still, he sometimes has had to take drastic action against dissenters, so has been easy to criticize as a tyrant behind a bureaucratic façade. But many among Russia's populace consider Putin a national savior.

No contemporary U.S. president can be compared to Putin. John F. Kennedy might have risen to that level. But the centrifugal forces annihilated Kennedy, and all of his successors have been controlled by globalist entities, preeminently the financial elite, especially the Rockefellers and the Deep State.

Background to Ukraine's Maidan Coup

U.S. intelligence agencies began plotting to take over Ukraine in 1946, even before the creation of the CIA. After World War II ended, U.S. intelligence, along with Britain's MI6, created a vast covert network in those parts of Europe occupied by the American and British armies. Its purposes were to thwart attempts by leftist parties to take over Western European nations and to launch operations against the Soviet Union.

After the CIA was formed in 1947, much of its activity was concentrated in Italy. Later, the European covert network morphed into Operation Gladio, which specialized in false flag events that could be blamed on communist "terrorists."

The CIA and MI6 also attempted to create underground networks in the nations of Soviet-controlled Eastern Europe, as well as component regions of the Soviet Union itself. These attempts had little success.

Regarding Ukraine, *Red Street Journal* has released declassified CIA-archived documents from November 1946 that originated with the U.S. Chief of the Field Installation in Munich, Germany. These lay out plans for "the collection of...counter-intelligence from the Ukrainian SSR and from other regions of the Soviet Union which is already available to or can be obtained by the Ukrainian underground movement."

Provisions included the use of underground personnel "as couriers and radio operators." Also included were night drops of equipment over western Ukraine, and creation of a "secret radio station" and "secret meeting places for agents." Plans also included:

> ...establishment of agents in cover jobs such as International Red Cross,...supplying necessary documents,...[and] equipping agents with personal weapons, money, and poison for suicide if caught.[15]

Originally called Operation Belladonna, the project morphed into Cartel, Androgen, Aecarthage, and Aeorodynamic. The purpose was to prepare for possible "open conflict between the United States and the USSR." Successors to the project continued on CIA books until 1991, when Ukraine gained independence after the Soviet Union collapsed. However, sources agree that the Soviets were still able to counter the creation of an effective Ukrainian underground and eliminate its agents by the 1950s.[16]

During the remainder of the Soviet era, Ukraine grew into a critical part of the Soviet Union's agricultural breadbasket and industrial infrastructure. Heavy industry, including the manufacture of military products like aircraft engines, became concentrated in the Russian-speaking eastern oblasts in the Donbass region. Ukraine's hydrocarbon resources were also concentrated in the Donbass.[17]

In 1954, the Soviet government under Khrushchev transferred the Crimean Peninsula from the Russian Soviet Federation of Socialist Republics to the Ukrainian Soviet Socialist Republic. Thoroughly Russian in its ethnicity, language, and culture, Crimea was home to the Soviet Black Sea fleet headquartered at Sevastopol. This transfer was viewed at the time as an administrative convenience.

Consequently, Crimea remained part of Ukraine after the 1991 Soviet collapse. By agreement with Russia, all the nuclear weapons the Soviets had stationed in Ukraine, including Crimea, were shipped out to the Russian Federation. Ukraine, now independent, renounced any claim to being a nuclear power.

Even after the break-up of the Soviet Union, Ukraine retained a close association with the Russian Federation. Much of Ukraine had been part of Russia for over three centuries. The only period of Ukrainian independence had been 1917–1922, during the Russian civil war. Ukraine had virtually no history, tradition, or infrastructure of modern statehood.

It's said that Boris Yeltsin, as president of Russia after 1991, took almost no interest in the affairs of Ukraine after the Soviet collapse. Like other former Soviet components, Ukraine faced a power vacuum which the U.S. began taking steps to fill.

Except for people of Polish, Hungarian, and Romanian ethnicity residing in the west, Russian culture continued to predominate, with Ukraine joining the informal association with Russia and other former Soviet republics through membership in the Commonwealth of Independent States, or CIS.

A small minority of largely western Ukrainian militants, operating as right-wing paramilitary groups, were immersed in an ideology dating from collaboration with the German invaders during World War II. These became the Azov Battalion, Right Sector, Svoboda Party, etc., that took part in the Maidan coup of 2014, with some units then incorporated into Ukraine's army. These paramilitary groups receive training from the CIA's "Ground Branch."[18]

Plans for the U.S. absorption of Ukraine into its sphere of influence had begun to develop during the 1990s, with ideas starting to circulate of Ukraine's joining the EU and NATO. Ukraine joined NATO's Partnership for Peace in 1994. In 2005 the "Orange Revolution" created a pro-Western undercurrent but without much support from the population. By 2014, Ukraine and Georgia were actively being targeted for future NATO membership, with Russia bitterly opposed.

The 2000s and beyond was a period of great uncertainty, shifting internal alliances, and confusion about Ukraine's identity and allegiances. Meanwhile, Ukraine's state-owned industries and large collective farms became subject to a similar process of privatization and looting to what was going on in Russia itself. As in Russia, an oligarchic class had emerged with a strong Jewish component.

The Ukrainian oligarchs began making their move toward political power even as Ukraine was gaining a reputation as the most corrupt nation in Europe. As late as 2021, the U.S. State Department warned American entrepreneurs against doing business in Ukraine. But by then, Hunter Biden, Joe

Biden's son, had become involved in Ukraine by membership on the board of Burisma, an oil conglomerate.

Igor Kolomoisky, media magnate and later governor of Dnipropetrovsk Oblast was a leading oligarch, heavily involved in the broadcast industry. His protégé was Volodymyr Zelensky, later president of Ukraine during the war against Russia.[19]

The Maidan Coup

In February 2014, the government of President Barack Obama overthrew the democratically elected government of Ukraine and installed a pro-American junta.

The coup was perpetrated through the combined efforts of the CIA, MI6, the National Endowment for Democracy, billionaire George Soros, State Department Neocon Victoria Nuland,[20] USAID and the U.S. embassy, cadres of Ukrainian Neo Nazis, and U.S.-paid demonstrators and terrorist snipers hired to cause panic by shooting police and bystanders in the streets of Kiev.[21] John Kerry had replaced Hillary Clinton as secretary of state.

Fearing the violence, elected President Viktor Yanukovych and his aides fled to Russia, having tried to steer Ukraine along a path of neutrality between Russia and the West. The coup was followed by declarations of independence by the pro-Russian oblasts of Donetsk and Lugansk in Donbass, armed skirmishes, and the routing of the Ukrainian army.

Russian efforts to stabilize the situation, leaving Donetsk and Lugansk as federated territories within Ukraine via the 2015 Minsk peace agreements between Russia and Ukraine with guarantees from France and Germany, was sabotaged by the U.S. Former German Chancellor Merkel and French President Hollande, along with former Ukrainian President Peroshenko, would declare in 2022 that there was never any intention to enact the Minsk agreements. It was just a smokescreen to deceive Russia while NATO was building Ukraine's military capacity for a future war.[22]

Based on a referendum garnering more than eighty-five percent popular approval, Russia then annexed Crimea, a culturally Russian province where they already had forces stationed, including legal and longstanding possession of the Black Sea naval base at Sevastopol. The Ukrainian Neo Nazis' response was to set alight the Trade Union Center in the Black Sea port of Odessa, burning alive 42 pro-Russian separatists trapped inside.[23]

After Russia annexed Crimea, U.S. President Obama signed an executive order declaring a state of "national emergency" in the U.S. Obama declared that the annexation presented:

…an extraordinary threat to the national security and foreign policy of the United States constituted by the actions and policies of persons that undermine democratic processes and institutions in Ukraine; threaten its peace, security, stability, sovereignty, and territorial integrity; and contribute to the misappropriation of its assets.[24]

The "national emergency" gave the Obama administration the authority to deal with any such threats to U.S. foreign policy. President Joe Biden later renewed the declaration. This declaration was a major step toward the U.S.'s proxy war against Russia.

For the next eight years the Kiev junta, now armed and trained by the U.S. and NATO, bombarded the pro-Russian separatist regions of Donbass. Over 14,000 were killed, mostly civilians. The ethnic cleansing was intended to drive Russian-speakers out of Ukraine, ensuring their removal from an electorate that stood in the way of U.S. hegemony via the Kiev-supported regime. Ukraine also built massive fortifications within the Donbass borders to prepare for a possible Russian armed attack.

Finally, with Putin's efforts to forestall the conflict having come to naught, Russia reacted by sending forces of its own army into Ukraine in February 2022. Russia called this invasion a "Special Military Operation," with the army supporting the Donetsk and Lugansk militias. Several million Ukrainians fled into Russia or adjacent EU nations. But there were already millions of Ukrainians working at jobs within the EU. The depopulation of Ukraine had begun and continues today as hapless undertrained troops are flung without air support against Russian defenses, failing to even reach their first lines. Such is the bitter fruit of the Obama administration's actions and policy in Ukraine.

History's Judgment on Obama

Besides the unprecedented bailouts of an outlaw financial industry, President Barack Obama escalated the murderous U.S. wars in Iraq and Afghanistan, became the "king" of drone assassinations,[25] then went on to destroy Libya. He failed at doing the same to Syria only by Russia's intervention. Add to that his numerous interventions and sanctions in Latin America and the expansion of the U.S. military presence in Africa and the Middle East. Though Obama did sign the JCPOA to prevent nuclear arms development by Iran, U.S. assent to the agreement was rescinded by President Donald Trump.

But it must be said that the peak of Obama's aggression was his subversion and takeover of Ukraine, leading to the possibility of World War

III against Russia. All this was done with virtually no participation from or approval by Congress. Congressman Dennis Kucinich said that Obama's attack on Libya without congressional approval was an impeachable offense. The same applies to Obama's Maidan coup in Ukraine.

But there was another disaster to reckon with. The financial crisis Obama dealt with by massive bank bailouts was the direct result of the systemic financial fragility brought into being by the privately owned and operated fractional reserve banking system based on usury that had been around since the Middle Ages and had never been effectively adapted to modern industrial conditions.

This system cannot be reformed. Change can only start with the abolishment of the Federal Reserve and rebuilding of the monetary system enabling sovereign control of money creation by the U.S. government to be used for the benefit of its population. The closest the U.S. ever got was the system presided over by presidents Lincoln through McKinley.

As stated in the previous chapter, Obama's "solution" to the financial crisis, other than massive bailouts that added to the U.S. national debt, was Fed Chairman Ben Bernanke's zero-percent interest rates combined with "quantitative easing." Under this policy, the Fed itself was forced to purchase Treasury bonds that were no longer marketable due to the zero-interest policy; these bonds were then loaned to banks as cheap reserves.

Since the banks could lend little money to a shaky private sector, they remained solvent but failed to benefit the economy at large. Growth of the government's debt promised inflation later. Obama's financial program was a classic case of "kicking the can down the road." The result is today's $73 trillion federal government debt.

Among the victims of the Bush/Cheney/Obama era of nearly sixteen years of wars-of-choice are the tens of thousands of U.S. veterans with PTSD or who have committed suicide, along with their suffering spouses, children, friends, colleagues, and society at large.[26]

The 2016 Election

Obama's secretary of state, Hillary Clinton, former first lady and U.S. senator, was handpicked by Obama, the Democratic National Committee (DNC), Wall Street, and the Deep State for the Democratic Party's presidential nomination in 2016. Clinton and the DNC undermined and subverted the opposition candidacy of Senator Bernie Sanders of Vermont who led a populist revolt. Sanders wanted a higher federal minimum wage and a universal prepaid health-care system. Obviously, such things were beyond the pale for the "world's greatest democracy" with its trillion-dollar war budget.

In a 2017 by-lined article in *Politico,* newly designated DNC chair Donna Brazile revealed how she had learned of the unethical takeover of the Democratic Party by Hillary Clinton. She found that she now must phone Senator Sanders to tell him the takeover was a *fait accompli* and that he had no choice but to "work as hard as he could to bring his supporters into the fold with Hillary, and to campaign with all the heart and hope he could muster."[27]

The takeover was a *quid pro quo* for paying the remainder of a massive $24 million debt foisted on the DNC by then-president Barack Obama. By agreement with then-chair Debbie Wasserman Schultz, Hillary for America and the Hillary Victory Fund had paid off eighty percent of Obama's remaining debt, about $10 million, and had placed the DNC on an allowance. The Clinton campaign continued to funnel money to the DNC up to the party's nominating convention.[28]

In return for the bailout, the DNC signed a Joint Fund-Raising Agreement with Clinton campaign official Robby Mook. The Agreement "specified that in exchange for raising money and investing in the DNC, Hillary would control the party's finances, strategy, and all the money raised. Her campaign had the right of refusal of who would be the party communications director, and it would make final decisions on all the other stuff. The DNC also was required to consult with the campaign about all other staffing, budgeting, data, analytics, and mailings."

Faced with such institutional power, the Sanders campaign battled but came up short, winning 1,847 of delegates pledged through primaries or caucuses vs. 2,204 for Clinton. But it was party politics that put Clinton over the top as she won the support of 560 unelected "superdelegates" vs. 47 for Sanders. Thus, it was Clinton money and influence that killed Sanders' populist insurgency.

In the months preceding the election, publication by *Wikileaks* of campaign manager John Podesta's emails, with the campaign falsely accusing Russia of a hack,[29] revealed a Clinton candidacy fraught with anxiety over her performance weaknesses, poor political instincts, and tendency to level insults at the supporters of, first, Sanders, and then Donald Trump. It could be said that given Clinton's prominence, the 2016 presidential election was hers to lose, and lose she did.

Black voters were the DNC's secret weapon. Sanders would have won the primary vote except for massive support for Clinton by southern blacks. It was assumed the same demographic would win the general election for Clinton against Donald Trump.

During the 1990s, the Democratic and Republican Parties had carried out a program of gerrymandering that created a number of "safe" districts that would assure the election of party incumbents. This included a number of guaranteed black congressional districts. The main job of these members of Congress has been to deliver the black vote for the Democratic Party in national elections. Needless to say, gerrymandering in their favor was a new experience for African Americans.

The same went to a lesser degree for the Hispanic vote. All this points to an attempt to portray the Republican Party as the party of "white supremacy." Of course, the Republicans had set themselves up by embracing Nixon's Southern Strategy.

And with U.S. demographics shifting in the direction of a majority non-white electorate, the objective was to consign the Republican Party to oblivion, especially with an open southern border continuing to funnel millions of undocumented immigrants into the country who are dependent on Democrat-controlled metropolises or secure "blue" states like California.

The Trump Presidency

Donald Trump won the 2016 election and became president, elected largely by white middle-class voters whom Hillary Clinton contemptuously called the "deplorables," a rather unique libel whose conception must have resulted from considerable cogitation. These were largely voters from states that had been robbed of their manufacturing jobs by her husband Bill Clinton's neoliberal outsourcing "reforms."

After the Bush/Cheney catastrophe, there was a power vacuum in the leadership of the Republicans. Trump easily wiped the slate clean in his march to the White House by trouncing non-entities like Jeb Bush, Marco Rubio, and Ted Cruz in the primaries.

Hillary Clinton and the DNC, supported by partisans within the FBI and elsewhere in the Deep State, immediately blamed her loss on Russian influence. Clinton herself authorized what came to be called "Russiagate" to further such claims. Eventually the allegations were disproven after dominating public discourse for years, but not before several Trump operatives, including Lieutenant General Michael Flynn, had their careers ruined or were thrown into prison for charges connected with their supposed collusion.

With conflict in Ukraine underway due to the Ukrainian government's disregard of the Minsk Agreements, the Deep State and mainstream media embarked on a project to demonize Russia and Putin before the expected confrontation. Hillary Clinton's "Russiagate" was facilitated by the media insofar as it played so well into this demonization.[30]

Russiagate and Jade Helm 15

There was a sidelight to the "Russiagate" story in connection with the time in 2015 when the U.S. military conducted the "Jade Helm 15" exercises in the Southwestern states, with the main focus being Texas. One of the locations was Livingston, Texas, where the army took over a Walmart store for several weeks. Local observers saw what looked like coffins being moved in and out.

Jade Helm 15 involved the U.S. Army Special Operations Command, along with several other military units. The Defense Department said the purpose was to "train soldiers in skills needed to operate in overseas combat environments, including maneuvering through civilian populations." This was nonsense, because the military units were sequestered at sealed locations with no possibility for such "maneuvering" evident.

Word of Jade Helm 15 caused an alert among observers. Governor Greg Abbott made a show of calling out the Texas State Guard to pretend he was keeping an eye on events. Civilian groups were out in force with binoculars and surveillance gear.

Rumors abounded that this was a drill for the government to declare martial law and round up dissenters. Rumors of FEMA detention camps were revived. The media responded as they were accustomed to do by mocking citizens' concerns as "conspiracy theories."

Then in 2018:

> Air Force General Michael Hayden, who served as NSA director and CIA chief under Presidents Bill Clinton, George W. Bush and—briefly—Barack Obama, told MSNBC's *Morning Joe* on Wednesday that Russians sought to "use dominance in the information space to directly attack the will of an adversary population and win that way."[31]

He went on to say that the conspiracy theories about Jade Helm 15 were in fact Russian disinformation. He said:

> At that point, I'm figuring the Russians are saying, "We can go big-time," Hayden said. "And at that point, I think they made the decision, 'We're going to play in the electoral process.'"

Hayden's absurd contention that citizens' concerns about Jade Helm 15 were actually a Russian dress rehearsal for "Russiagate" gives a clear picture of how the Deep State was using Russia to warp opinion and cover their own tracks, no matter how improbable the means.[32] And yet—and yet—large sectors of the American public—primarily Democrats—bought it!

Trump Under Attack

"Russiagate," Hayden's claims about Jade Helm 15, the later suppression of the Hunter Biden laptop story, etc., were clearly part of a designed long running "hate and fear Russia" plan perpetrated by the Deep State to prepare public opinion for the planned U.S. proxy war against Russia in Ukraine.

When the story about Hunter Biden's compromising laptop surfaced just before the 2020 election, the FBI, with the laptop in custody, lied to social media by claiming the story was "Russian disinformation" and succeeded in suppressing it. In May 2023, a House committee disclosed that the CIA and the Biden presidential campaign had colluded in writing the open letter signed by fifty-one former intelligence officials also claiming the story was "Russian disinformation." Biden bragged about this letter in his last debate against Trump, claiming that the laptop story was "garbage."

As his presidential term began in 2017, Trump's critics from the "liberal" media were able to dig up plenty of dirt on him with respect to his relations with women and his business practices, including stiffing contractors and selling luxury condos to alleged *Russian mafia* [!] money launderers.

Trump also said he wanted the U.S. to get out of nation-building abroad. And that European nations should pay their fair share of NATO's costs. He made overtures to North Korea, spoke of improving relations with Putin, and disrupted trade with China with new tariffs. He pulled out of the Trans-Pacific Partnership to save American jobs, and he suspended negotiations on the Transatlantic Trade and Investment Partnership.

Trump also made plans to exit Afghanistan, and reduced troops in Iraq. He did send a few soldiers into Syria but refrained from bombing Syrian cities. He stopped funding ISIS and ordered elimination of the U.S. combat role in Syria. Even worse, Trump was starting to question U.S. NATO participation.[33] Finally, he had set his sights on breaking up the big Wall Street banks by restoring Glass-Steagall to divide deposit banking from investment operations.[34] Nothing came of this, but it did show the direction of his thinking.

On the other hand, Trump made extraordinary concessions to Israel, announcing a plan to move the U.S. embassy to Jerusalem and ordering the assassination of Qasem Soleimani, head of the Iranian Quds Force.[35] At the instigation of Deep State operatives within his administration, he also conducted the major provocation of canceling the INF Treaty with Russia. He also pulled out of the Paris Climate Accords.

But Trump was nowhere near starting a big war against anyone, much less Russia. However, he had little control over appointments to policy-making positions, including his own White House. Otherwise, where had "We Lie-Cheat-and-Steal" Mike Pompeo[36] come from? Not to mention Neocon retreads like John Bolton whom Trump taunted by asking if had invading Ireland in mind.[37] When Pompeo, Bolton, and the Joint Chiefs presented Trump with a plan to bomb Iran, he refused.[38] Finally, when Trump ordered U.S. troops out of Syria, his aides and the military refused to comply, even boasting about their refusal.[39]

Trump often had a deer-in-the-headlights look when faced with blustering through difficult questions. But he had the helicopter salute on the White House lawn down pat. As far as the big questions of war-and-peace were concerned, he was probably the least dangerous president in modern U.S. history.

Hillary Clinton had said she wanted the U.S. to do even more "nation building" in Syria, Ukraine, Afghanistan, and elsewhere. What this meant, of course, was more U.S. terrorism and killing people of questionable friendliness to the U.S., unless we needed to pay them to do some of our own dirty work in obscure places. It was the U.S., after all, who hired al Qaeda to fight against Russia in Afghanistan and who bankrolled ISIS.

From the start, what we shall now call the Shadow Men wanted to get rid of Trump ASAP, whereas "We Came-We Saw-He Died" Hillary Clinton[40] was their kind of gal.

The Shadow Men

What we shall call the "Shadow Men" are those who sit atop the Deep State pulling strings on behalf of the ultimate heavies—the financial controllers and those who rule them. In Europe, the biggies include the Rothschilds, like those who run the *Economist*. In the U.S., it remains the Rockefellers, or at least their successors, like Jamie Dimon, the King of Wall Street.[41] Another name that frequently pops up is Bill Gates. One of the chief string-pullers, Henry Kissinger, like Obama a Nobel Peace Prize winner, is still around at age 100, at least at this writing.

On January 22, 2017, a curious article appeared in the *Sunday Guardian-India* entitled, "Shadow Men Work to Remove President Trump."[42] The author was a highly placed Indian academic and journalist, Madhav Nalapat. Nalapat states that his sources were knowledgeable individuals "based in Chicago, Washington, New York, and London." He writes, only *two days after Trump's inauguration:*

If the plans activated since November 8, 2016 [election day], by the Shadow Men succeed, the 45th President of the United States will not last in office beyond a thousand days from his swearing-in on January 20, 2017. The term Shadow Men refers to officials and policymakers operating in a stealthily coordinated manner to ensure the furtherance of specific agendas unrelated to the public interest.... They represent the hitherto ubiquitous and dominant Wall Street-Atlanticist alliance that has devised and implemented policy in Washington for several decades. These Shadow Men form an informal club of intelligence operatives, businesspersons, officials, and politicians whose relevance to policymaking and whose monetary wealth depend on the continuation of policies helpful to the interests they support, even though these may be harmful to the country they belong to.[43]

The existence of a cabal like the "informal club" described by Nalapat has long been suspected by observers of the American scene. A certain amount of flesh was put on the bones of suspicion early on by Professor Carroll Quigley in *The Anglo-American Establishment.* Nalapat continues:

Individuals with direct knowledge of the "1,000-day plan"...warn that January 20, 2017, marked not just Donald J. Trump's first day in office as the 45th U.S. President, but the acceleration of an ongoing campaign that has been designed to ensure that President Trump "does not continue in the world's most powerful job for more than a thousand days."

Nalapat describes the planned mechanism for getting rid of Trump:

According to individuals revealing details of the 1,000-day ouster plan, "the preferred route is a steady increase in public pressure, which would lead to the 45th President's impeachment by the U.S. Congress on the basis of presumed misdemeanors. These would be played up by media persons, who regularly get briefed by officials active in the shadow network.... It was explained that the Shadow Men are apprehensive that the strong-willed billionaire may refuse to get house-trained in the manner that Hillary Clinton so transparently was. In their view, the role of an elected head of state is in many aspects ceremonial, and on matters of national security and strategy, he or she should, in essence, follow the agenda set for him by the interests represented and protected by the Shadow Men."

President Donald Trump was impeached by the U.S. House of Representatives on December 13, 2019. This was *1,057 days* after Trump's inauguration, *almost exactly* on the Shadow Men's timetable that Nalapat revealed. Like so much else that was going on at the time, the impeachment also revolved around Ukraine and Russia, with Trump supposedly soliciting Ukrainian President Zelensky in a single phone call for adverse information on Biden when he was vice president.

Trump fought back by going on Twitter. Trump called the impeachment inquiry "a coup." He said it was intended "to take away the power of [the] people, their vote, [and] their freedoms" and that the Democrats were "wasting everyone's time and energy on bullshit." He said: "All Republicans must remember what they are witnessing here—a lynching! But we will WIN!"

In House Intelligence Chairman Adam Schiff's opening argument at Trump's Senate impeachment trial, he chastised Trump for abusing the U.S. relationship with Ukraine by citing a statement from a witness at the House impeachment inquiry: "The United States aids Ukraine and her people so that they can fight Russia over there, and we don't have to fight Russia here." Thus, the impeachment attempt had a close relationship to the fearmongering of "Russiagate."

But the impeachment failed. Trump was acquitted in the Senate on February 5, 2020, and continued in office.

Nalapat continues:

> Although media reports claim that U.S. intelligence agencies are hostile to the just sworn-in President, "in actual fact most within the middle layers (of these agencies) are supportive of his (Trump's) views," with only the politically-connected higher levels signing on to the 1,000-day agenda now in the process of being implemented. The sources say that the new head of state of the world's most powerful country "has an astonishing amount of goodwill within the operational level of the investigative and intelligence agencies."

From Nalapat's final paragraph:

> Given the pervasive influence of businesses dependent on China and the Middle East in Washington, a key objective of the Shadow

Men is to ensure that the Enemy Number One slot remains with Moscow and its allies such as Iran and not migrate to Beijing or to Saudi Arabia. Another is to ensure that the interests of Wall Street and the Atlantic Alliance continue to be given primacy in U.S. policy. The worry of the Shadow Men is that a U.S. President "who has yet to be house trained by the bureaucracy the way Barack Obama was in his very first week as President of the United States, and who has over three decades developed strong and consistent views on geopolitics and on economics over decades" of careful cogitation may succeed in shifting U.S. policy away from the Wall Street-Atlanticist embrace that has been the norm since the 1980s. "Trillions of dollars are at stake, so there is nothing to get surprised about that tens of millions have been spent these past months on ensuring that the agenda of the Wall Street-Atlanticist alliance continues to be official U.S. policy," a source said, adding that "President Trump represents the most potent threat to such interests in two generations."[44]

So far, they haven't done to Trump what an earlier generation of Shadow Men did to John F. Kennedy. At least not yet.

On the other hand, we may be seeing a slow-motion assassination attempt against Trump by Democratic prosecutors carrying out their agenda during the run-up to the 2024 presidential election. This includes, of course, the second Trump impeachment after the January 6, 2021, "riot" at the U.S. Capitol, the indictment of Trump by New York district attorney Alvin Bragg, the two federal indictments by special counsel Jack Smith, and the indictment of Trump on racketeering charges by Georgia prosecutor Fani Willis.

Cumulative sentencing of Trump if convicted on all charges would exceed 700 years. Never mind that this growing weaponization of the judicial process could eventually destroy the rule of law in the U.S. altogether.[45] The Democrats' assault on Trump has coincided with growing indications of financial corruption by President Joe Biden in league with his son, Hunter, of Ukrainian Burisma fame. House Republicans are inching toward a possible vote on impeachment. On August 21, 2023, Washington, DC, attorney Robert Barnes explained on a *Duran* program with Alexander Mercouris and Alex Christoforou that while the charges against Trump are without merit, it will likely be up to the Supreme Court to sort out the question of whether an incumbent party can utilize "lawfare" to destroy a political appointment.

But the surprising thing was that so many voters formerly characterized as "liberal" have flipped to become gung-ho supporters of the Democratic

Party's adoption of the Bush/Cheney posture of "forever wars" and of the Democrats' fanaticism in marching lockstep into the various levels of societal shutdowns and dictated Big Pharma "solutions" to the COVID crisis. Word is that with the alleged appearance of a new COVID variant, masking and lockdowns may resume in the fall of 2023.

The Democrats' current alignments have historical roots. In the 1930s, the Rockefeller banking/industrial dynasty broke with what was then the Morgan/Rothschild alliance with the conservative-minded Republican Party by embracing Roosevelt's New Deal. As recounted earlier, the Rockefeller-funded Council on Foreign Relations became the primary booster of U.S. post-World War II financial and military hegemony. Roosevelt and most other Democrats fully embraced these measures.[46]

The Rockefellers also became the dominant force behind the liberal wing of the Republican Party, with Nelson Rockefeller a chief advisor to Eisenhower, later becoming VP under Ford, with his brother David the architect of global finance and chief sponsor of Democrat Jimmy Carter. Thus the Rockefellers controlled both parties while also performing as the godfathers of the CIA which was entrenched as the enforcers of the national security state.

But after Republicans Bush and Cheney implemented the disastrous "War on Terror," the power of the liberal/globalist/totalitarian mind set gravitated back to the Democratic Party via Obama, Hillary Clinton, and now Biden. Obviously, Trump, once the Democrats' New York City darling, became the outlier. Now in 2023 Trump has pledged to destroy the Deep State and end the Ukraine war in twenty-four hours by bringing Zelensky and Putin together to negotiate.

We have been watching as the Shadow Men have been trying to destroy Trump for the last seven years. Thus far he has proven a stronger adversary than they may have expected.

COVID-19

At almost exactly the time President Donald Trump was being tried and acquitted in the Senate, marking an initial failure of the Shadow Men's vendetta, the virus causing COVID-19 was identified in an outbreak in the Chinese city of Wuhan.

Attempts to contain the virus in Wuhan failed. The World Health Organization declared the outbreak a "public health emergency of international concern" on January 30, 2020, and a pandemic on March 11, 2020.

As of this writing, almost four years after the initial outbreak, the official death count of the World Health Organization is 6,955,141. The U.S. has the highest death count of any nation at 1.1 million.

Amazingly, no official body has been able to identify definitively the source of the COVID-19 virus. In March 2023, WHO Director-General Tedros Adhanom Ghebreyesus said the origin of COVID-19 was unknown and that, "If any country has information about the origins of the pandemic, it's essential for that information to be shared with the WHO and the international scientific community."[47]

Early claims, including statements made by Trump, that the source was an animal market in Wuhan, are no longer taken seriously. Such claims evidently motivated Trump to call the infection the "China virus."[48] Trump is continuing to use that term in his campaign speeches for the 2024 presidential election. But the claims have shifted.

Allegations have now been made that the virus escaped from the Wuhan Institute of Virology. This has been vigorously disputed by Chinese researchers and the Chinese government. On February 27, 2023, the Chinese Foreign Ministry released an official statement that "certain parties should stop rehashing the 'lab leak' narrative, stop smearing China, and stop politicizing origins-tracing."[49]

Allegations have also come forth that the virus was engineered through "gain-of-function" research, where the potency of a pathogen is artificially increased. At the center of the growing controversy over gain-of-function research is Dr. Anthony Fauci, the recently-retired director of the National Institute of Allergy and Infectious Diseases (NIAID) and the U.S. pandemic "czar" under both Trump and Joe Biden. Early on, Fauci was among those pushing the animal market theory.

Expert testimony before the U.S. Senate has stated that there is *no civilian use* for gain-of-function experiments.[50] Such statements focus suspicion on the development of pathogens for use as bioweapons, which is how gain-of-function research has been identified by presidential candidate Robert F. Kennedy, Jr.

From the Amazon.com introduction to Kennedy's new book, *The Wuhan Cover-Up:*[51]

> Gain-of-function" experiments are often conducted to deliberately develop highly virulent, easily transmissible pathogens for the stated purpose of developing preemptive vaccines for animal viruses before they jump to humans. More insidious is the "dual

use" nature of this research, specifically directed toward bioweapons development.

The *Wuhan Cover-Up* pulls back the curtain on how the U.S. government's increase in biosecurity spending after the 2001 terror attacks—facilitated by Dr. Anthony Fauci, director of the National Institute of Allergy and Infectious Diseases (NIAID)—set in motion a plan to transform the NIAID into a de facto Defense Department agency.

While Dr. Fauci zealously funded and pursued gain-of-function research, concern grew among some scientists and government officials about the potential for accidental or deliberate release of weaponized viruses from labs that might trigger worldwide pandemics. A moratorium was placed on this research, but true to form, Dr. Fauci found ways to continue unperturbed—outsourcing some of the most controversial experiments offshore to China and providing federal funding to Wuhan Institute of Virology's (WIV's), leading researchers for gain-of-function studies in partnership with the Chinese military and the Chinese Communist Party.

But is Wuhan a red herring? Another possible source that has been mentioned for the COVID-19 virus is the U.S. military lab at Fort Detrick, Maryland. Former CIA analyst Philip Giraldi states:

> The argument that it originated…at Ft. Detrick, Maryland, and was deliberately weaponized and released in China to weaken that country's economy and military is somewhat compelling. [52]

Obviously, such speculation provides a new twist to the Wuhan lab leak theory, which is why some view the Wuhan allegations as a Deep State false flag cover story. Was this why, in a February 2023 *Fox News* interview, FBI Director Christopher Wray said:

> The FBI has for quite some time now assessed that the origins of the pandemic are most likely a potential lab incident in Wuhan. Here you are talking about a potential leak from a Chinese government-controlled lab.

Wray went on to say that the virus has killed "millions of Americans," which is not true, and "that's precisely what that capability was designed for,"[53] for which he gave no evidence. This is obviously irresponsible discourse and big-time China-bashing. Is it part of a government campaign to soften up public opinion for a possible future U.S. war against China?

Raising suspicions is the fact that Ron Unz of the *Unz Review* has documented that the presence of COVID in China was known to the Defense Intelligence Agency even before COVID began to be identified among the Chinese population in late 2019.

Notably, the initial outbreaks were in China, Iran, and northern Italy. Commenting on writings by Ron Unz, Dr. Kevin Barrett has suggested that COVID may have been developed as a military weapon against China and Iran, with blowback carrying the infection back to the U.S. and around the world. Barrett also points to massive infections of chickens and pigs in China as additional possible biowarfare attacks.[54]

As shown by an interview on *Redacted* with Sasha Latypova, a former pharmaceutical executive, indications are that the Pentagon was in charge of planning for the COVID pandemic, possibly starting as early as 2013, and that another pandemic may be in the planning stages. Allegations of the pandemic as a U.S. military action have also been made by Robert F. Kennedy, Jr.[55]

Although no one has demonstrated publicly where COVID-19 came from or whether it was released accidentally or deliberately, the one thing that is certain is that the COVID pandemic became the single most impactful event of the Donald Trump presidency. With the decision by authorities to conduct massive shutdowns of businesses, schools, restaurants, and public events, the U.S. economy crashed from mid-2020 onwards. Almost six million employees were laid off, with many more being allowed by their employers to work from home. Similar disasters took place in other countries.

The economic collapse that resulted was met by a type of "money-printing" beyond that which was undertaken previously by the Federal Reserve in 2008–2009. This time, instead of bailouts going to the banks, direct cash payments were made to U.S. taxpayers to help people ride out the crisis, along with support for extended unemployment benefits and loans to businesses. For many of these loans, repayment was waived.

The U.S. stock market went into steep decline, while continued low interest rates resulted in another housing bubble reminiscent of 2006-2007. Employees now allowed to work at home could buy a house anywhere they wanted. As the pandemic eased, inflation began to take off, with the Federal Reserve now courting recession by raising interest rates to levels not seen for fifteen years. The higher rates were intended to draw in capital from abroad for T-bond investments but rendered worthless the low-rate bonds already held by banks using them as reserves. The integrity of the entire U.S. banking industry was now placed at risk, demonstrating yet another fatal flaw in the Federal Reserve system.[56]

Meanwhile, with Dr. Anthony Fauci leading the U.S. response, Big Pharma reaped a bonanza. Over his fifty-year career as head of NIAID, Fauci had overseen the explosion of vaccines as the response of choice for infectious diseases at the same time the U.S. was seeing a huge increase in chronic and degenerative diseases like diabetes, obesity, asthma, and, among children, autism, ADHD, learning disabilities, and other serious developmental conditions. The government now recommends that children receive over sixty vaccine inoculations!

We have also seen explosive growth of iatrogenic illness, with adverse drug reactions becoming the fourth leading cause of death after heart disease, cancer, and stroke.[57] Relaxation of drug approval rules by the FDA has resulted in poorly tested drugs reaching the market, including children's vaccines.[58] Almost entirely ignored by officials and Big Pharma are issues of wellness, nutrition, and environmental stressors like air and water pollution.

Robert F. Kennedy, Jr. came out with his monumental book, *The Real Anthony Fauci: Bill Gates, Big Pharma, and the Global War on Democracy and Public Health* in 2021. Commentators who criticize Kennedy as being "anti-vaxx" have obviously never read this landmark volume which exposes the tremendous harm Fauci, Gates, and others in the health policy establishment have done.

During the pandemic, Fauci became a "beloved" figure as he seemed to be calmly guiding the nation through an existential crisis.[59] But many believe the lockdowns punished strong, healthy children and adults for an infection inflicted mainly on the elderly and people with pre-existing conditions.

Fauci's own preferred patent medicine, remdesivir, proved hugely expensive, dangerous, and largely ineffective. But it made Big Pharma a lot of money. Most prescribed medications, other than Paxlovid, are intended only for people already hospitalized or on ventilators.[60]

The focus of Dr. Fauci and most of the U.S. medical establishment was on the mRNA vaccine. But given that the Pfizer and Moderna vaccines granted emergency authorization by the FDA were virtually untested, a major controversy has arisen over their safety, due to large numbers of reports of adverse reactions, particularly myocarditis. As with other vaccines, Big Pharma was indemnified against lawsuits.[61]

Health authorities have failed in their responsibility to provide the public with information needed to make an informed decision about whether to submit to the vaccine. Neither the WHO, NIAID, CDC, FDA, nor employers mandating vaccination, have provided data on whether the vaccines are effective and actually prevent COVID—indications are that they are not[62]—or what exactly is the incidence and severity of adverse reactions. Thus we are seeing a total failure of the public health system at the federal, state, local, and international levels. Why is no one being held accountable for this?

Additional controversies exist about the recommended treatment modalities for COVID, though, since no proven cure exists, the pressure to step up use of still-unproven vaccines is overwhelming. Xavier Becerra, the California lawyer who heads the Department of Health and Human Services which oversees NIAID, CDC, and FDA, has recommended a COVID booster every two months![63]

One treatment modality that some doctors have used is ivermectin. But its use is ridiculed, even though more than eighty clinical studies are ongoing, mainly in "third world" countries. Being a generic medication, long in use and already approved for certain conditions, ivermectin as a COVID treatment would not make Big Pharma a lot of money. Fauci has demonized the use of medications like ivermectin and hydroxychloroquine that many believe saved Africa from the ravages of the pandemic.

Is there a more sinister agenda? Robert F. Kennedy, Jr. contends that the COVID vaccine was military-produced though marketed through Pfizer and Moderna. Dr. Kevin Barrett suggests that "…the mRNA rollout, and perhaps the pandemic itself, amounted to a large-scale test of new military technologies."[64]

F. William Engdahl writes, "Some organizations have suggested that the true aim of the vaccinations is to make people sicker and even more susceptible to disease and premature death." A comment supportive of population control by Bill Gates, billionaire head of the Bill and Melinda Gates Foundation, has been widely cited in support of this notion. In a talk at an invitation-only TED2010 Conference, Gates spoke of the need to reduce the earth's population. He said:

> First we got population. The world today has 6.8 billion people. That's headed up to about 9 billion. Now if we do a really great job on new vaccines, health care, reproductive health services, we lower that by perhaps ten or fifteen percent.[65]

"New vaccines" and "health care" *lowering* the population? Reproductive health services? One possible concern that has been raised with regard to the COVID vaccine is miscarriages among pregnant women. My personal physician lost *two successive babies* after receiving shots. She successfully carried a third baby to term after refusing another shot. Another concern has been the apparent frequency of healthy young people, often athletes, dropping dead from heart failure after being vaccinated.[66]

These concerns have become so widespread that they prompted an August 11, 2023 statement by Senator Ron Johnson to *Fox News* that, "This was all pre-planned by an elite group of people.... We're up against a very powerful group of people.... We are going down a very dangerous path, but it's a path that is being laid out and planned by an elite group of people that want to take total control over our lives, and that's what they're doing bit by bit."

For a glimpse at what may be the core of an ongoing plan for planetary population reduction, see Dr. Meryl Nass's analysis: "The WHO's Proposed Amendments Will Increase Man-Made Pandemics." *Brownstone Institute*, August 17, 2023. The article can also be found at merylnass.substack.com, August 17, 2023.[67]

The U.S. has also created a new Office of Pandemic Preparedness and Response Policy (OPPR) for "leading, coordinating, and implementing actions related to preparedness for and response to, known and unknown biological threats." Heading the office will be Major General (ret.) Paul Friedrichs. This office can be expected to work closely with WHO in its destructive undertakings.

The 2020 Election

President Donald Trump ran for reelection in 2020 but lost against former Vice-President Joe Biden. As was the case four years earlier, Senator Bernie Sanders was poised to win the Democratic Party presidential administration. Again, the party snatched the nomination away from Sanders. This time the DNC got Rep. Jim Clyburn of South Carolina to endorse Biden and pressured candidates Amy Klobuchar and Pete Buttigieg to drop out of the Democratic primary the weekend before Super Tuesday.[68] Biden won the nomination on the strength of the party establishment's "safe" constituency of black politicians.

During the campaign, hatred of Donald Trump became the dominant theme of Democratic Party politics. This attitude was exemplified by Jill Biden's advice to Bernie Sanders' supporters that they should switch allegiance and vote for her husband solely in order to defeat Trump.[69]

We are all well-versed in the objections President Donald Trump raised to the vote counting in the 2020 election. "Stop the Steal" is a memorable slogan. My own view is that Trump would have been better off to concede the election and wait for his next opportunity in 2024. But he allowed himself to be sucked into a quagmire that culminated in the "riot" at the Capitol on January 6, 2021.

The Deep State is expert at fomenting civil strife to make targeted groups look guilty, and there are certainly indications in the video footage that the Capitol police were allowing members of the crowd to move freely around the Capitol complex, even indicating, once inside, where and where not to go, resulting in some reaching Nancy Pelosi's office. Were there provocateurs inciting the Capitol violence? Were there provocateurs among Trump's inner circle egging him on in the days and weeks beforehand? We may never be allowed to know.

What we do know is that when Steven Sund, the chief of the Capitol police, requested National Guard assistance on January 6 to hold back the surging crowd, he was stonewalled for seventy-one minutes before his chain of command and the Defense Department gave him the help he needed to restore order. Sund also says he had never been told that the FBI was aware of the expected presence of identified domestic terrorists so was unable to take precautions ahead of time. Tucker Carlson, when interviewing Sund on his Twitter channel, said, "This sounds like a setup to me."[70]

An overlooked point is that it is impossible for *anyone* to verify the results of a U.S. presidential election. How can this be done with fifty states and several territories, each with its own voting rules, no uniform national system, and no impartial watchdog? How can voting machines be considered free from manipulation and accurate without a paper trail? In any event, only about half of registered voters even bother to vote. Yet the U.S. destroys other nations purportedly to make them more "free" and "democratic."

Russia Invades Ukraine

After Joe Biden was inaugurated, events moved inexorably toward the Russian invasion of Ukraine. In a 2019 study, the Rand Corporation wrote that "providing lethal aid to Ukraine" was a prime means for the U.S. to weaken Russia. A proxy war might do this, though Rand cautioned about the danger of the U.S. being drawn into a larger conflict.

Since the start of the Biden administration, the U.S. had been preparing a comprehensive array of meticulously targeted economic sanctions intended to take down the Russian economy once the conflict was ignited. The 2019 Rand study had identified sanctions as a prime means of attack, though it cautioned that "their effectiveness will depend on the willingness of other countries to join in such a process." It also noted that "sanctions come with costs and, depending on their severity, considerable risks."

When Russia invaded Ukraine, the sanctions were ready. They would backfire, weakening the EU, especially Germany, by loss of low-cost Russian energy imports.

Russia's Pre-Conflict Demands

Russia had its own perspective on what was about to break out. On December 15, 2021, about two months before sending forces into Ukraine, and having failed to forestall conflict via the Minsk agreements, Russia sent a list of last-ditch demands to the U.S. and NATO that was released publicly two days later.[71] The proposals were extremely broad, having two primary components. One was to return the borders of NATO to where they were before the Clinton administration began to move eastward by incorporating the nations of Eastern Europe. The second would require an equally profound shift in Western foreign policy. It called for the U.S. and Russia to pull all nuclear weapons out of any other nation where they were stationed and to confine them within their own boundaries.

The proposal would immediately eliminate the most aggressive features of the new Cold War the U.S. had launched and would have it pull back from the pressure it had been exerting against Russia over the last quarter-century.

The U.S. and NATO ignored the proposals. It was Russia's last attempt to get the attention of the West before resorting to arms.

The U.S. Started This War

By the winter of 2021–2022, the U.S. had its pieces in place as the NATO-trained Ukrainian army dug into their carefully prepared fortifications in western Donbass. The war in Ukraine had begun, even as Russian troops stood watching at the border. The trap against Russia had been laid. Three decades of hard work by the U.S. Deep State and its military and political subsidiaries would now pay off.

Though the Russians had moved their military forces to the borders of Ukraine over the weeks preceding the war, they refused to take immediate action when Ukraine's renewed bombardment of the Donbass began on February 16, 2022. According to Swiss military analyst Jacques Baud, *this date marks the actual start of the war.* Colonel Baud states that President Biden knew of this bombardment which preceded the Russians crossing the border eight days later and that if Germany and France had honored their commitment to guarantee the Minsk Agreements, the war never would have happened.[72]

Russia launched its "Special Military Operation" on February 24, 2022, explaining its attack by referring to the UN Charter. The UN Charter outlaws aggressive war but supports legitimate self-defense. It does not support U.S. global military hegemony, the Wolfowitz Doctrine, the "War on Terror" as defined by George W. Bush, or Full-Spectrum Dominance. These U.S. policies effectively are in contradiction to both international law and the UN Charter.

But this reminder: the U.S. acted to start the war by overthrowing the democratically elected government of Ukraine in 2014. The coup could not have taken place without U.S. planning and support. Victoria Nuland, Undersecretary of State for Political Affairs, said the U.S. spent $5 billion preparing for it. So from this perspective, when the Russians crossed the border, the U.S. proxy war against Russia was already *almost eight years old.*

Maintaining Control of Europe

Russia's invasion of Ukraine marks the end of the historical narrative contained in this book. A description of all that has happened since then would require a book of its own—in fact innumerable books.

One additional topic should be mentioned, however. This is the assertion by Dr. Michael Hudson and others that the U.S.'s real target of the war is Germany and, by extension, the rest of the EU. This makes sense in light of the narrative we have been following of how Britain, and then the U.S., saw as their primary objective to prevent the domination of the European Continent by any competing power. Hence the much-repeated refrain about the purpose of NATO being "to keep the Americans in, the Russians out, and the Germans down."

Some say that the worst nightmare for Britain was always a German-Russian alliance.[73] Now, the likelihood of German-Russian cooperation was emerging via the economic bonds being created through the construction of the Nord Stream 1 and 2 pipelines allowing Russia to supply gas for German industry without having to transport it across any East European nation. At a 2021 press conference with German Chancellor Olaf Scholz standing next to him, President Joe Biden said that Nord Stream would be terminated if Russia invaded Ukraine. When the pipelines were blown up on September 26, 2022, responsible commentators realized immediately that the act was likely carried out by the U.S., even as the usual media propagandists were making the absurd claim that Russia blew up the pipelines of which it was a partial owner—when if it had wished to exert pressure, it could simply have turned off the gas.[74]

The likelihood that the pipelines were destroyed by the U.S. was confirmed by the now-celebrated article by Pulitzer-prize winning journalist Seymour Hersh published on February 8, 2023. It was clearly an act of war perpetrated, Hersh claimed, by the U.S. and its collaborator Norway against Russia and the U.S.'s own European "allies." Then in April 2023, when Scholtz visited Washington, DC, the U.S. announced that Germany had been "asked" to move its armaments industries to the U.S.

The failure of Europe to protect its own economic interests with regard to the pipelines or protest their destruction demonstrated the extent to which the EU is a U.S. vassal. It was the U.S. and Norway that gained the most, with U.S. higher-priced liquified natural gas exports now finding a market in Europe and Norway's oil and natural gas exports to Europe reaping huge windfall profits.

Immigration

The huge number of immigrants unleashed due to America's "forever wars" and U.S. dollar imperialism that have combined to destroy civilized life in nations around the world has also facilitated U.S. domination of Europe. Millions of refugees from Iraq, Afghanistan, Syria, Libya, the Balkans, Central America, and elsewhere have poured into Europe, the U.S., and Canada both legally and illegally. European immigration was embraced by former German Chancellor Angela Merkel. To all of these have been added millions of Ukrainians fleeing poverty or escaping the U.S. proxy war against Russia.

Sweden, for example, has almost reached the point of having more immigrants than native-born persons. There has been little planning in any country for the assimilation of immigrants. The stress on public services and infrastructure, including education, housing, transportation, and public safety has ballooned out of control around the world. Jobs for most immigrants do not exist, leaving them dependent on state welfare. Immigrants are often jammed into high-rise apartments with multi-family occupancy, essentially slums. Under such conditions, France is exploding and Britain reeling, while other countries like Hungary seek to exclude immigrants seeking asylum altogether.

Young people with nothing to do, no hope, no income, and no future are a recruiting ground for criminal gangs which are growing exponentially within Western nations. Police and judicial systems cannot keep up, and crime rates are soaring. The same thing has happened in the U.S. that has afflicted other countries. Violent drug and human trafficking gangs from Mexico and Central America are ubiquitous. Everywhere nationalist-oriented political parties are gaining support as citizens rise up to oppose the social and cultural chaos.

Full-Spectrum Dominance Has Failed

U.S. action in fomenting the failing proxy war against Russia in Ukraine *may* be the last-gasp attempt at Full-Spectrum Dominance. The financial kingpins have so stripped the U.S. of its manufacturing base by shipping production to cheap-labor countries that, when faced with an industrial-level war as in Ukraine, the U.S. is no match for industrial powerhouses like Russia and its allies in the production of armaments. Going to war against a nuclear power like Russia is a far cry from taking down countries like Iraq, Afghanistan, or Libya.

The war in Ukraine is also generating an economic competitor and alternative to the West by speeding the growth of BRICS—the Russian-led consortium consisting of Brazil, Russia, India, China, and South Africa, with other nations now joining. At the BRICS summit in South Africa in August 2023, Argentina, Egypt, Iran, Saudi Arabia, the UAE, and Ethiopia were admitted. Alternatives to the U.S. dollar and SWIFT payment system are in the process of being created even as we speak. Putting out fires by attacking each of these nations individually or in combination is less possible by the day.

Meanwhile, the Federal Reserve and its subordinate central banks in the West have panicked. The current program of rapid and steep interest rate increases aims to suck in the resources of U.S.-controlled or still-influenced nations in order to compensate for the military setback in Ukraine. Statements that the Federal Reserve is "fighting inflation" is a smokescreen, as pulling surplus cash out of the economy can easily done by raising taxes. Interest rate increases may slow some types of economic activity, but they put upward pressure on prices by adding an interest surcharge to every transaction. *The actual intent of the interest rate increases is to ensure continued dollar hegemony.*

The result will be to crash what remains of the producing economies of the U.S. imperial sphere. It's the ultimate result of fractional reserve banking and compound interest usury. Insofar as no one benefits but the top layer of the financial elite, eventually the parasite kills the host. That is now happening. The screams of the host—earth's suffering human population—can be heard into the night.

Full-Spectrum Dominance has failed. The world is no longer willing to be victims of U.S. banking usury and hedge fund predators backed by military control.[75] *Now the U.S. must find a different basis for its existence.*

The rational choice is to become a responsible member of a multipolar world. This can only start with U.S. transition to a self-sustaining monetary system. The Appendix at the end of this book will attempt to explain what such a system might look like.

A major player standing in the way of such epochal change is Britain, the primary cheerleader for the 2003 U.S. war against Iraq and the present proxy war against Russia in Ukraine. When the Ukrainian government was close to ending the war with Russia through the Istanbul negotiations of March 2022, it was British Prime Minister Boris Johnson who traveled to Kiev to order Ukrainian President Zelensky to discontinue the peace talks, even though a signed agreement between Ukraine and Russia had already been reached. As a stunning two-part documentary by Scott Ritter titled *Agent Zelensky* reveals, the Ukrainian president has been under direct management by Britain's MI6.

Over a year and more than 300,000 Ukrainian dead soldiers later, on May 19, 2023, British Defense Minister Ben Wallace warned "that the UK could enter a direct conflict with Russian and China in the next seven years and has called for an increase in military spending to counter the potential threat." Speaking to the *Financial Times,* Wallace said "a conflict is coming with a range of adversaries around the world." [76] But insofar as retired British general Richard Barrons "claimed that the UK would run out of ammo within just hours in case of a major conflict,"[77] who would do the fighting? As with World Wars I and II, Britain must be counting on the U.S.

As of this writing, Ukraine's heavily touted spring 2023 counteroffensive has failed. Will the U.S./UK/NATO double down and escalate? Will other nations be drawn into the active fighting? Will Russia try to end the war with its own terminal counterattack? We don't know. We may hope that this is where the plan to "recover" the U.S. for the British Empire ends.

ENDNOTES FOR CHAPTER 17

1 See Jeremy Kuzmarov, *Obama's Endless Wars: Fronting the Foreign Policy of the Permanent Warfare State* (Clarity Press, 2019).

2 Ibid., 180.

3 Wayne Madsen, *The Manufacturing of a President: The CIA's Insertion of Barack H. Obama Into the White House* (independently published, 2012).

4 The DNC formula worked with Obama's two terms, but flopped when Hillary Clinton lost the presidency to Donald Trump in 2016. But Joe Biden came roaring back with his own "identity" campaign in 2020. By this time, Democratic Party identity politics had morphed into "wokeism" and "cancel culture."

5 I learned of this meeting after my retirement from the U.S. Treasury Department while writing my book, *We Hold These Truths: The Hope of Monetary Reform*.

6 This controversial depiction of the U.S. "drawdown" in Iraq has been promulgated by Colonel Douglas MacGregor.

7 Hillary Clinton said of the assault on Gaddafi, "We came. We saw. He died." She said this on TV while chuckling. For sheer cold-heartedness, her statement ranks with Madeleine Albright's televised remark that the death of 500,000 Iraqi children from the 1990s sanctions was "worth it."

8 Stephen Lendman, "Why Libya Was Attacked," in Cynthia Mckinney, ed., *The Illegal War on Libya* (Clarity Press, 2012), 105.

9 "Gold Dinar: the Real Reason Behind Gaddafi's Murder," *Millenium State* [blog], May 3, 2019. https://millenium-state.com/blog/2019/05/03/the-dinar-gold-the-real-reason-for-gaddafis-murder/

10 As of March 2023, the U.S. has at least 900 troops in Syria, deployed without permission of the Syrian government. These troops are engaged in removing oil from Syria's producing oil fields and transporting it into Iraq. According to the Syrian government, this oil, amounting to eighty percent of total Syrian production, is being removed illegally. The U.S. is reportedly also raising a new terrorist jihadi force to attack Syria's legitimate government. Such actions are glibly referred to as "foreign policy" and "soft power."

11 While I believe freedom of speech should apply to whistleblowers, I am not opposed to censorship in matters of morality and incitement to crime. For instance, I believe the government has failed miserably in censorship of pornography and prosecution of cybercrime.

12 Snowden's family roots go back to colonial days with ancestors serving in the Revolutionary War. NSA headquarters at Fort Meade sits partially on land once owned by the Snowden family.

13 Fred Weir, "War-ravaged Chechnya shows a stunning rebirth - but at what price?" *The Christian Science Monitor*, March 21, 2012. https://www.csmonitor.com/World/Europe/2012/0321/War-ravaged-Chechnya-shows-a-stunning-rebirth-but-at-what-price

14 Commentators have pointed out the resemblance of 21st century Russia and China to the U.S. under the "American System" in the late 19th century. See especially: Matthew Ehret, "Mendeleyev, Witte and the Revival of Russia's Lost Revolutionary Potential of 1905," *Matthew Ehret's Insights* [blog], November 14, 2021 – https://matthewehret.substack.com/p/mendeleyev-witte-and-the-revival; and Cynthia Chung, "WHY Russia Saved the United States: The Forgotten History of a Brotherhood," *Through A Glass Darkly* [blog], July 6, 2023 – https://strategic-culture.org/news/2020/12/25/why-russia-saved-the-united-states

15 Field Project Outline, Redbird/CARTEL, Case Officer to Chief of Field Installation, Munich, Germany, November 1946, Document MGMA-2167, Declassified and Released by the Central Intelligence Agency, 2007. Published by *Red Street Journal*.

16 Ukrainian Neo Nazi leader Stepan Bandera was assassinated by the KGB in Munich in 1959.

17 The Chernobyl nuclear power plant that melted down in 1986 is located in northern Ukraine.

18 Fabio G.C. Carisio, "CIA's Ground Branch is Training Ukrainian Paramilitaries against Russia," *GOSPA News*, January 15, 2022.

19 At most, about one percent of Ukraine's pre-war population was Jewish. Much of Ukraine's Jewish population had emigrated to the U.S. and Canada decades earlier.

20 Nuland was part of the Neocon cabal that directed the Project for a New American Century (PNAC). She is a top official in Biden's State Department and, according to Seymour Hersh, was part of the group that blew up Nord Stream 1 and 2. See note below.

21 For details on the Maidan snipers, see Kit Klarenburg, "'Rigorous' Maidan Massacre Exposé Suppressed by Top Academic Journal?", *ScheerPost*, March 18, 2023.

22 At a NATO summit press conference in Vilnius, Lithuania, on July 12, 2023, British Defense Secretary Ben Wallace said that the UK, the U.S., Canada, and Sweden had been building up Ukraine's military capabilities before the start of the conflict with Russia in February 2022.

23 "How did Odessa's fire happen?" *BBC News*, May 6, 2014. https://www.bbc.com/news/world-europe-27275383

24 *Redacted* newsletter, March 6, 2023. Already the U.S. was ramping up its propaganda barrage.

25 U.S. military drone attacks against "terrorist suspects" continue to this day. They are sometimes called "extrajudicial murder" and by the UN are termed "extrajudicial executions" (UN Office of the High Commissioner of Human Rights, https://www.ohchr.org/en/special-procedures/sr-executions). They are certainly unconstitutional as acts of war without a congressional declaration. See *Redacted* newsletter, May 19, 2023, for a particularly egregious killing of a civilian sheep herder.

26 USO.org reports that "30,177 active-duty personnel and veterans who served in the military after 9/11 have died by suicide—compared to the 7,057 service members killed in combat in those same 20 years."

27 Donna Brazile, "Inside Hillary Clinton's Secret Takeover of the DNC," *Politico*, November 2, 2017.

28 Controversy has arisen over whether individuals from Ukraine or the Ukrainian government donated money to the Clinton campaign. While Ukrainian oligarch Victor Pinchuk donated at least $8.6 million to the Clinton Foundation between 2009 and 2013, there is no record of direct campaign contributions. See Bill McCarthy, *PolitiFac.com* in *Austin-American Statesman*, February 24, 2022.

29 WIlliam Binneyo and Ray McGovern, "The Dubious Case on Russian Hacking," *Consortium News*, January 6, 2017. https://consortiumnews.com/2017/01/06/the-dubious-case-on-russian-hacking/

30 On May 15, 2023, Special Counsel John Durham issued an official Justice Department report concluding that the FBI should never have launched an investigation into the Trump campaign, as there was never any evidence of collusion with Russia in the 2016 election. The report stated that CIA Director John Brennan briefed President Obama and Vice-President Biden on the "alleged approval by Hillary Clinton on July 26, 2016, of a proposal from one of her foreign policy advisers to vilify Donald Trump by stirring up a scandal claiming interference by Russian security services." Thus, Obama and Biden were knowing parties to "Russiagate."

31 Matt Largey, "'Jade Helm' Conspiracy Theories Were Part Of Russian Disinformation Campaign, Former CIA Chief Says," *KUT 90.5*, May 3, 2018. https://www.kut.org/politics/2018-05-03/jade-helm-conspiracy-theories-were-part-of-

russian-disinformation-campaign-former-cia-chief-says

32 In the May 21, 2023 *ScheerPost,* Chris Hedges wrote: "The cynical con the Democratic Party and the FBI carried out to falsely portray Donald Trump as a puppet of the Kremlin worked, and continues to work, because it is what those who detest Trump want to believe. If Russia is blamed for Trump's election, we avoid the unpleasant reality of our failed democratic institutions and decaying empire.... The myth allows us to believe that Democratic politicians, like the establishment Republicans who have joined them, are the guarantors of a democracy they destroyed."

33 For a detailed look at the long list of Trump's decisions and behaviors that made the Deep State want to get rid of him, see Christian Parenti, "Trump against Empire: Is that why they hate him?" *The Grayzone,* February 15, 2023.

34 "Trump is considering new Glass-Steagall-style bank rules," *BBC,* May 1, 2017.

35 Kevin Barrett, "When the U.S. Government Murdered Qassem Soleimani: License to Kill?", *VT,* March 30, 2023.

36 Pompeo bragged in a speech at Texas A&M University that as Trump's CIA director, "We lied, we cheated, we stole."

37 Parenti, op.cit.

38 Interview with Col. Douglas MacGregor by Judge Andrew Napolitano, *Judging Freedom,* March 21, 2023.

39 Alexander Mercouris, *The Duran,* March 26, 2023.

40 "Libya's ongoing destruction belongs to Hillary Clinton more than anyone else. It was she who pushed President Barack Obama to launch his splendid little war, backing the overthrow of Muammar Gaddafi in the name of protecting Libya's civilians." *American Conservative,* January 10, 2020.

41 So called by Nomi Prins.

42 M. D. Nalapat, "Shadow men work to remove President Trump," *The Sunday Guardian,* January 22, 2017. https://www.sundayguardianlive.com/news/8132-shadow-men-work-remove-president-trump. As far as I can tell, this article, published in India, has never appeared in the Western media. It is still available, however, on the internet.

43 Some of the Shadow Men are congressional staff members. During my frequent treks up Capitol Hill as a government analyst, I also learned that every member of Congress has a Deep State "minder" to be sure they don't get any funny ideas about independent action. All politicians have to be "house broken" by the Deep State. Jesse Ventura has told some amusing stories about his contacts with Deep State operatives in the basement of the Minnesota state house. We have met others in this book in their contacts with Arkansas Governor Bill Clinton at the Mena airport.

44 Regardless, Trump's administration took numerous actions hostile to Russia, including sanctions and enhancement of the U.S. bases in Poland and Romania, along with covert aid to Ukraine in the military build-up after the Minsk Agreements.

45 On the theme of a "Shadow Men" cabal, see my own *ScheerPost* article from June 25, 2023, "A Wild, Conspiratorial, Fantastical View of World Politics: Might It Be True?"

46 The most notable exception was Henry Wallace, Roosevelt's vice-president from 1941–45. Wallace favored post-war peace with the Soviet Union, with Roosevelt moving in the same direction prior to his death. But Wallace was pushed aside in his own presidential aspirations by backers of Harry Truman who was used by Rockefeller-backed forces to create the CIA and national security state.

47 *RT.com*, March 4, 2023.

48 Deep State idiocy is boundless. In March 2023, *The New York Times* came out with a story that a "raccoon dog" in China was now suspected as being the source of the virus.

49 "Stop 'smearing' China with lab leak narrative, Beijing tells U.S.," *RT News*, February 27, 2023.

50 At an August 3, 2022, hearing of the Senate Homeland Security and Governmental Affairs Subcommittee on Emerging Threats, Dr. Richard Ebright, Rutgers University chemistry professor and Waksman Institute of Microbiology laboratory director, said that although "gain-of-function research can advance scientific understanding, it has no civilian practical applications." Further, in a statement to the House Select Subcommittee on the Coronavirus Pandemic, Dr. Robert Redfield, former director of the Centers for Disease Control and Prevention (CDC), said he had "no doubt" the National Institutes of Health (NIH) and Dr. Anthony Fauci funded gain-of-function research that likely resulted in the creation of COVID-19 and its subsequent leak. *Veteranstoday.com,* March 10, 2023.

51 Review by Robert Malone, "The Wuhan Cover-Up by RFK Jr.," *Brownstone Institute*, August 9, 2023.

52 Ekaterina Blinova, "Wuhan Lab 'Leak' Yarn Returns: How U.S. Reporter Pours Gasoline Into Washington's War Machine," *Sputnik News,* February 27, 2023.

53 *RT.com*, March 1, 2023.

54 Personal communication with Dr. Barrett.

55 Jeremy Kuzmarov, "Was the Pentagon and CIA Behind the COVID-19 Pandemic?" *CovertAction Magazine*, February 27, 2023. Kuzmarov writes further, "In a new book, *The Truth About Wuhan: How I Uncovered the Biggest Lie in History* (New York: Skyhorse Press, 2022), Huff claims that his boss at EcoHealth Alliance, Dr. Peter Daszak, was working with the CIA and that beginning in 2012, he oversaw the development of the biological agent known as SARS-CoV-2 that results in the disease COVID-19."

56 In March 2023, the Silicon Valley Bank in Santa Clara, California, closed its doors for this reason and was taken over by the Federal Deposit Insurance Corporation. Time will tell whether this signaled a larger collapse of the U.S. banking system similar to the 2008–2009 crisis.

57 Center for Drug Evaluation and Research, 2021.

58 In 1976–79 I worked as an analyst in the Office of the Commissioner at FDA. There was immense political pressure coming from Big Pharma to weaken FDA's drug approval process. In the 1980s after Reagan became president and Dr. Anthony Fauci became head of NIAID, the floodgates were opened with streamlined approvals becoming routine.

59 Fauci is "a beloved figure among Democrats," *Krystal Kyle and Friends*, July 22, 2023.

60 See the Mayo Clinic's "COVID-19 drugs: Are there any that work?" for a glimpse of the completely inadequate collection of treatments.

61 In his book on Fauci, et. al., RFK Jr. recounts that the test which FDA used to approve the Pfizer vaccine showed that vaccine recipients had more deaths from all causes than the placebo control group and four times as many cardiac arrests. See Table S4, p. 78.

62 My wife and I both came down with COVID after three shots. People say without evidence that the illness would be worse without them.

63 *Redacted* newsletter, August 9, 2023.

64 Dr. Keven Barrett, "Are COVID Vaccines—Like COVID Itself—a 'Pentagon Project'?" *Substack.com,* February 26, 2023.

65 F. William Engdahl, "Bill Gates talks about 'vaccines to reduce population,'" *Geopolitics-Geoeconomics* [blog], March 4, 2010. http://www.engdahl.oilgeopolitics.net/Swine_Flu/Gates_Vaccines/gates_vaccines.html

66 Despite the government's shielding vaccine manufacturers from liability, a class-action lawsuit has been filed by the families of COVID victims. See Monica Dutcher, "Families of COVID Victims Sue EcoHealth Alliance Alleging Gain-of-Function Research Caused 'Undue Risk' and 'Harm,'" *The Defender*, Children's Health Defense News and Views, August 14, 2023. "The families of four people who died from COVID-19 and one person injured by the virus are suing EcoHealth Alliance and the international nonprofit's president, Peter Daszak, PhD, and a cohort of government and elected officials, hospitals, military personnel, and others." The lawsuit cites a DHHS Inspector General's report described in a November 3, 2021 article by Sharon Lerner in *The Intercept*, titled, "NIH Officials Worked with EcoHealth Alliance to Evade Restrictions on Coronavirus Experiments."

67 According to Robert F. Kennedy, Jr., the World Health Organization is controlled by Bill Gates.

68 There were reports that Barack Obama made the calls to Klobuchar and Buttigieg.

69 Millions of people have allowed themselves to be goaded by propaganda into an irrational hatred of Donald Trump. I have never voted for Donald Trump but don't hate anybody, and I am a registered Democratic voter in the state of Maryland. I plan to vote for RFK Jr. in the 2024 Democratic primary on the basis of his position on the COVID pandemic and his opposition to the U.S. proxy war against Russia in Ukraine. I believe that Biden's actions in fomenting that war leading to the useless sacrifice of the lives of hundreds of thousands of people disqualifies him for reelection. Also see my *ScheerPost* article, "An Argument for the Relevance of RFK, Jr.," scheerpost.com, September 3, 2023.

70 Twitter.com/TuckerCarlson, Episode 15, August 10, 2023. Also see Steven Sund's new book, *Courage Under Fire: Under Siege and Outnumbered 58 to 1 on January 6.* Former White House economic adviser Peter Navarro "has said candidly in interviews that the plan among Trump loyalists was to use a peaceful protest at the Capitol to pressure Congress and Vice President Mike Pence to reject the certification of Electoral College votes from states where the results were being contested by the Trump campaign. Contrary to the claims of the January 6 [congressional] committee, Navarro said that the violence that erupted interfered with the plan and ensured its failure." Jordan Dixon Hamilton, "Peter Navarro Convicted of Contempt of Congress, *breitbart.com,* September 7, 2023.

71 Israel Shamir, "To Make Sense of War," *Veterans Today,* December 17, 2022.

72 Jacques Baud, "The Military Situation in the Ukraine: Update," *Labour Heartlands,* April 15, 2022.

73 This has historical precedence in the Molotov-Ribbentrop Pact of 1939 that preceded the outbreak of World War II. This short-lived German-Soviet non-aggression treaty that Britain succeeded in breaking up, threatened to undermine the project of enticing Germany and the Soviet Union into destroying each other. The possibility of a German-Russian alliance as a key to European solidarity was reflected in the aspirations of French President Charles de Gaulle for a united Europe "from

Portugal to the Urals." A similar idea was expressed by Mikhail Gorbachev in the late 1980s when speaking of "our common European home."

74 At a January 26, 2023, U.S. Senate hearing, Undersecretary of State for Political Affairs Victoria Nuland said, "…I think the administration is very gratified to know that Nord Stream 2 is now…a hunk of metal at the bottom of the sea." Previously, Secretary of State Antony Blinken described the incident as a "tremendous opportunity" for the U.S.

75 The hedge fund benefiting most from the Ukraine disaster is said to be BlackRock.

76 Reported in *RT.com,* "'Conflict is coming,' UK Defense Secretary Warns," May 19, 2023.

77 Ibid.

CHAPTER 18

Betrayals & Challenges

Thus says the Lord: Observe what is right, do what is just; for my
salvation is about to come, my justice about to be revealed.
Isaiah 56: 1, 6-7

Don't worry about anything; instead, pray by telling God
what you need, and thank him for all he has done for you.
St. Paul's Epistle to the Philippians 4:6

The son of Adam will not pass away from Allah until he
is asked about five things: How he lived his life, how he utilized
his youth, by what means he earned his wealth, how he spent
his wealth, and what he did with his knowledge.
Prophet Muhammad

For those who have an intense urge for Spirit and Wisdom,
it sits near them, waiting.
Patanjali, The Yoga Sutras

I have always prayed, and I believe that the Almighty
has always protected me.
Geronimo (1829–1909), Apache warrior

Everything on the earth has a purpose,
every disease an herb to cure it, and every person a mission.
This is the Indian theory of existence.
Mourning Dove Salish (1888–1936), Salish Indian Nation, Montana

All those who love money or the glory of the world,
adore and do homage to the devil.
Saint Anthony of Padua (c. 1190/91–1230)

Wirkliche Freiheit ist die Frucht erfüllter Notwendigkeit
und soll dazu dienen, Höheres als Freiheit zu erreichen!
Bô Yin Râ (Joseph Anton Schneiderfranken, 1876–1943)[1]

The Recurrent Anglo-American Devastation of Germany

In the pages of this book I have had a particular focus on aspects of history to which I have a personal or family connection. With Native Americans, specifically the Flathead tribes of Montana, the connection is to my place of birth in Missoula and my family ties to that region. In the case of Germany, I also have strong personal ties, have traveled often to the German-speaking countries, and am studying the German language. With the U.S. military's post-World War II occupation of Germany, hundreds of thousands of other U.S. citizens have spent time in that country or worked there, many marrying German nationals and raising German-American children, and some staying as retirees.

Native Americans and the people of Germany have both been subject to intense assault from the Anglo-Americans going back generations. Similar assaults have been perpetrated on African Americans and many others in the world due to the Anglo-American drive for world conquest under their Machiavellian "might makes right" obsession. Numerous nations have been attacked, exploited, or destroyed by the U.S. and their British allies since World War II. Many of the victims of Anglo-American aggression and their descendants are still suffering from serious psychological disturbances as a result. But human beings are resilient, though recovery takes time.

In their hubris, the Anglo-Americans now seem to have designated Russia, Iran, and China as the next candidates for destruction. This preposterous project stems from their panic at losing their grip on the world.

But the law of karma is inexorable. Today, as the populations of both the U.S. and Britain collapse into despair and growing poverty, the Anglo-Americans are now visiting on themselves what formerly they did to others. *I don't think it's a stretch to say that the Anglo-American elites are perpetrating what may prove to be genocide against their own populations.* Pharmacoterrorism, destruction of the quality of the food supply, open borders, weaponization of the judicial system, censorship of free speech, and the costs of education, homes, cars, and travel beyond the reach of ordinary people are directly tied to their policies.

Prior to the huge influx of immigrants into the U.S. around the period of the Civil War, and before the arrival of more immigrants from Ireland, Scandinavia, Eastern Europe, Russia, and Italy, the second most populous ethnic group after those of British origin were the Germans. The house I live in was built by Germans in the 1820s. German as a living language was spoken in parts of the U.S. into the 20th century. Much of the religion practiced in America derives from German sects.

Betrayals and Challenges

The German-speaking nations have been one of the world's intellectual and spiritual centers, giving us figures like Luther, Dürer, Bach, Leibniz, Mozart, Goethe, Immanuel Kant, Karl Barth, Bô Yin Râ, and many others. Going back to medieval days, we have Meister Eckhart, Johannes Tauler, "the Frankfurter" (author of the *Theologia Germanica*), Angelus Silesius, the Rhineland Mystics, and more.

But when the U.S. banking elite joined forces with the British to stamp out Germany as a commercial rival by waging what became World War I, the Germans were absurdly demonized as the "Hun." This demonization continued through World War II, with the Anglo-Americans showing every intent of wiping Germany off the map. Arguably the most important geopolitical events of the 20th century concerned Britain's extended takedown of Germany as its main continental European rival and its competitor in securing control of the world's energy resources. In order to succeed in this objective, Britain was able to commandeer U.S. financial and industrial might.

After World War II, a section within the U.S. government wanted to obliterate Germany as a modern state by reducing its economy to farmland.[2] A contrary view prevailed—that a strong industrial Germany, but without a military arm, would bolster U.S./British deterrence against the Soviet Union. Once again, Germany, now as part of NATO, was set up for potential conflict with a Russian enemy. This was despite a German affinity toward Russia going back centuries.

Meanwhile, without the wasteful burden of a military machine, Germany showed what a peacetime economy can accomplish by performing its postwar "economic miracle." Germany was able to rebuild its devastated cities, where approximately fifty percent of all structures in major German urban areas had been destroyed by British/American bombing,[3] producing obliterated city landscapes similar to those that now litter the Middle East due to the "War on Terror." A million German civilians had been killed in the bombing. Another million died after World War II from disease, starvation, and forced relocation.

Forbidden to rearm, the Germans were able to bestow their manifold technical talents and spirit of innovation on creating a consumer-oriented economy that became a world powerhouse and has formed the core of the productive infrastructure of the European Union. By 1973, the German economy began to run on cheap gas brought by pipeline from Russia. The gas also heated German homes, with the surplus being sold to other European nations. This demonstrated that cheap energy is a key to modern industry.

Today we see the challenges facing Germany as that nation struggles to cope with the current crisis marked by relentless pressure from its U.S.

overlord to support the U.S. proxy war against Russia in Ukraine. Many view the relationship between Germany and the U.S. as that of a vassal state to its feudal master. I personally have no doubt that the leading German politicians and media magnates are on the U.S. government payroll or under their control, by whatever other means.

Today is the third time in a little over a century that Germany and Russia have been manipulated into conflict, though Germany claims it is not a "party" to the Ukraine war—it's just supplying military equipment and munitions. But Germany is also fully in compliance with all the EU sanctions against Russia. These sanctions are backfiring and leading to Germany's economic decline, which, as Michael Hudson and others have averred, may be an underlying intent of the Anglo-Americans.

Today, Germany's role is playing itself out in the face of what much of the world acknowledges as the September 2022 U.S. attack on the Nord Stream pipelines intended to sever Germany from reliance on Russian natural gas. Nord Stream 1 was built to transit gas beneath the waters of the Baltic Sea and began to operate in 2011. When Nord Stream 2 was being built and readied for action, the U.S. began to balk at the implications of closer economic relations between Germany and Russia.

As outlined in the previous chapter, the U.S. role in the Nord Stream attack reached the news via reporting by Seymour Hersh. After Hersh broke the story of how the U.S., with Norway's help, perpetrated the crime, Hersh subsequently reported on a March 2023 White House meeting between President Joe Biden and German Chancellor Olaf Scholz, where the two discussed the pipeline exposé, and:

> ...as a result, certain elements in the Central Intelligence Agency were asked to prepare a cover story in collaboration with German intelligence that would provide the American and German press with an alternative version for the destruction of Nord Stream 2.[4]

Such a story did come out, claiming implausibly that Nord Stream 2 was blown up by a six-person attack using the rented yacht Andromeda and was done by elements friendly to Ukraine. Journalist Alex Christoforou likened the claimed assault to an attack by the "SS Minnow."

With the CIA cover story failing to fly, Seymour Hersh next reported that President Biden had decided to pull the trigger on Nord Stream in retaliation for Germany's being less than enthusiastic about providing weapons to Ukraine. In other words, it was both *blackmail* and *collective punishment* against the German nation. Some, including Hersh's sources, called it an act of war by the U.S. against its supposed German ally. The warmongering government of Poland joined in the pressure on Germany.

The prospect has now been raised of Germany and the rest of Europe becoming deindustrialized by the U.S. proxy war. Germany has also been pressured to deplete its small military arsenal by shipping tanks, guns, and munitions to Ukraine.

The German Green Party, which is part of the ruling coalition with the SPD under Chancellor Olaf Scholz, has been the most vociferous group in Germany in support of the proxy war. The Greens appear to dismiss the prospects of deindustrialization as not incompatible with their own agenda.

I could see clearly in my recent travels to Germany that dissension is growing against further participation in the war as the German economy has gone into recession. As the U.S. continues to state that it will support the government of Ukraine in its war with Russia "for as long as it takes," statements that are echoed by German leaders, *there is no clear definition by anyone in Germany* as to what the expected "end state" of its participation in the war is to be.

No one seems to have a clear vision of what Germany will look like when the war ends, what its future relations with Russia may be, or how Germany will cope in an environment where energy costs and inflation are exploding. The German government, along with other EU nations, has used subsidies to mitigate the impact of rising fuel prices, but some German factories are starting to shut down, such as those engaged in fertilizer and glass production. Bankruptcy rates have increased dramatically. U.S. hostility toward China may also cause reductions in German imports of cheap Chinese consumer products. More inflation may ensue.

While in Germany, I spoke with Germans about the economic situation. I saw anger over the government's embracing of the Green agenda which, along with the Ukraine war, is creating an energy crisis that may result in the national decline into deindustrialization long advocated by those who hate Germany. The Green Party's role in shutting down Germany's nuclear power plants is spoken of with contempt. BMW is converting to electric automobiles. Since these will run on electricity generated by coal-powered electrical plants, the electric cars are now being derided as "coal-powered" automobiles.

Germany has been a highly-organized economic dynamo based on manufacturing and export of well-designed consumer and industrial products. Germany also has a public banking sector that keeps mortgages within reach for ordinary people, though the inflation has also affected housing prices. Most people in Germany are making a living—there has been no impoverished underclass, except among the growing number of refugees, to which hundreds of thousands of Ukrainians have now been added.

But the German *stasis* is starting to tremble from the fallout from the U.S. proxy war in Ukraine. If this war devolves into a major economic breakdown, the results could be explosive, and it will not be the Russians who will be blamed. It will be the German pro-American government, above all the Greens who shut down the nuclear power plants.

Let me add that in spite of its national guilt complex often commented on, I do not hold Germans responsible for World War I or II. Those wars were foisted on them and on the people of other countries through circumstances no one proved capable of averting. The "war guilt" the Western allies imposed on Germany over World War II was no more valid than the one they imposed at the Treaty of Versailles. The Holocaust, an occurrence challenged by many for which some have been prosecuted, has enabled this reality to be overshadowed. The real war guilt belongs to the Anglo-American elite who seem convinced they'll also win the next world war, forgetting that it was really the Russians who decisively prevailed over Germany in World War II.

I have immense respect for the German people, regret the suffering inflicted upon them through the wars of the last century, and hope for a future of peace, where the Germans can find their place in the multipolar world that is arriving. Germany and the German-speaking nations of Switzerland and Austria have educated, resourceful, restrained, law-abiding populations. There is also an underlying spiritual vitality in these nations that is hard to define but exists nonetheless.

Eventually, the U.S. military machine will have to vacate Germany, as its presence is only an incitement to instability that a decreasing number of people in Germany want and no one needs.

The Americans claim NATO exists as a defensive alliance to counter Russia's "imperial ambitions." But what Russia really wants is for NATO, with its U.S.-provided nuclear weapons, simply to back off. Sensible Europeans understand this. Russia today is not a military threat either to Germany or the rest of the EU, unless the EU sticks with the Anglosphere through the coming crisis. Then all bets are off.

Meanwhile, it's Poland and tiny Lithuania that are beating the war drums most loudly against Russia. But they are doing this to further their own ambitions, not for European security.

I'll close this section with comments from author and German factory owner Dr. Fadi Lama, who recently wrote an important book: *Why the West Can't Win: From Bretton Woods to a Multipolar World* published by

Clarity Press. In a personal email, Mr. Lama writes of his own recent visit to Germany:

> …it was clear from closures of big industries in the region, and laying off hundreds of workers such as at Bosch Rexroth plant in the nearby small town of Michelstadt among others, that the German economy is heading downwards.

On Germany's recent deal with the U.S. for purchase of Liquified Natural Gas (LNG), he pointed out that:

> U.S liquefied natural gas (LNG) developer Venture Global LNG on Thursday said it had signed a twenty-year deal to provide Germany's Securing Energy for Europe GmbH (SEFE) with 2.25 million tonnes per annum (MTPA) of LNG.
>
> With this deal, Venture Global would become Germany's largest LNG supplier, with a combined 4.25 MTPA of LNG, the company said.
>
> From *Statista:* Up to May 2022 Germany was importing 5,000 million cubic meters or five billion cubic meters per month. From BP, converting 5 billion cubic meters to tons of natural gas = 3.675 million tons of LNG monthly or 44 MTPA.
>
> So the widely heralded deal with Venture Global amounts to less than 10% of what Germany used to get from Russia.
>
> This deal will in no way compensate for Russian gas. More importantly, industry needs cheap reliable energy to be competitive. LNG is significantly more expensive than pipeline gas, and shipping it thousands of miles makes it even more expensive. Furthermore, long term availability of sufficient gas for export from the U.S. is not guaranteed. Thus unfortunately, German deindustrialization will proceed at an accelerated rate.

N.G. Brown of *The Duran* on-line community writes:

Germany. Somebody please stage an intervention here. This is one sick puppy, who is pretty much slashing his wrists and arms and legs every single day and nobody does anything to stop it? Germany is bleeding profit and prosperity, jacking up its bills and cutting back on income, isolating itself from anything "scary" or "mean" per U.S. rules.

How long the German people will tolerate all this is the question of the hour for the future of Europe. As of July 2023, a poll by the Allensbach Institute for Public Opinion Research shows that the Olaf Scholz "Traffic Light" coalition, which includes the Social Democratic Party, the Greens, and the Free Democratic Party, is polling at twenty-one percent satisfaction vs. sixty-two percent a year ago.

This is a coalition that is collapsing, but what will replace it? Showing rapidly growing strength in the polls is the nationalist AfD Party—Alternative für Deutschland—but inclusion of the AfD in any manner has been anathema to German mainstream politicians and, by extension, their American masters. After being crushed in two world wars and lifted off the mat by American loans, the German mainstream today—including the press—has been thoroughly Atlanticist. Is this about to change?[5]

The Flathead Reservation Tribes' Survival and Adaptation

We have seen in earlier chapters how the Native American tribes of Montana's Flathead Reservation—the Salish, the Kootenai, and the Pend d'Oreilles—have occupied western Montana for thousands of years—since prior to the last Ice Age, when megafauna such as mastodons still roamed North America. The tribes belong to the hunter-gatherer culture of ancient North America, related to the similar cultures of northern Eurasia.[6]

Over the centuries, the tribes evolved a complex culture based on sustainable utilization of plant and animal resources. They organized themselves under a system of hereditary chiefdoms and, as they tell us, followed a religion based on spiritual affinity with the universe and its ruling spirits in an attitude of continual prayer. It is a bedrock principle of Native American religion to seek the aid of helping spirits through such practices as the vision quest, rituals, sacred dances and music, and observance of taboos.[7]

The Flathead tribes of Montana are among the most successful Native American nations. Their chiefs decided early on that they would not make war against the U.S. but would do their best to adapt, so they were spared some of the trauma the whites visited on American Indians elsewhere.

The tribes have taught themselves to be adept at modern self-government, including upholding their rights and interests in dealing with federal, state, and local governments. Due to an active program of purchasing non-tribal reservation land as it comes up for sale, the tribes now own a majority of reservation property, after losing much of it to whites under the federal government's Allotment decrees of the early 20th century.

Two active culture committees act as custodians of tribal history, language, and traditions. From the CS&KT website:

> The Séliš-Qlispé Culture Committee was first created in 1974–75 in response to the urgent concern of many traditional elders that we needed to take strong action to ensure that our culture would be carried on by the younger generations, and by the generations yet to come. Since that time, we have worked hard in many areas to ensure that both our language and way of life will always survive and flourish.

The Kootenai Culture Committee was created in 1975:

> ...the mission has been to preserve and perpetuate the traditional language and culture of the Ksanka people. This includes identifying, gathering, preparing, and storing all the traditional foods and medicines as well as carrying on all the ceremonial practices and all of the worldview values that go along with it.

A central task of the culture committees is to preserve the tribal religion. Some tribal members are also members of Christian churches and may practice both the native and Christian teachings. Some still attend Mass at St. Ignatius Catholic Church on the reservation, formerly the church of the St. Ignatius mission.[8] But the preservation and promotion of the ancient pathways remain central, particularly with the decline of vitality of modern secular American culture and its corruption with the idea that everything and everyone is for sale.

Tribal Government

The Flathead Reservation's tribal government includes agencies for finance, health, education, housing, economic development, police and courts, and natural resources. In 1976 the tribes established Salish Kootenai College, with its own TV station and commercial press. The tribes operate Two Eagle River School, where tribal languages and history are taught to school-age children. Tribal education also places strong emphasis on the visual and performing arts. Athletics also receive strong emphasis, with tribal youth excelling at basketball and football.

The tribal government is the reservation's largest employer, with 1,200 full-time and 600 seasonal employees. The tribes own the Seli'š Ksanka Qlispe' Dam on the Flathead River—the first Native American nation to own and operate a hydroelectric facility. They have also taken over ownership of the Flathead Reservation Bison Range from the federal Fish and Wildlife Service.[9]

With its own bank and lending program, the tribes seek to minimize economic "leakage," where tribal money or earnings of its members are sucked out of the community by businesses whose ownership lies elsewhere.

Native Americans, including tribal and non-tribal members on the reservation, continue to be outnumbered by whites. This goes back to Allotment days when "surplus" reservation land was opened to white settlement. Allotment also created a process of racial blending that has resulted in many mixed-race marriages.[10] Today some descendants of white Allottees are voluntarily willing their property to the tribes.

Since 1923, the jurisdiction of Lake County, Montana, and several other counties have been superimposed on the reservation, with Polson the Lake County seat and public schools serving both white and Indian pupils. The Catholic Indian boarding school at St. Ignatius closed in the 1970s.

Many challenges face the reservation tribes today. One is pressure from the rapid growth in population of cities and counties to the south and north of the reservation. The population of the region is now over a quarter million and growing. Many of the newcomers moving to Montana are retired people or independently wealthy. Many are work-at-home tech employees from California. These tend to drive up housing prices, exacerbating the lack of affordable housing in the region.

Also raising the price of properties is the growing practice of by outside investors of buying lower-end houses for use as Airbnbs, VRBOs, or second homes. It's local people who may have lived in a place their entire lives who are priced out of the market. No one at the political level in Montana or nationally has taken ownership of this growing scandal.

Health

The Flathead Reservation has been under the constant threat of endemic rural poverty, less so perhaps than some other Indian reservations, but the impact on health, including incidence of obesity, diabetes, and heart conditions, is serious. So is crowding in homes and limited access to health care facilities.

During the COVID pandemic, tribal members suffered the same traumas as rural populations elsewhere, with impacts especially on the elderly and those with pre-existing conditions, and isolation and mental health issues with people under lockdowns and quarantines.

Indian reservations also have a higher than average incidence of suicide, drug and alcohol addiction, and depression, with the Flathead tribes faring better than many but not immune. Early on, alcohol was outlawed on the Flathead Reservation, though readily available in adjacent towns and frequently brought onto the reservation by whites and Indians from outside. After the start of white homesteading around 1910, alcohol flowed freely. Today every town along U.S. 93, which runs the length of the reservation, has liquor stores, along with AA groups. The tribal police have the job of dealing with frequent DUI violations.

Fentanyl in its medical form is used in anesthesiology, but it has also become the greatest contributor to a national toll of overdose deaths now exceeding 100,000 per year. Fentanyl is so potent that Montana's drug authorities use "One pill can kill" as a slogan. Fentanyl is the substance of choice for drug-running Mexican gangs that work out of Washington, Oregon, British Columbia, and California and have penetrated all of Montana, including the state's seven Indian reservations. Tragedies often result, including overdose deaths that usually affect younger people.

Federal and state authorities in Montana have busted both large and small drug networks and are obtaining convictions of pushers arrested with stashes of meth and fentanyl, along with guns and rolls of cash. Unlike heroin and cocaine, fentanyl can be made in laboratories without plant-based substrates. The Drug Enforcement Administration reports that large quantities of chemical substrates for fentanyl enter Mexico from China.[11]

The tribes run a six-part suicide prevention and intervention program called Reason to Live Native that includes free suicide screenings, prevention and intervention activities, therapy and referrals, support groups, healthy activities that promote resiliency and a sense of belonging, and access to cultural activities that offer a sense of connectedness and wellness.

Affordable Housing

As noted previously, lack of affordable housing on the reservation is a serious problem, as it is throughout the U.S. Housing inflation on and around the Flathead Reservation has run in the range of two to three hundred percent, or greater, in the last two to three years. Tribal residents have been priced out of the market, particularly lower income members. Inflation of prices for lumber and building materials has made a bad situation worse. This has led to overcrowding and increased homelessness, with the attendant health risks.

It is increasingly difficult for tribal members to buy, build, or rent homes. The same goes for people moving to the reservation to work for the tribal government.[12] It's a growing crisis in adjacent areas as well.[13] At the same time, tribal members have an advantage in being able to lease land from the tribal government on which they can build homes.

The pressure on the tribal Housing Authority, led by Director Jody Cahoon Perez, has become immense, with staff working to find new options. She says that the "sweet spot" for an affordable house, given tribal income levels, is $125,000, which today is impossible to find. Options being explored include turning Amish sheds into "tiny houses," building a tribally-owned small-home manufacturing facility, and setting up a Habitat for Humanity group.

The Housing Authority administers several programs to help lower-income tribal members, including management of over 500 rental units, operation of two trailer parks, and making plans for a new multi-family rental project. Nine new single-family homes are being built for lower-income applicants. The Housing Authority is also purchasing a forty-acre tract near Pablo for a new home ownership project.

A recent study of the tribal housing crisis by a consultant documented the problems facing the reservation but gave no real solutions. No one at the political level has made any real attempt to explain the *causes* of the housing inflation or why the authorities do nothing about it. As in many instances in the past, the tribes are stuck with being part of a larger social and political system destructive to their way of life.

Natural Resources

Population growth and industrial development cause increased pollution and pressure on natural resources and the water supply. This especially applies to the regional growth of urban areas, where commercial facilities and upscale housing developments proliferate. Availability of water is a particular issue in the western states, including western Montana, where rainfall is less than fifteen inches per year. A growing problem is the aging of residential septic systems where waste leaches into streams and groundwater. This applies especially to Flathead Lake, where non-tribal members own most of the lakefront property.

The tribes have created a vigorous program of environmental monitoring, enhancement, and protection that includes restoration of bull trout habitat, restriction of wilderness areas to tribal use, protection of the grizzly bear population on the Mission Mountain range, and resolution of longstanding water rights issues in cooperation with county, state, and federal authorities.

Use and ownership of water rights was contested recently when a coalition of citizen groups was able to stop approval of an application by a local corporation to extract water from the Flathead basin aquifer and sell it at retail. The application would have allowed *two billion* units of water per year to be bottled and sold.

A major environmental issue is the spillover of pollution from areas adjacent to the Flathead Reservation. In December 2022, Rich Janssen, Jr., director of natural resources for the tribal government, traveled to Washington, DC where he delivered a statement to members of the Biden Administration and the Canadian Embassy about the threats posed to downstream waters by coal mining in British Columbia's Elk Valley. Janssen was part of a delegation of indigenous tribal leaders who cited the Boundary Waters Treaty of 1909, arguing that the Canadian government must do more to regulate the politically-powerful Teck Coal company.

The most serious environmental hazard is the toxic concentration of selenium in mining residues which then leaches into the watershed, with elevated levels having been detected. Responding to pressure, U.S. President Biden and Canadian Prime Minister Justin Trudeau issued a joint statement saying the two governments "intend to reach an agreement in principle by this summer to reduce and mitigate the impacts of water pollution in the Elk-Kootenai watershed, in partnership with Tribal Nations and Indigenous Peoples."[14]

Economy

On a topic reflecting reservation economics, an organization working to make a difference is the Mission Mountain Food Enterprise Center in Ronan. The center operates as part of Mission West Community Development Partners. The purpose is to support regional food sovereignty by offering facilities and training to gardeners, farmers, and ranchers in processing of garden produce and livestock for local marketing and consumption.

The tribes are also planning to construct a 4,000 square foot facility in Ronan for processing and selling meat from wild game such as bison, elk, deer, and moose, as well as cattle. The facility will utilize Ronan city water and septic systems.

Much of the leakage of tribal earnings and savings off the reservation is due to shopping for products produced elsewhere. A Walmart super-center sitting within the reservation at Polson illustrates the point.

Any contribution that tribal members can make to food sovereignty is a significant boost to individual and community sustainability, along with restoring a healthy diet. Another possible area of exploration mentioned for the reservation has been the development of a local scrip currency.

As have many other tribes, they have taken advantage of the economic opportunities that have come their way, including the establishment of casinos under the Indian Gaming Regulatory Act of 1988.

Among the economic issues facing the tribes is obtaining affordable sources of credit in order to buy or build a home or start a business. This is no different from the dilemmas facing people in the rest of the U.S. or worldwide where credit is controlled by private banking and financial institutions who lend to those already well-off. Publicly owned banks are desperately needed everywhere but are anathema to the financial elite. The tribal bank on the Flathead Reservation is such a public bank, but it can only lend to people with ability to repay. Among the native population, poverty rates exceed that of the population at large, both within Montana and the nation.

Thus an increased availability of lending resources may not be much of a help to individuals, which makes communal tribal enterprise so important. Historically, the federal government tried to exert control over the Indians by breaking down tribal culture into atomized individual and family economic units. Today this policy is being challenged by efforts by the tribal government to create a sustainable economic community.

Our 2023 Visit

Having paid two separate visits to Montana and the Flathead Reservation in 2022, I returned in 2023 with my sister Christine to attend the annual powwow at Arlee. Several thousand people congregated, mostly Native Americans, including Flathead tribal members and Indians from the Western states and Canada. The rest were white spectators, though the powwow is not heavily advertised as a tourist attraction. There were numerous children, many of mixed race, playing on the grounds.

Surrounding the open-air performance pavilions were extensive camping grounds with teepees, tents, and campers interspersed with portable toilets and a tent mall of vendors selling food, beverages, clothing, and souvenirs, including valuable hand-made jewelry.

This was the 123rd powwow at Arlee. During the late 19th century, when the government was suppressing traditional ceremonies, the Flathead government agency banned their performance. In his 1886 annual report, reservation agent Peter Ronan, in other respects a friend to the Indians, asked headquarters in Washington for funding to build two new jails to enforce "the Code of Indian Offenses." The first offense was performing the traditional dances. Another was activities of the medicine men.[15]

But the tribal chiefs reasoned that if they billed the event as a Fourth of July Independence Day celebration, little could be done to stop it. The powwow has gone on ever since, except for two COVID years during government-enforced lockdowns.

At times during the powwow's history, alcohol and drugs were a problem, sometimes instigating violence among both Indian and white attendees. Now at the entrance to the powwow complex a large electronic sign read,

"No Alcohol; No Drugs." Admission was free, though the reservation's armed tribal police look you over when you drive in.

During the five days we attended, what we saw was a joyful and intense expression of traditional dancing, drumming, and songs by 500 or so performers, most dressed in traditional costumes, celebrating a culture thousands of years old. Each session began with thanks to the Creator and ended with prizes to the winners of numerous dancing and drumming competitions.

There were also intervals when we spectators could circle around the dance floor with the costumed participants. The announcers spoke both in native languages and English. To experience the power of the chanting and drum circles with names like Blacklodge and Wild Rose was an unforgettable experience, one that my father also enjoyed when he would travel to the reservation in the 1940s with his homemade teepee.

The best source of information on the Arlee Powwow comes from Salish teacher and spiritual advisor Johnny Arlee, who was present and spoke at the 2023 event. A book containing his teachings is *Over a Century of Moving to the Drum,* published by the Salish Kootenai College Press and the Montana Historical Society Press. The word "powwow" means "war dance." In his book, Arlee explains many of the traditional dances, including the war dance, the scalp dance, the scout song, the prairie chicken dance, and others. The book also contains commentaries by participants, including "The Powwow for Me is a Very Religious Experience," an interview with Salish/Northern Cheyenne powwow dancer Bryan Brazill. He says that, "All the powwow dances came from spiritual beginnings."

The formation and cultivation of dancing and drumming groups form an important part of Native American life today. Powwows have proliferated and are performed at schools, community celebrations, public events, and competitions. Children start learning to drum, sing, and dance at an early age and continue through their life at school and beyond. Many of the performers make the circuit of Western powwows throughout the summer months.

Future of the Flathead Tribes

What is most impressive about the Native Americans of the Flathead Reservation is the age and resiliency of their culture. They have fought for what they have today—not by going to war against the whites, but by their determination and ingenuity in forging a unique way of life that combines their traditional culture with the demands of living in the modern world. Their determination in preserving the beauty, health, and diversity of the reservation ecosystem is remarkable, as is their compassion and sense of responsibility in caring for each other, including their children, in the face of so many challenges. The whites, including the churches, tried to turn the

Indians into individualists, competent imitators of the white man's way of life, but neglectful of any path resembling the communal values of the traditional Indian culture.

My own sense is that whatever the Indians today can do to elevate themselves above being subject to the pressures of inflation, food degradation, pollution, and congestion coming from the outside world, the better chance they will have of facing the future. Of course the same applies to all of us.

Conclusion: The U.S. Has Been Hollowed Out[16]

Control of the U.S. by an irresponsible financial elite has resulted in the following:

- growing poverty with pervasive household and business debt;
- lack of meaningful work amid increasing automation;
- wage stagnation;
- disappearance of affordable housing[17] and rampant homelessness;
- inadequate and unaffordable healthcare;
- inflationary profiteering by banks and big corporations;
- destruction of family farming, pervasive chemical pollution, GMO degradation of the food supply.[18]

We also face:

- Falling life expectancies with engineered addictions (opioids and fentanyl) and pandemics (COVID) combined with destruction of the family and of normal social and community relationships.
- Mass shootings, racial divisions, explosion of cybercrime, large and dangerous criminal gangs.[19]
- Private ownership of the money supply contrary to the Constitution, vast debt pyramids based on non-productive derivatives, financial exploitation by big banks and hedge funds, pervasive money laundering.
- Weaponized corporate-owned media monopolies monitored and controlled by the banks, hedge funds, and Deep State and a judicial system where destruction of individuals through state-sponsored "lawfare" has become the norm.
- Foreign policy based on wars of aggression, huge profits for weapons manufacturers, domination by political crime families and

politicians, corruption of the political process by wealthy donors and compromised legislators, worldwide CIA-sponsored terrorism, a revolving door between big business and government, a weaponized FBI at war with dissidence, illegal mass surveillance of citizens.

- With the collapse of the U.S. global financial empire, a breakdown in order can be seen everywhere. Criminality is epidemic. We see in particular the criminalization of the world of the internet and smart phones, with scams and larceny everywhere. Governments seem either incapable or unwilling to intervene.

- Our young people are being decimated. CDC reports that 2022 saw the highest rates ever of youth suicides and homicides. These are youngsters who have lost their reason to live.

- Some have suggested, with good reason, that the chaos we are witnessing, not just in the U.S. but throughout the world, is a campaign of genocide by the global elites against working class and lower income people, including indigenous. There is a repugnant term in circulation: "useless eaters."[20]

And on it goes. This disorder has been triggered by betrayals of America's founding principles by the financial elite and their political cronies.

Betrayals

We have betrayed the principles upon which our country was founded: as a republic based on law with a social contract between government and "We the People." These betrayals have been the result of many historical events:

- The strange coincidence of the assassination of three Republican nationalist presidents within the space of thirty-five years: Abraham Lincoln, James Garfield, and William McKinley, all of whom put the national interest at the forefront;[21]

- Formation of a "secret society" by Cecil Rhodes and Nathaniel Rothschild to "recover America for the British Empire" and collusion in that project by J.P. Morgan, the Rockefellers, the Schiffs, the Warburgs, and countless others;

- Surrender by Congress of its Constitutional authority over the nation's monetary system by passage of the Federal Reserve Act of 1913, followed by takeover of the nation's newspapers by the banking interests, entrance into World War I, then triggering the Great Depression by shipping U.S. gold to Europe;

- The decision by the Franklin D. Roosevelt administration to adopt the plan presented by the Rockefellers' Council on Foreign Relations for the U.S. to use World War II as a springboard to attain global military dominance, a plan reaffirmed by the Wolfowitz Doctrine of 1991 and today's military doctrine of Full-Spectrum Dominance;
- Creation of the national security state after 1946 with the Rockefeller-founded CIA being given unconstitutional powers to make covert war against anyone deemed a threat, with the CIA still today at the center of a huge organized crime syndicate;
- The 1963 *coup d'etat* that took place when President John F. Kennedy was assassinated, followed by assassinations of Dr. Martin Luther King, Jr., Robert Kennedy, JFK, Jr. and Senator Paul Wellstone;
- The neoliberal Rockefeller plan to remove the gold peg, implement the petrodollar, and make U.S. dollar hegemony the globalist tool for world dominance;
- Allowing the Neocons to infiltrate then take over the U.S. foreign policy apparatus in league with the Rhodes Scholar elite that today runs the Biden administration;
- The 9/11 false flag attack, now suspected by many to have been perpetrated by the government/Neocons, that ushered in the Bush/Cheney/Obama "War on Terror," which in reality is effectively U.S.-sponsored state terrorism against anyone in the world not a compliant vassal;
- Spending $19 trillion on bailing out the banking industry after the fraud-triggered crash of 2008–2009 without any substantive reform of the debt-based financial system;
- Use by the Deep State of electronic surveillance and manipulation of media to create a totalitarian propaganda and surveillance environment as shown by "Russiagate," "the Twitter Files," etc.;
- The COVID pandemic, likely resulting from gain-of-function research on viruses subsequently released at selected locations, followed by draconian lockdowns and enforced implementation of untested and dangerous vaccines.
- U.S./Neocon action in advancing NATO to Russia's borders and overthrowing the elected government of Ukraine in the coup of 2014, followed by provocations intended to draw Russia into the

current proxy war, with China/Taiwan next and World War III possibly to follow.

Our Predatory Financial System

The financial system continues to prey, not just on Americans, but on the entire world. As of April 2023, with their current program of steep interest rate increases, the Federal Reserve and other Western central banks are now conducting a massive raid to suck in the resources of nations still under U.S. control to prop up what is left of the producing economy of the U.S.

All this is part of the Great Reset, the final assault by the banking elite on the world's people—the terminal black hole of fractional reserve banking and compound interest usury.

The Federal Reserve claims the purpose of interest rate increases is to bring inflation back to their two percent target. But the inflation is largely a corporate hedge against supply chain disruptions, unrelated to "too much money pursuing too few goods," a situation that could easily be remedied through adjustments to tax policy. The same lie was propagated when the Federal Reserve ignited the Volcker Recession in 1979 on the heels of U.S.-instigated oil price shocks. That too was a banking elite power play.

Meanwhile, as warnings of recession loom, the Federal Reserve interest rate hikes create windfall bank profits. In April 2023, JP Morgan Chase reported record revenues and profits for the first quarter of 2023. Profits per stock share were up fifty-two percent from the previous year. The bank's net interest income grew forty-nine percent to $20.8 billion, with revenue for the consumer and community banking unit soaring by eighty percent.[22]

The *same day* that the report of profits at JPMorgan Chase came out, the media reported a mass shooting at an Alabama *teen birthday party* where four were killed and twenty-eight wounded. The assailant was still at large. These two events are connected at a deep level. "No man is an island." Do you hear that, Jamie Dimon et al?

Let's take one last look at how things actually work.

JPMorgan Chase is the flagship of the Rockefeller financial empire, headed by David Rockefeller until his death in 2017 and now by his heirs and successors. As an example of how things work, look at the world's largest microchip manufacturer, Taiwan Semiconductor Manufacturing Company (TSMC), one of the largest and most critical tech businesses on the planet.

While the Taiwanese government owns a small stake in TSMC, most of its stock is owned by U.S. banks and investment hedge funds, including JPMorgan Chase, Sanders Capital, Bank of New York Mellon Corp., Morgan Stanley, Capital World Investors, Massachusetts Financial Services, Fisher

Asset Management, and American Balanced Fund. Another major shareholder is Royal Bank of Canada. For a time, Warren Buffet owned stock. These are the firms that control TSMC and name its board of directors.

Banks could only hold these investments after the Bill Clinton administration repealed Glass-Steagall in 1999. Bill Clinton, with his banking deregulation program, including the appointment of Rockefeller protégé Alan Greenspan as Fed Chairman, was one of the best friends the Rockefellers and their ilk ever had.

Today, the U.S. is preparing to fight a war against China, with the main issue being the Chinese claim on Taiwan to which the U.S. agreed in 1979 and is also approved by the UN. *One of the main drivers toward World War III is to ensure control of TSMC and other assets on Taiwan by U.S. finance.* Indeed, Nancy Pelosi's provocative trip to Taiwan had a side agenda of shoring up her own Taiwanese assets.

I have a good friend with much wisdom who has no doubt that a nuclear holocaust will soon be triggered. Those in the Deep State probably believe they have deep enough bunkers to survive it. That's what "continuity of government" is about. Indeed, Anthony Blinken just averred on *60 Minutes Australia* that nuclear war is not worse than climate change. And why not? *The U.S. military has wanted to wage nuclear war against Russia since 1950 and has toyed with the notion against other countries as well, ever since.* And we know whose interests the U.S. military serves.

As I've stated earlier, one reason Russia and China are able to lead much of the world away from U.S. hegemony is that they possess *government-owned central banks* with the ability to direct public and private investment into socially beneficial domestic economic policies as defined by the central government. The U.S., by contrast, is saddled with the privately-owned Federal Reserve System whose *primary function is to make money for the already very rich.*

It does not matter what other changes or reforms are made if the Federal Reserve System is not abolished. Calls for economic reform by politicians like Bernie Sanders and Marianne Williamson complain about inequities *but never mention the Federal Reserve.* Calls by others like Ron Paul to "audit the Fed" are a tiny step in the right direction but again address only the tip of the iceberg. Demands by Occupy Wall Street to "Abolish the Fed" were more germane, which is why the movement was crushed by the police in the streets of New York City. The only 2024 presidential candidate to mention a review of Federal Reserve power is Ron Desantis.

So first, get rid of the Federal Reserve System. Let the U.S. financial industry go into a controlled bankruptcy. Build a new financial system based on Greenback-type currency, as Lincoln did. The NEED Act introduced in Congress by Representative Dennis Kucinich in 2011 would accomplish this. Such legislation could only be implemented by a national unity government. See the Appendix for further explanation of the NEED Act.

Closing Thoughts

I wrote this book from our home in Maryland, two miles from the Antietam Battlefield. The place is off the beaten path and retains many memories of our country's past, back to Native American and pioneer days, even further to the Ice Age and the raising of the Appalachian Mountains. Remnants of Indian life abound, like the rhyolite quarries for making spear points and arrowheads, and I've been able to discover an Indian burial ground along the Appalachian Trail, four miles away on South Mountain. The region has a strong German heritage, with annual German-American festivals. There's a living spiritual energy emanating from the many Christian churches, Protestant and Catholic, some dating to the 18th century, and from the presence of traditional Mennonite culture arising from churches, farms, and markets. The area is an oasis from the prevailing U.S. social chaos.

I have tried to interweave two particular story lines into this discussion of *Our Country, Then and Now*. One has been my family's history, including my own personal experiences. The second has been an appreciation of the history of the Native American tribes whose ancient homeland surrounds my birthplace in Montana. I believe these narratives are representative of American lives. Readers will have their own personal and family stories to relate. These are important to remember and share.

The conclusions I have drawn about the world crisis are obviously similar to those of many other commentators. But I also think I have elucidated some fresh perspectives on possible solutions at least to the financial crisis. Obviously, a narrative that covers 400 years of history in a single volume cannot include everything of importance. I know I have given short shrift to many important themes. One is the labor union movement. Another is the adoption by many black Americans of the Islamic religion through movements like the Nation of Islam. Another is the Latino culture involving vast numbers of recent Hispanic immigrants. Another is "climate change."[23]

In the face of all the problems we have been examining, we must now learn to move forward in a positive way, without hatred. To do so, we must embrace *both our diversity and our traditional values*. As expressed by the title of a famous book of the 1950s and 1960s, we need to rediscover and value *The Family of Man.*

This book is my hymn to The Family of Man, and a call for us to embrace each other in kindness and cooperation as we progress toward the future. The divisions within the U.S. today are appalling. So is the manifest greed, violence, and exploitation.

This must change.

ENDNOTES FOR CHAPTER 18

1 Translated: "True freedom is the fruit of necessity and serves to attain greater things than freedom." For those who wish, I would like to recommend the writings of German Bô Yin Râ as a guide to humanity's future. His books may be found in English translation on the website of the Kober Press of Berkeley, California. The works of Bô Yin Râ have provided guidance to thousands in the German-speaking world for over a century. Germany will survive the present world crisis, and I firmly believe that the time will come when the teachings of Bô Yin Râ will provide a basis for the next phase of Western spirituality. Particularly recommended is the book *The Meaning of This Life.*

2 This was the Morgenthau Plan proposed by U.S. Secretary of the Treasury Henry Morgenthau in a 1944 memorandum entitled *Suggested Post-Surrender Program for Germany.* The plan was enthusiastically supported by Prime Minister Churchill's top scientific adviser, Frederick Lindemann.

3 The leading advocate for saturation bombing of German cities during World War II was the aforementioned Frederick Lindemann. He wrote that in order to maximize civilian casualties, "Bombing must be directed to working class houses. Middle class houses have too much space round them, so are bound to waste bombs."

4 Seymour Hersh, "The Cover-Up," *Substack.com,* March 22, 2023.

5 Former French president Nicolas Sarkozy says that Europe is "dancing on the edge of a volcano." *RT.com,* August 24, 2023.

6 See Ake Hultkrantz, *Native Religions of North America: The Power of Visions and Fertility* (Long Grove, Illinois: Waveland Press, 1998).

7 The best available account of Montana native spirituality is Verne Dusenbury, *The Montana Cree: A Study in Religious Persistence* (University of Oklahoma Press, 1998).

8 Telling "Coyote Stories" is an important means by which the tribes socialize and pass on life lessons and traditional knowledge. These may be told at public events. Liz Dempsey, "Telling Coyote Stories passes culture and knowledge to the next generation," *Char-Koosta News,* March 2, 2023. http://www.charkoosta.com/news/telling-coyote-stories-passes-culture-and-knowledge-to-the-next-generation/article_c96e3d22-b951-11ed-b63e-97b3f9c0a385.html

9 The Bison Range with tribal commentators is featured in Ken Burns' four-hour documentary, *The American Buffalo,* due to be released by PBS in October 2023. The range has played a critical role in the preservation of the nearly-extinct North American bison, now numbering over 500,000. Bison are also being raised on numerous commercial ranches. For a detailed account of how the Bison Range was restored to the Flathead Reservation tribes, see Micah Drew, "Return of the Buffalo," *The Flathead Beacon,* May 25, 2022.

10 Of course even before Allotment there were mixed marriages, including the common practice of white traders marrying Indian wives.

11 It is not only Mexican drug gangs responsible for the drug epidemic. In April 2023, Joanne Marian Segovia, the head of the San Jose, California, Police Officers' Association, was charged with spending nearly a decade of her tenure trafficking in synthetic opioids from half a dozen countries and shipping them to U.S. customers out of her official office. Busted by the Department of Homeland Security, she faces a possible twenty years in prison.

12 As of 5/19/2023, Zillow lists only one house in the St. Ignatius area on the reservation for under $200,000; it has 918 square feet and costs $197,000. Another house has 846 square feet and costs $239,000. One with 1,032 square feet costs $255,000. These are the three cheapest houses in the area, completely out-of-reach for area wage earners.

13 On June 16, 2023, the *Flathead Beacon* reported that "Janice Hensel, a Missoula resident…was recently displaced from her longtime [Missoula, MT] apartment after the property management company that owns the building raised her rent from $1,200 a month to $1,930 a month, a rent increase of roughly 60%. Hensel, a retired paralegal who now lives on disability income, and her husband Bob Burt, who recently suffered a brain aneurism, are two of many Missoula residents forced to navigate housing prices that have far outpaced wages in the area over the past decade." Comments by readers point not only to housing inflation and shortages but to tax increases by local government as a factor. One commenter alleges that one of the major property owners raising rents in Missoula is a convicted drug dealer. Local and state legislatures are doing nothing to ease the crisis or stop profiteering, even as elderly disabled people are driven from their homes.

14 Tristan Scott, "Indigenous Leaders Take Transboundary Pollution Concerns to Washington," *Flathead Beacon,* March 29, 2023.

15 Peter Ronan, edited by Robert J. Bigart, *"A Great Many of Us Have Good Farms": Agent Peter Ronan Reports on the Flathead Indian Reservation, Montana, 1877–1887* (Pablo, Montana: Salish Kootenai College Press, 2014), 364.

16 National Security Adviser Jake Sullivan used this phrase in an address of April 27, 2023: "America's industrial base has been hollowed out," similar to statements made by President Donald Trump for years. Sullivan stated that the cost of rebuilding U.S. industry is $3.5 trillion, a figure much too low. Sullivan offered no answer to where the investment capital would come from.

17 U.S. cities are being made unlivable by the inflation of housing prices, both purchase and rental. The government has no explanation or solutions for this inflation. In fact, governments promote inflation, as inflation drives up tax revenues, including local real estate taxes.

18 One of the most dangerous and ubiquitous pollutants is the herbicide atrazine, which Robert F. Kennedy, Jr., says can alter the sex of frogs and that exposure among young people should be a cause for concern.

19 In a discussion with a Washington, DC, police officer I was told that within that metropolitan area Latino gangs are conducting massive amounts of human trafficking, drug, gambling, and car theft operations while being ignored by politicians. The same is happening in many other U.S. cities. The gangs are connected with Mexican cartels which operate with impunity in controlling the huge influx of illegal immigrants from all over the world on the southern border.

20 See Michel Chossudovsky, *Global Research #5742626*, "'Billionaires Try to Shrink World's Population': Secret Gathering Sponsored by Bill Gates, 2009 Meeting of 'The Good Club,'": "Is Worldwide Depopulation Part of the Billionaires' Great Reset?": "For more than ten years, meetings have been held by billionaires described as philanthropists to Reduce the Size of the World's Population culminating with the 2020–2022 COVID crisis. Recent developments suggest that 'depopulation' is an integral part of the so-called COVID mandates including the lockdown policies and the mRNA 'vaccine.'" According to the *Wall Street Journal:* "In May 2009, the billionaire philanthropists met behind closed doors at the home of the president of the Rockefeller University in Manhattan. This secret gathering was sponsored by Bill Gates. They called themselves 'The Good Club.' Among the participants were the late David Rockefeller, Warren Buffett, George Soros, Michael Bloomberg, Ted Turner, Oprah Winfrey, and many more. In May 2009, the *WSJ* as well as the *Sunday Times* reported (John Harlow, Los Angeles) that 'Some of America's leading billionaires have met secretly to consider how their wealth could be used to slow the growth of the world's population and speed up improvements in health and education.'" The emphasis was not on population growth (i.e. Planned Parenthood) but on the reduction in the absolute size of the world's population. Many support there being a reduction in the absolute size of the world's population. The question is: by what means? The lack of governmental care for the wellbeing of their domestic populations has had many manifestations, including government failures to protect their own populations as with the Katrina hurricane disaster and the 2023 Hawaii wildfires. Another is the U.S. proxy war against Russia in Ukraine, where the Biden administration is determined to fight to the last Ukrainian, with the death toll on both sides already close to half a million. Will the global elites simply try to save the system while sacrificing the masses, as suggested by William I. Robinson in *Can Global Capitalism Endure?* Or could their plans be more pernicious yet?

21 These American assassinations were mirrored by assassinations of similar progressive political figures in Russia, which saw itself as an ally of the U.S. from the time of the Civil War until well into the 20th century.

22 JP Morgan posts record profits," *RT.com,* April 17, 2023.

23 See Josh Mitteldorf, "Yes, Ecosystems Are Collapsing. No, It Has Nothing to Do With CO_2," *Children's Health Defense,* August 7, 2023. https://childrenshealthdefense.org/defender/ecosystems-collapsing-not-co2-emissions/

APPENDIX

Monetary Reform

It's one thing to call for the end of the Federal Reserve. But it's also essential to define what should be put in its place. In 2011 Congressman Dennis Kucinich introduced his National Emergency Employment Defense (NEED) Act. The NEED Act was created through research and proposals originated by Stephen Zarlenga of the American Monetary Institute.

- The NEED Act would abolish the Federal Reserve System and replace it with a Monetary Authority within the US Department of Treasury. The Monetary Authority would serve as a central bank, a depository for federal funds, and a point of origin for direct issuance of US currency. The Federal Reserve infrastructure would be absorbed by the Monetary Authority.

- The NEED Act would restore to Constitutional government the sovereign power to create money. The private banking system could only lend money beyond its deposit base by borrowing it first from the Monetary Authority according to established guidelines. Bank issuances would also be in US currency.

- The NEED Act is not socialism; private enterprise would be the backbone of the economic system where money is controlled by a responsible public agency.

- Through the NEED Act, the federal government would provide direct funding for infrastructure projects, for paying down the national debt, and for interest-free loans to state governments.

- Capitalism would be restored to its proper place, which is to finance production, not as bank-imposed debt parasitical to the working economy.

- Legitimate earning of profits by productive enterprise would remain.

- Speculative trading in derivatives and all forms of money laundering would be banned.

- A citizen's dividend could be applied as needed that would allow workers and families to flourish through income adequate to meet everyday needs. The dividend would be flexible, reflecting actual economic production and would meet the intent of proposals for a basic income guarantee.

- Savings would be encouraged through a responsible banking system that treats credit as a public utility. The move by the Federal Reserve to control every individual's funds through a Central Bank Digital Currency would be halted.

- Federal expenditures would be generated through direct spending, authorized by law, by the Monetary Authority, as was done through the Greenbacks. The national debt would gradually be paid off, with no further need for federal government borrowing. Taxation could then be adjusted downward with no need to pay interest on the national debt.

The catastrophic weakness of the bank-centered financial system was exposed during the 2008–2009 collapse and the Great Recession that followed. The experience was repeated in the economic meltdown that came with the COVID pandemic of 2020–2022. In both cases, the government drove the nation into unpayable debt to forestall collapse of the U.S. economy. The national debt currently stands at $31 trillion. The only way it can ever be eliminated under present policies is through hyperinflation that is destroying the finances of every US citizen.

The NEED Act stands as a ready alternative. Similar ideas of direct payment of government obligations through self-generated currency have been circulating for generations. The NEED Act brings these ideas to a form suitable for immediate implementation. The NEED Act is part of the platform of the Green Party.

The following resources are recommended for further study and action:

Congressman Dennis Kucinich's 2011 NEED Act: https://www.congress.gov/bill/112th-congress/house-bill/2990

Stephen Zarlenga, *The Lost Science of Money: The Mythology of Money, The Story of Power* (American Monetary Institute) https://monetary.org/misc/buy-the-book/

Stephen Zarlenga, *Presenting the American Monetary Act* (American Monetary Institute, 2010). Includes the text of the act introduced in 2010. https://monetary.org/images/pdfs/32-page-brochure.pdf

Matthew Ehret, *Rising Tide Foundation*, https://www.risingtide-foundation.org, and *Canadian Patriot Review*, https://canadianpatriot.org

Alliance for Just Money, https://www.monetaryalliance.org; Alliance for Just Money Board Member Howard Switzer blogs at *Howard's Newsletter*, https://howardswitzer.substack.com

Ellen Brown, *Web of Debt: The Shocking Truth About Our Money System and How We Can Break Free*, 5th rev. ed. (Third Millennium Press, 2012), https://www.worldcat.org/title/778886199. Brown is founder and co-chair of the Public Banking Institute: https://publicbankinginstitute.org/

Christine A. Desan, "How To Spend a Trillion Dollars: Our Monetary Hardwiring, Why It Matters, and What To Do About It," Harvard Public Law Working Paper No. 22-04, March 12, 2022. https://papers.ssrn.com/sol3/papers.cfm?abstract_id=4056241. Desan is managing editor of the online monetary reform project, Just Money: https://justmoney.org.

Also see my own writings:

Richard C. Cook, *We Hold These Truths: The Hope of Monetary Reform* (Denver, Colo.: Tendril Press, 2009). Explains the gap between GDP and national income that is filled by bank lending. https://www.worldcat.org/title/733234833

"A Bailout for the People: Dividend Economics and the Basic Income Guarantee," presented at the 8th Congress of the U.S. Basic Income Guarantee Conference and the 2009 Eastern Economics Association Annual Conference in New York on February 27, 2009. https://dandelionsalad.wordpress.com/2009/02/03/bailout-for-the-people-the-cook-plan-by-richard-c-cook-2/

Credit as a Public Utility: Solution to the Economic Crisis, 6-part video series [144:11 total RT], YouTube, last updated on July 4, 2011: https://www.youtube.com/playlist?list=PLFB9A12F6D128A4C6

"Failure of the U.S. Monetary System and the Solution," *ScheerPost,* June 7, 2023. https://scheerpost.com/2023/06/07/failure-of-the-us-monetary-system-and-the-solution/

For more information, email the author at: monetaryreform@gmail.com

Acknowledgements

First and foremost I want to acknowledge the help of my cousin Johnny Lathrop, our family archivist, without whose unfailing support and encouragement this book could not have been written. Johnny read the entire manuscript at least three times, offering suggestions, reactions, and corrections along the way. He also furnished copies and originals of family documents that opened windows on key aspects of American history. A particular pleasure was our tour with our wives Karen and Christie at the Antietam Battlefield. Johnny is a true man of faith in the wilderness of this world!

Looking back, the professionals with whom I served during the first four years of my federal government career at the U.S. Civil Service Commission taught me what honest, compassionate public administration can accomplish if allowed to do its work.

Much later, the seven years I interacted with visitors while working with knowledgeable colleagues for the Maryland Park Service in the Blue Ridge Mountains gave me valuable grounding in the natural and human history of one of the crossroads of American history and taught me how to see through the eyes of a scout.

When it came time to write this book, Dr. Lewis Coleman, a medical pioneer in the study of the mammalian stress mechanism, was a companion, sounding board, and mentor through the entire writing and publication project. Particularly rewarding were our California meetings and our dialogue on the U.S. proxy war against Russia in Ukraine.

Fellow Clarity author Dr. Fadi Lama read an early draft thoroughly and critically and offered invaluable insight. Support, insight, and encouragement also came from Howard Switzer, Dr. Kevin Barrett, Ellen Brown, Matthew Ehret, F. William Engdahl, the *ScheerPost,* and others.

Rich Janssen, Jr., was a key source of help and inspiration in gaining access to the culture of the Confederated Salish and Kootenai Tribes of Montana, as were Jody Perez and other tribal members. I am especially grateful to those who make the Bison Range and Hot Springs mineral bath accessible to us visitors and to the staffs of the Three Chiefs Center and the Ninepipes Museum.

Nothing can replace the experience of my sister Christine and myself attending the 2023 Arlee tribal powwow and taking part in timeless Native

American ceremonies, music, and dances. We were also able to attend the 2023 Arlee Independence Day parade and rodeo. Regular reading of the *Char-Koosta News* and the *Flathead Beacon* provided an essential background narrative to the life of the Flathead Reservation, the Northwest Montana region, and the Montana Native American community.

My cousin Judy Laughrun and her husband Steve helped me regain my Montana roots, as did the staff of the Seeley Lake Historical Museum, near where my mom grew up in the Montana wilderness. Thanks also to the staff of Travelers' Rest State Park, where Lewis and Clark camped with their Corps of Discovery, the Historical Museum of Fort Missoula, and the clergy and staffs of the St. Mary's Mission in the Bitterroot Valley, the St. Ignatius Mission on the Flathead Reservation, and the St. Francis Xavier Catholic Church in Missoula.

I was born a block from St. Francis Xavier at St. Patrick Hospital. Thanks also to the University of Montana from which my dad graduated, where our family lived in student housing, and where I paid several visits during our 2022–2023 journeys.

Authors Dr. Donald Gibson and Peter Janney, both in their amazing books and personal contact, gave me essential insights into the life and death of my first political hero, President John F. Kennedy.

A regular diet of videos by Robert F. Kennedy, Jr., Scott Ritter, Colonel Douglas MacGregor, Alex Christoforou, Alexander Mercouris, Tucker Carlson, Tulsi Gabbard, and others kept me informed while trying to write during these days of geopolitical crisis and U.S. social chaos.

Toward the end of the writing came a reading of Dennis Kucinich's *Division of Light and Power,* bringing an important boost of morale and clarity in formulating key issues. Since I first met Dennis twenty years ago, his insight, integrity, and courage have always helped me feel I was on the right track. Thanks also to Dennis's wife Elizabeth and her help.

My publisher and editor Diana Collier of Clarity Press meant everything in getting what I hope is a coherent, readable document into the hands of the public. Her support was rock solid. Diana, I am a huge fan of Clarity Press and the many indispensable volumes you have published. Joining your roster of authors is a true honor.

Above all, my wife Karen has helped open my inner doors to the often-elusive peace of mind and heartfelt joy, also aiding the ongoing sharing of life with our beloved extended families. Her putting up with my absorption in the book for over a year was amazing, as was her companionship on book-related journeys to Montana, Germany, and California. Thanks too to all the hospitable people we met in the U.S. and Europe on these trips.

Services and friendships at the church we attend in Boonsboro, Maryland, have also meant much.

Finally, a special thanks to friends at the German Bô Yin Râ Foundation and the Kober Press in Berkeley, who have taught me that real freedom starts with meeting the requirements of necessity—*notwendigkeit*. America faces collapse today because we have ignored this principle and because such ignorance incurs potentially catastrophic guilt.

Let me add that my book is essentially a compilation of true stories reflecting the American experience. It would be foolish to deny that many of these stories are heart rending. In order for me to keep my balance and sense of humor I have had to share these stories with others, both in writing them and in offering them to you, the reader.

At the end of the day, all I can say to *everyone,* readers included, is "Thank you."

Index

9/11 263, 290, 315, 320, 327, 331, 341–349, 354, 361, 422
   Commission 343
   prelude to 321–324
   truth movement 343

A

ABM Treaty 354
Acoma Pueblo 330–31
Adams, John Quincy 48, 65
Adelman, Ken 352
adjustable rate mortgages 358
affordable housing 414–15, 420
Afghanistan 106, 119, 162, 210, 230, 253, 263, 287, 290, 292–94, 303, 321, 322, 324, 327, 332, 341, 346–50, 353, 355, 363, 364, 366, 368, 376, 381, 382, 396, 397
African American 85, 278, 311, 366
Agee, Philip 290, 308
Agent Orange 278
Agnew, Spiro 271, 286
Agrarian Republicanism 32
Aguinaldo, Emilio 137
AIG 359
Al 284
Alamo 50
Alaska 47, 51, 66, 71, 83, 107, 247
al-Assad, Bashar 368
Albright, Madeleine 319, 323, 334, 399
Aldrich, Nelson 183–85
Aldrich Plan 186, 188
Aldrich-Vreeland Bill of 1908 185–86
Allende, Salvador 281
Alliance for Just Money 431
allotment 117, 127, 141, 145, 178, 179, 180, 225, 227, 261, 262

All the President's Bankers (book) 183, 356
al Qaeda 294, 303, 321–322, 347–349, 351, 382
Alternative für Deutschland (AfD) 412
Al Wertheimer 223–24
America First Committee 237
American Bankers Association 188
American Israel Public Affairs Committee (AIPAC) 315, 351
American Monetary Institute 313, 429, 430
"American System" 53, 55, 57, 90, 399
The American Whig Review 64
Anaconda Copper 166, 226
Anaconda Lumber 221
Andreessen, Marc 336
The Anglo-American Establishment (book) 182, 203, 205, 383
Anglo-Saxons 153
Antietam, Battle of 88, 95, 96, 97–99, 100, 102, 126, 425, 432
Antifa 339
Appalachian Mountains 46, 55, 95, 425
Appomattox Courthouse 89
Arabia 194, 207, 245, 280, 294, 315, 322, 340, 341, 343, 353, 385, 397
Arab Spring 366
Aristotle 27–28
A River Runs Through It (film) 114
Arlee, Johnny 419
Arlee, Montana 147–48, 227, 433
Arlee Powwow 148, 227, 418–19, 432
Armstrong, Hamilton Fish 236–237
Arthur, Chester A. 110, 133
Ashcroft, John 349
Ash Hollow, Battle of 120
Ashkenazi Jews 108

Assange, Julian 368
assassinations 133, 136, 139, 154–55, 161, 193, 237, 264–65, 268, 269, 274, 285–288, 304, 314, 332, 337, 351, 367, 381, 385, 421
Astor, John Jacob 34–35
Atlantic Council 309, 339
atomic bomb 257, 262–64, 270
Attu Island 246, 248–49
Australia 129, 151, 160, 177, 352, 424
Austria-Hungary 108, 151, 192, 208, 231
Azov Battalion 374

B

Baghdad 46, 352–53
Baker, James 325
Balfour Declaration 199, 205, 245
Baltic Republics 240
Baltimore, Maryland 8, 29, 45, 54, 95, 97, 127
Bank for International Settlements 283
bank holiday 213
Bank of England 31, 157, 184, 190, 192, 207, 211, 236
Bank of North America 25
Bannock Indians 128
Baptiste, Catherine 146
Baring, Alexander 14
Barings Bank 14, 33–34
Barrett, Kevin 389, 391, 401, 432
Barrons, Richard 398
Baruch, Bernard 251, 273
basic income guarantee 284, 430
Basotho people 165
BATF 345
Batista, Fulgencio 258
Baud, Jacques 394, 403
Bay of Pigs 259, 262, 269
Bear Stearns 359
Becerra, Xavier 391
Begin, Menachem 295
Belgium 77, 79, 139, 156, 162, 163, 175, 193, 194, 237
Belmont, August 57, 86, 87, 88, 105, 134, 151
Benjamin, Judah 62, 88

Berlin to Baghdad railroad 163
Bernanke, Ben 357, 359, 364, 377
Bevin, Ernest 244
Bezos, Jeff 336
Bible 44, 45, 78
Biddle, Nicholas 58
Biden, Hunter 374–375, 381
Biden, Joe 2, 18, 19, 37, 209, 320, 332, 334, 336, 344, 348, 350, 351, 352, 362, 363, 364, 366, 374, 375, 376, 381, 384–87, 392–95, 398, 400, 403, 408, 417, 422, 428
Big Oil 315, 340
Big Pharma 339, 340, 345, 386, 390, 391, 402
Big Six 208, 356, 357
Bilderberg Group 283, 316
Bill and Melinda Gates Foundation 340, 391
Bill of Rights 10
Bimetallism 86
bin Laden, Osama 263, 321, 322, 347, 348
Birney, James G. 67
von Bismarck, Otto 94, 163
Bison Range, National 72, 149, 260, 414, 426, 432
Bitcoin 25
Bitterroot Valley 73, 77, 79, 81, 82, 115, 116, 117, 433
Blackfeet Indians 18, 73, 78, 82, 114, 116, 261, 330
Black Hawk War 40, 43, 102
Black Hills 24, 47, 118, 119, 120, 121
"Blackrobes" 76, 77, 78, 81
BlackRock 283, 316, 361, 404
Blacks 48, 49, 85, 117, 124, 125, 126, 131, 172, 175, 187, 284, 319, 333, 378
Black Sea 106, 162, 176, 192, 373, 375
Black Thursday 208, 210
"Black Wall Street" 126
Blair, Tony 347, 363
Blankfein, Lloyd 337
"Bleeding Kansas" 69
Bliss, Betsy Elizabeth 43
Bliss, John F. 41, 43, 44

Bliss, Thomas 1, 3, 4, 40, 91, 140
blitzkrieg 235
BLM 339
Bloomberg, Michael 336, 428
Blue Ridge Mountains 432
BMW 409
Boer War
  First 136, 167
  Second 168, 173, 174
Bolsheviks 200, 232
  Bolshevik Revolution 138, 158, 194, 200, 205, 232
Bolton, John 335, 382
Bonners Ferry Kootenai Indians 178
Boonsboro, Maryland 95, 97, 98, 434
Booth, John Wilkes 85, 94, 105
bootlegging 92, 135, 209
B&O Railroad 46, 95, 127
Borodino, Battle of 162
Boundary Waters Treaty of 1909 417
Bowman, Isaiah 237
Bowman, Robert 312
Boxer Rebellion 138
Bô Yin Râ 405, 407, 426, 434
Bozeman Trail 76, 119, 120
Braddock, Edward 95
Bragg, Alvin 385
Brazile, Donna 378, 400
Brazos River 50, 69
Bremer, Arthur 285
Brennan, John 400
Bretton Woods Agreements 240, 251, 281, 282, 410
Britain 3, 9, 10, 13, 24, 31, 33, 34, 39, 48, 54, 56, 57, 59, 64, 65, 67, 68, 76, 84, 85, 86, 87, 88, 97, 105–108, 110, 113, 114, 129, 131, 132, 136, 137, 139, 140, 151–71, 173, 174, 176, 177, 181, 182, 184, 185, 189, 191–210, 216, 217, 228, 231, 232–45, 250, 252, 257, 288, 292, 327, 337, 339, 341, 346, 347, 352, 368, 372, 395, 396, 398, 403, 406, 407
  Battle of 234
British Commonwealth 203
British East India Company 157, 162
British Empire 39, 65, 107, 139, 150–54, 157, 167–70, 173, 176, 181, 182, 194, 202, 203, 232, 237, 240, 337, 398, 421
British North West Fur Company 78
British "poodle" 240
Brooklyn Navy Yard 197, 218
Brown, Ethel 197, 218, 431, 432
Brown, John 69, 85
Bruton Parish Church 272, 273
Bryan, William Jennings 113, 130
  Cross of Gold speech 113, 130
Brzezinski, Zbigniew 283, 293, 339, 341
"bubble-ization" 359
Buchanan, James 60, 62, 87
Buchanan, Patrick 231, 318
Buffett, Warren 237, 268, 269, 274, 286, 357, 424
Buford, John 99
Bull Moose Party 179, 187
Bureau of Indian Affairs (BIA) 117, 127, 147, 177, 180, 225–27, 301, 308, 329, 330
Burnside, Ambrose 98
Burns, William 355
Bush, George H.W. 229, 273, 298, 299, 304, 306, 307, 311, 313, 314, 318, 324, 327
Bush, George W. 229, 272, 273, 286, 320, 322, 327, 332, 335, 336–364, 366, 380, 395
Bush, Prescott 229, 256, 299, 314
Business International Corporation 365
Butler, Smedley 228
Buttigieg, Peter 392, 403

C

Cabin Fever (book) 221, 224
Caldwell, Kansas 142, 217
Calhoun, John C. 53
Cambodia 271, 277, 278
Camp David 295, 334
Camp Fortunate 74
Camp Peary 2, 246, 249, 290

Canada 6, 7, 9, 11, 15, 39, 64, 66, 68, 71, 72, 73, 77, 84, 96, 107, 108, 128, 129, 146, 149, 151, 179, 293, 396, 400, 418, 424
Cape Colony 165, 166, 167, 174
Capitol building 135
Carey, Henry C. 56, 90, 107
Carlson, Tucker 339, 393, 433
Carnegie, Andrew 182
Carnegie Foundation 339
Carnegie Institute for International Peace 235
Carter, Jimmy 2, 283, 291–301, 303, 308, 386
Casey, William 300–304, 309, 324, 333
Cash-and-Carry 235
Castro, Fidel 258, 262, 272, 314
Catherine the Great 160–61, 176, 372
Cayman Islands 156, 326
CBS News 334, 335, 363
Cecil, William, 1st Baron Burleigh 154
Central Asia 158, 176, 192, 292, 327, 341
Central Powers 151, 191, 231
Challenger space shuttle 2, 300, 302, 311–313, 328
charcoal 46
Char-Koosta News 426, 433
Charles II
 restoration of 157
Charles V 156
Chase Manhattan Bank 236, 267, 282, 298
Chase National Bank 183, 252
Chase, Salmon P. 89, 93
Chechnya 371, 399
Cheney, Dick 286, 287, 298, 315, 317, 321, 327, 333, 335, 336–364, 340, 341, 344, 346, 349, 350–57, 361, 365, 377, 379, 386, 422
Cherokee Indians 7, 11, 39, 49, 142, 143
Cherokee Strip Livestock Association 142
Chicago 36, 45, 121, 126–27, 141–42, 269, 271, 361, 365, 382

Chief Bushyhead 142
Chief Charlo 116–17, 178, 180
Chief Eneas 178
Chief Victor 81, 116
Chile 49, 252, 273, 280, 281, 283, 288, 307
Chillicothe, Ohio 43
China 9, 35, 47, 112, 138, 158, 162, 170, 184, 235, 238, 241, 244, 279, 280, 288, 291, 294, 299, 320, 332, 334, 340, 355–56, 369–71, 381, 384, 387, 388, 389, 397–99, 402, 406, 409, 415, 423–24
"Chopsticks" 144
Christianity 7, 45, 83, 87
Christoforou, Alex 385, 408, 433
Church Committee 287–290, 310
Church, Frank 287, 289
Churchill, Randolph 167
Churchill, Winston 167, 196, 238, 245, 252, 303
 Iron Curtain speech (1946) 245
Church of England 3, 166
CIA 2, 210, 230, 237, 242, 243, 246, 251–65, 268–74, 276–78, 281, 285–89, 290–91, 293–94, 299–300, 303, 304, 306, 308, 309, 310, 311, 314, 315, 318, 319, 320–22, 324, 325, 327, 328, 332–34, 338, 339, 343, 348–50, 352, 353, 365, 366–67, 368, 371–75, 380–81, 386, 388, 398–402, 408, 421–22
 drug trafficking 290, 311
 torture chambers 277, 287, 349–350
Citibank 296
Civil War 2, 10, 18, 22, 36, 44, 47, 48, 51, 53–57, 60–62, 69, 83, 84–105, 106–112, 116–18, 120, 123, 125, 127, 130–32, 135, 140, 147, 151, 163, 164, 181, 183, 184, 195, 196, 200, 214, 218, 232, 337, 406, 428
Claessens, Brother 81
Clark Fork River 71, 73, 74, 83, 114, 115, 147, 226
Clark, Wesley 353
Clark, William 15, 75, 77–78, 83

Index

Classical Economics 56
Clay, Henry 53, 55–56, 59, 66, 90
Cleveland, Grover 110, 127, 130, 132–34, 137, 184, 328
Clifford, Clark 304
climate change 72, 424–25
Clinton, Bill 190, 311, 318–19, 333, 365, 371, 379–80, 401, 424
    Clinton administration 369, 371, 394
    impeachment 319, 324
Clinton, Hillary 351, 366–67, 375, 377–79, 382, 383, 386, 398–401
Clyburn, Jim 392
CNN 312, 339, 345
Coalition Provisional Authority 352
Cobell, Elouise 330, 335
"Cobell" project 329
cocaine 311, 415
C&O Canal 46, 55, 95
Code of Indian Offenses 418
Coinage Act of 1837 130
Coinage Act of 1873 113, 114
COINTELPRO 287
Cold War 156, 242, 244, 251, 252, 255–56, 273, 280, 315, 319, 320, 324, 355, 364, 370, 394
Coleman, Lewis 432
collateral damage 366
College of William and Mary 2, 12, 271–72, 276
Collier, John 227, 229
Colony Club 223
color revolutions 230, 310, 332, 354, 371
Columbia River 68, 71, 73, 74, 77, 226
Columbus, Christopher 5
Cominform (Communist Information Bureau) 252
Committee on the Present Danger 299–300
Commonwealth in Christ 135
Commonwealth of Independent States (CIS) 374
compound interest 28, 53, 86, 93, 131, 132, 183, 340, 371, 397, 423
Compromise of 1877 125

concentration camps 137, 175
Confederate States of America 47
Confession of Faith 168–170, 171
Confessions of an Economic Hit Man (book) 273
Connally, John 268
Consolidated Salish & Kootenai Tribes (CS&KT) 115–16, 260–61, 413
    Housing Authority 416
Continental Congress 21, 30, 198
Continental currency 21, 24
Continental Divide 13, 71, 74, 76, 114–16, 219, 225, 261
Continental System 161–62
Continuity of Government (CoG) 321, 424
Contras 310
Cooke, Jay 92, 111, 114, 132, 196
Cook, Frederick Steele Fitts 51, 197, 222
Cook, Richard C. 308, 431
Coolidge, Calvin 206, 210, 235
Cordilleran ice sheet 71–72
Côté, Jean-Marc 172
Council on Foreign Relations 204, 235, 236, 254, 269, 298, 309, 316, 318, 323, 337, 355, 356, 386, 422
cover-ups 268, 269, 274, 285, 312, 353
COVID-19 227, 332, 339, 360, 386, 387, 388, 389, 391, 402, 403, 414, 418, 420, 422, 428
Coxey, Jacob 135
Coyote Stories 426
Crazy Horse 8, 18, 120, 122, 124
Credit Mobilier 111
Crimea 85, 355, 373–75
    Crimean Peninsula 373
    Crimean War 106, 162
Crime of 1873 106–124, 130
critical infrastructure 346, 356
cryptocurrency 25
Cuba 65, 133, 137, 202, 258, 259, 262–64, 274, 288, 290, 349, 355
Cuban Missile Crisis 262–264
Cuban Revolution of 1960 133, 258
Currency Act of 1767 20–21

Custer, George Armstrong 8, 18, 100, 120–25, 193, 343
Custer's Last Stand 121–123
Czar Alexander II 106
Czechoslovakia 239, 252
Czolgosz, Leon 138

D

Dahlgren, Madeline 96
Dardanelles 194
Darwin, Charles 153
Das Kapital (book) 57, 110
Davis, Reuben 65
Dawes Act 127, 149, 178, 179, 224, 261, 329
D-Day 239, 248, 256
death squads 310, 350, 353
Debs, Eugene 188, 192, 198
debtors' prison 27
debt slavery 38
Dee, John 154, 171
Deep State 1, 18, 253, 289, 315–16, 320, 324, 338–39, 361, 362, 369, 372, 377, 379, 380, 381, 382, 386, 388, 393, 394, 401, 402, 420, 422, 424
de Gaulle, Charles 240, 264, 403
Dehio, Ludwig 152, 155, 171, 203
Democratic National Committee (DNC) 285, 366, 377–79, 392, 398, 400
Democratic Party 57–58, 63, 86–88, 122–23, 125–26, 130, 271, 276, 284–85, 293, 352, 366, 377–79, 385–86, 392, 398, 401, 412
Democratic Review 63
"deplorables" 379
Desan, Christine A. 431
Desantis, Ron 424
Desert Storm 230, 314
De Smet, Pierre Jean 79
détente 279, 291–93, 303, 370
Deutsche Bank 359
Dien Bien Phu 258
Dimon, Jamie 336, 359, 360, 382, 423
Dixon, Joseph 179–80

Dnieper River 233
Dobrynin, Anatoly 264
Dole, Robert 308
dollar diplomacy 184
dollar hegemony 282, 338, 371, 397, 422
dollar peg 241
domestic dependent nations 12, 302
domino theory 265
Donbass 355, 373, 375–76, 394
Doolittle, James 259
dot.com bubble 298, 357
Douglass, Frederick 85
Douglas, Stephen A. 87
Dow Chemical 2, 246, 249, 267
Drake, Edwin L. 46
Drake, Sir Francis 153
drang noch osten 327
Dred Scott decision 69, 85
drone assassinations 332
Drug Enforcement Administration (DEA) 415
Dukakis, Michael 314
Dulany, Daniel 95
Dulles, John Foster 257
Dunham, Ann 365
The Duran 364, 411
Durant, William C. 134
Dutch East India Company 157

E

Early, General Jubal 100
Earned Income Tax Credit 284
Eastern establishment 337
Eastern State Hospital 278
The Economist 339, 382
Edison, Thomas 133
Edmunds, George F. 148
Edward VII 167
Egypt 106, 160, 167, 174, 295, 315, 367, 397
Eisenhower, Dwight D. 238, 239, 254, 256–60, 262, 265–66, 302, 386
farewell address 259
Elbe River 239

Electronic Funds Transfer (EFT) 26, 295, 328, 329, 336, 346
Elizabeth I 153–156
Emancipation Proclamation 106
Emerging Economic Stability Act of 2008 360
Empire State 55
empresarios 50
Engdahl, F. William 205, 235, 244, 250, 273, 307, 391, 403, 432
English Channel 3, 152, 154, 238
Entente Cordiale 164
Environmental Protection Agency (EPA) 284
"Era of Good Feeling" 55
Erie Canal 45, 55
Espionage Act 198, 368
eugenics 153, 364
Eurodollar 241, 282, 337
European Union (EU) 264, 339, 374, 376, 394, 395–96, 407–410

F
Facebook 339
Fair World Order 340
Falwell, Jerry 320
The Family of Man 425
Family Assistance Plan 284
Family Jewels 287
The Farm 246
Fauci, Anthony 387, 388, 390, 402
Federal Bureau of Investigation (FBI) 210, 269, 273, 275, 278, 287, 311, 320, 321, 324, 334, 338, 339, 347, 348, 379, 381, 388, 393, 400, 401, 421
Federal Deposit Insurance Corporation (FDIC) 26, 214, 305, 402
Federal Drug Administration (FDA) 390, 391, 402
Federalist Party 53
Federal Reserve 23, 25, 27, 38, 61, 83, 88, 93, 112, 134, 172, 182, 184, 186, 188–91, 195, 196, 206, 207, 210–14, 228, 236, 241, 282, 283, 295–98, 300, 305, 308, 313, 329, 356, 357–60, 377, 389, 390, 397, 421–24, 429–30
  Act of 1913 23, 61, 189, 206, 421
  Bank of New York 206, 236, 297, 360
  Notes 25, 61
Feith, Douglas 299
Fentanyl 415
Ferdinand, Archduke Franz 193
Fetterman Fight 120
Fifth Republic 264
Files, James E. 274
Financial Management Service (FMS) 295, 312, 328, 346
Fink, Larry 336
"FIRE" economy 183
First Bank of the United States 28–32, 34–35, 55
The Flathead Beacon 426, 427, 433
Flathead Allotment Act of 1904 149, 179, 302
Flathead Culture Committee 302, 413
Flathead Indian Irrigation Project 149, 180
Flathead Reservation 2, 70, 73, 82, 92, 114–118, 146–48, 177–81, 190, 224–25, 260–62, 274, 302, 412–19, 426, 433
Flynn, Michael 379
Food and Drug Administration (FDA) 291, 390, 391, 402
food stamps 149, 283
Ford Foundation 339
Ford, Gerald 254, 278, 286–288, 291, 294, 308, 314
Ford, Henry 133
Forster, William 101, 218
Fort Connah, Montana 146
Fort Detrick, Maryland 351, 388
Fort Duquesne 95
Fort Eustis 271, 272
Fort Missoula, Montana 83
Fox News 339, 388, 392
fractional reserve banking 25, 26, 29, 47, 53, 59, 61, 131, 183, 306, 312, 329, 340, 371, 377, 397, 423

France 3, 6, 12–14, 31–33, 54, 57, 65, 76, 77, 83, 85–87, 97, 106, 107, 131–32, 136–37, 151, 153, 156–64, 177, 185, 189, 191–94, 197, 201, 207, 212, 217–18, 228, 231, 233–240, 243, 254, 257, 264, 282, 375, 394, 396
Franco-Prussian War of 1870 201
Free World 257
French and Indian War 10, 95
French-Canadians 1, 74, 218
French Revolution 58, 160
Friedman, Milton 112, 124, 284, 359
Full-Spectrum Dominance 369, 395, 397, 422

G
Gabbard, Tulsi 433
Gaddafi, Muammar 367, 401
Gagarin, Yuri 266
gain-of-function research 387–88, 402, 422
Gallatin, Albert 14, 32–34
Games People Play (book) 159
Garfield, James 39, 116, 133, 136, 421
Garland, Judy 224
Gates, Bill 336, 382, 390–91, 403, 428
Geithner, Timothy 357, 359
General Colonization Law 349
General Electric 134, 209
genocide 133, 190, 270, 406, 421
George, David Lloyd 193
George V 175
German Empire 159, 163
German Green Party 409
Germany 33, 44, 57, 86, 108, 114, 132, 136, 140, 151–53, 158–59, 163–64, 173, 176, 177, 181, 184, 189, 191, 192, 193, 194, 196, 197, 200, 201, 202, 203, 207–209, 211–12, 216–17, 228, 231–40, 242–43, 252, 255, 257, 280, 288, 293, 324, 327, 331, 349, 373, 375, 394–96, 399, 403, 406, 407–411, 426, 433
Geronimo 405
gerrymandering 378–79

Gettysburg, Battle of 95, 99–100, 102, 120, 258
Ghani, Ashraf 348
Gibson, Donald 267, 268, 274, 433
Gilded Age vii, 109, 125, 207
Glacial Lake Columbia 71
Glacial Lake Missoula 71–72
glasnost 313
Glass, Carter 188
Glass-Steagall 316, 319, 357, 381, 401, 424
Glenn, John 266
gold dinar 367
Golden Triangle 278, 332
Goldman Sachs 283, 316, 334, 337, 356, 357, 359, 361, 366
gold peg 282, 298, 422
Gold Rush
   California 1848 25, 60
   Montana 1864 116
Goldschmid, Harvey 358
gold standard 86, 113–14, 126, 129, 151, 181, 192, 195, 201, 208, 211–13, 216, 228, 233, 241, 282
Goldwater, Barry 270, 272, 294
"The Good Club" 428
The Good War 228–31
Google 309, 339, 358
Gorbachev, Mikhail 313, 315, 324, 404
Gore, Al 332–333, 335
Grant, Orville 121, 122
Grant, Ulysses S. 39, 66
Great Britain 9, 13, 24, 31, 39, 56–57, 59, 64–65, 68, 76, 85, 86, 108, 110, 113, 131–32, 139, 151–53, 156, 164, 167, 169, 170, 173, 176, 184, 194, 201, 216–17, 228, 231, 234, 236, 288
Great Depression 25, 54, 113–14, 132, 190, 210, 212, 214, 216, 220, 235, 422
"Great Game" 162, 292
The Great Gatsby (book) 210
Great Railroad Strike of 1877 127
The Great Rapprochement 181–182
Great Recession 356, 360, 430

Great Reset 364, 423, 428
Greece 240, 243, 244, 338, 360
Greeley, Horace 93
Greenback Party 111, 123, 129
Greenbacks 22, 61, 91, 93, 109–110, 130, 132, 134, 215, 312, 430
Greenspan, Alan 283, 313, 338, 356, 357, 358, 359, 424
Greenwald, Glen 368
Gross Domestic Product (GDP) 306, 326, 357, 431
Guam 133
Guantanamo 258, 349, 350
Guatemala
  coup 258, 259
Guiteau, Charles 136
Gulf of Tonkin 270, 343
Gulf War 298, 307, 314–16, 319, 327

H

Habsburgs 231
Halliburton 298, 327
Hamilton, Alexander 14, 21, 24, 28–33, 36, 55–57, 69, 109, 243, 337
Hammarskjold, Dag 258
Harding, Warren 40, 188, 204, 206, 210
hard money 20, 58, 60, 90, 130
Hard Rock International 11
Harper's Ferry, West Virginia 98
Harrison, Benjamin 40, 133
Harrison, William Henry 59, 66
Harvard Law School 365
Haskell Institute 144
Hayden, Michael 380
Hayes, Rutherford B. 39, 98, 125–26, 128, 149
hedge funds 132, 283, 295, 329, 336, 420, 423
"Helicopter" Bernanke 359
Hellgate Treaty of 1855 82, 114–16, 147, 261–62, 301
Henry, Patrick 273
Heritage Foundation 339
heroin 270, 278, 348, 415
Hersh, Seymour 255, 274, 307, 335, 396, 400, 408, 426

Hill, A.P. 98
Hill, Carolyn 222
Hill, General D.H. 98
Hill, John Clark 140–41, 217–18
Hinckley, John, Jr. 304
Hiroshima 230, 242, 250, 263
Hitler, Adolf 162, 171, 212, 228, 232–34, 250, 327
Hobbes, Thomas 10
Ho Chi Minh 258, 278
Hoecken, Adrian 82, 115
Holland 156–158
Hollande, François 375
Holy Roman Empire 156, 163
Hong Kong 177, 238, 280
Hooker, Joseph 98
Hoover Dam 215
Hoover, Herbert 207, 210, 212
Hoover, J. Edgar 210, 275, 338
House, Edward 197
House of Windsor 153, 165
House Select Committee on Assassinations 269
House Select Subcommittee on the Coronavirus Pandemic 402
house slaves 47
Houston, Sam 50–51
Howard, Admiral Charles 153
Hoyt, Charles 178
Hubbard, Colonel G.E. 143, 145, 222
Hudson, Michael 28, 340, 362, 395, 408
Hudson's Bay Company 15, 73, 77, 146, 227
Hughes, Charles Evans 197
Humphrey, Hubert 271, 276, 285
Huns 165
Hunt, E. Howard 269, 286
Huntington, Samuel 339
Hussein, Saddam 298, 315, 340, 350, 351–52
hydrogen bomb 300
hydroxychloroquine 391
hyperinflation 207, 242, 430

I

ICBM 280
Ice Age 40, 71, 72, 119, 152, 412, 425
Illinois 11, 39, 40–45, 55, 87, 95, 102, 140–42, 217, 286, 365, 426
immigration 47, 50, 55, 108, 152, 171, 396
Imperial Guard 162
income tax 92, 110, 134–35, 188, 191, 194–95
   1913 Income Tax Amendment 188, 191, 304
Independent Treasury 57–61
Indian Claims Commission 115, 301, 308
Indian Gaming Regulatory Act of 1988 417
"Indian New Deal" 227, 261
Indian Removal Act 49, 68
Indian Reorganization Act of 1934 227, 261
Indochina 177, 238, 258
Indonesia 177, 270, 274
Industrial Revolution 57, 160
inflation 21, 24–25, 28, 30–31, 34, 38, 59–60, 91–92, 129, 206, 211–12, 214, 241, 282–83, 296–98, 300, 305, 307, 340, 356, 360, 377, 389, 397, 409, 415, 416, 420, 423, 427
INF Treaty 313, 324, 381
Inside the Company: CIA Diary (book) 290
The International Center for 9/11 Justice 343, 362
International Monetary Fund (IMF) 30, 241, 283, 296, 306, 338
Iran 258, 259, 293, 296, 299, 301, 303, 304, 310–11, 315, 327, 333, 340–41, 353, 355–56, 364, 376, 382, 385, 389, 397, 406
Iran-Contra 303, 310–11, 333
Iran coup 258–59, 315, 327
Iranian hostages 300
Iraq 119, 138, 194, 201, 207, 230, 263, 272, 287, 299, 304, 307, 314–15, 318, 320, 322, 327, 331–32, 340, 349, 350–53, 355, 358, 362–66, 368, 376, 381, 396–99
Iraq War 350–354
Ireland 2, 43, 95, 101, 108, 125, 153, 196, 218, 360, 382, 406
Irish Potato Famine 101
Iron Curtain 245
Iroquois Indians 75
ISIS 303, 352–53, 368, 381–82
Islamic Revolution of 1978–1979 258
isolationism 228
Israel 108, 199, 245, 269, 274, 280, 295, 299, 315, 320, 339, 343, 351, 353, 381, 403
Italy 57, 77, 81, 108, 125, 132, 136, 148, 151, 158, 163, 191, 208, 236, 238, 239, 240, 243, 251, 372, 389, 406
ivermectin 391

J

"Jackals" 252
Jackson, Andrew 23, 35, 48, 54, 57–58, 64, 66–67, 140
Jackson, Henry M. "Scoop" 299
Jackson State University 278
Jackson, Stonewall 97, 99
Jade Helm 15 380, 381
Jamaica 156
James Blair High School 268, 271
Jamestown 2, 8
Janney, Peter 274, 433
Janssen, Rich, Jr. 417, 432
January 6, 2021 385, 393, 403
Japan 87, 170, 177, 233–36, 238, 240, 247, 249, 251–52, 283, 293, 334, 369
JCPOA 376
Jefferson, Thomas 2, 12, 13, 15–18, 31, 52, 140, 161, 337
Jekyll Island, Georgia 185
Jesuit Order 7, 77–79, 146, 147, 261
JINSA 315

Index

Johnson, Andrew 107, 109, 132
Johnson, Boris 398
Johnson, Lyndon 126, 270–72, 276, 280, 283, 285, 293, 343
Johnson, Ron 392
JPMorgan Chase & Co. 229, 356–57, 359, 360, 423

K

Kaiser Wilhelm II 152, 164, 177
Kalispell, Montana 224, 246
Karzai, Hamid 348
Kellogg, Mark 122
Kelly, Gerald L. 77
Kennan, George 257, 322
Kennedy, John F. 126, 237, 259, 262, 266, 267–68, 274, 287, 372, 385, 422, 433
   American University Speech 266
Kennedy, John F., Jr. 328, 335, 422
Kennedy, Robert F. 264
Kennedy, Robert F., Jr. 307, 387, 389–91, 403, 427, 433
Kent State University 278
Kerr Dam 226
Kerry, John 351, 355, 375
Keynes, John Maynard 214, 240
Khalid Sheikh Mohammed 350
Khmer Rouge 277
Khrushchev, Nikita 259, 263
Killer Angels (book) 99
Kimberley diamond mine 166
Kingfisher Free Press 217
Kingfisher Journal 143
Kingfisher, Oklahoma 141, 143, 145, 217, 222
King, Martin Luther, Jr. 126, 267, 271, 275, 284, 422
King Phillip's War of 1675–1678 6
Kissinger, Henry 254, 255, 271, 277, 279, 280, 293, 294, 323, 339, 382
Klobuchar, Amy 392, 403
Knauer, Virginia 305
Kober Press 426, 434
Koch brothers 336
Kolchak, Admiral 232

Kootenai Indians 73, 77–78, 115–16, 124, 145, 150, 177–78, 190, 224, 226–27, 261, 274, 301, 302, 412, 413, 417, 419, 427, 432
Korea 9, 177, 230, 242, 265, 277, 369
Korean War 244, 261
Kristol, William "Bill" 299, 320
Kucinich, Dennis 215, 313, 377, 425, 429, 430, 433
Kucinich, Elizabeth 433
Kuhn, Loeb & Company 181, 183, 185, 186, 200
Ku Klux Klan (KKK) 126
Kulturkampf 163
Kuwait 272, 315, 318
Kuzmarov, Jeremy 275, 320, 333, 362, 398, 402

L

Lake County, Montana 414
Lama, Fadi 410, 432
Lamont, Thomas 208, 229
Lamour, Mirabeau Buonaparte 50
Land Run of 1892 140–145
Lansky, Meyer 245
Laos 253, 270, 274, 277, 278
Lathrop, Johnny 62, 101–102, 432
Laud, Archbishop 3
lawfare 385, 420
Lebanon 201
Lee, Robert E. 89, 97, 99–100
legal tender 22, 25–26, 34, 68, 91, 130
Lend-Lease 217, 235
Lenin, Vladimir 200
Less Developed Countries (LDC) 296, 306
Lewinsky, Monica 324
Lewis and Clark Expedition 6, 15, 16, 17, 68, 70, 73–77, 83, 96, 433
Lewis, Bernard 339
Lewis, Cotesworth Pinckney 272
Lewis, Meriwether 6, 15, 18, 83
"liars' loans" 358
Libertarianism 56, 105, 344
Liberty Loan program 196

Libya 230, 303, 315, 332, 353, 367–68, 376, 377, 396, 397, 399, 401
Liddy, E. Gordon 286
Lincoln, Abraham 39, 56–57, 66, 69, 84–85, 87–94, 99, 101, 104, 106–107, 109, 120, 130, 132, 136, 140, 181, 188, 215, 312, 372, 377, 421, 425
Lindbergh, Charles A., Jr. 218–19, 237
Lindbergh, Charles A., Sr. 186, 189, 202
Lindemann, Frederick 426
Liquified Natural Gas (LNG) 411
Little Big Horn, Battle of 122
Lobengula 173–74
Lolo Pass 116
London Times 93–94
"Lost Order" 98
Louisiana Purchase 12, 14, 33, 49, 76
Louis XIV 157–159, 160, 231
Lower Pend d'Oreille Indians 73, 116, 178, 302
Loyalists 9, 11
LSD 278, 287
Lumumba, Patrice 288
Lusitania 196–98, 232, 343

M

MacGregor, Douglas 250, 364, 399, 401, 433
MacKenzie, Charles 15
Madison, James 32, 35
"magic bullet" theory 268
Magna Carta 10
Maidan Coup 372–76
Malaysia 177
Malcolm X 126, 274
Manassas
  Second battle of 97
  Battle of First (First Bull Run) 88, 90
Manchuria 177, 235
Mandan Indians 15
Manhattan Project 242
Manifest Destiny 49, 52–53, 63–83
Manning, Chelsea 368

Mansfield, Mike 259, 265
Mao Tse Tung 244
Marjorie Virginia Peilow 197, 218
Marshall, George C. 239, 243
Marshall Plan 243, 251–52, 273
Marx, Karl 110
Mary Queen of Scots 153
mastodon 72
Mauser rifles 167, 174
Maximilian von Hapsburg-Lothringen 83, 107
Maxim machine gun 163, 173–74
Mayan ruins 8
McCaffrey, Barry 349
McCain, John 323, 355, 361, 366
McCarthy, Joseph 252, 257
McClellan, General George 87, 97, 99, 120
McCloy, John J. 253
McCone, John 262, 269, 281
McDonald, Angus 75, 145, 146, 227
McDonald, Charlie 227
McDonald, Duncan 146–48, 150, 227
McDonald, Frank 271
McGovern, George 284, 285, 308
McGovern, Ray 364, 400
McGuffey's Eclectic Reader 45
McKinley, William 40, 98, 133–38, 140, 187, 190, 337, 421
McNamara, Robert 263, 266
Medicare 283
Medvedev, Dmitri 370
megafauna 72, 119, 412
Mein Kampf (book) 232
Mellon, Andrew 208
"menagerie" 139
Mennonites 96
Mercouris, Alexander 385, 433
Merkel, Angela 375, 396
Merrill Lynch 304, 359
Mexican drug gangs 427
Mexican War 2, 44, 50, 66, 68–69, 84, 107, 197
MI6 171, 252, 273, 363, 372–73, 375, 398
Michel Chossudovsky 428

Milken, Michael 306
Milner, Alfred 174, 175, 176, 181–82, 193, 197, 199, 203–205, 232, 236
Milner's Kindergarten 176
Minsk agreements 375, 394
Mission Mountain Food Enterprise Center 417
Mission Mountains 147, 149, 224, 227, 260, 302, 416
Mississippi 47, 49, 65, 68, 88, 222, 285
Mississippi River 7, 11–14, 16–17, 29, 40, 45, 47, 52, 67, 97, 118, 260, 285
Missoula, Montana 70–74, 77, 83, 114, 117, 147, 178–79, 219, 220–21, 224, 246, 249, 260–61, 406, 427, 433
Missouri 13, 47, 73, 74, 76, 78, 245
MKULTRA 287
Molotov-Ribbentrop Pact of 1939 403
Monetary Authority 429, 430
Monetary Commission 61, 185
money printing 215, 241, 340, 360
Money Trust vii, 88, 172, 182–83, 189, 195, 206, 360
Monocacy, Battle of 95, 100
Monroe Doctrine 65–66, 107, 136, 140, 151, 163, 232
Montreal 6, 94, 96, 146, 293
Mook, Robby 378
Morgan, J.P. 88, 108, 129, 134, 172, 175, 181–84, 186, 190–91, 421
Morgan, J.P. "Jack" Jr. 191, 194–95
Morgan, Junius 181
Morgenthau Plan 426
Mormons 119
mRNA 390, 391, 428
MSNBC 380
muckrakers 184
Mujahideen 294, 303, 347
Mullins, Eustace 190, 212, 228, 250
Munich Security Conference 355
Murrah Federal Building 324
Muscle Shoals 215
Musk, Elon 336, 339
Mussolini, Benito 208

N

Nagasaki 230, 242
Nalapat, Madhav 307, 382
Napoleon 13, 14, 15, 33, 34, 97, 107, 122, 160–63, 165, 231
Napoleonic Wars 86
NASA 284, 300, 311–12
Nass, Meryl 392
Natal 166
National Banking Act of 1863 60, 92
National Bison Range 149
National Endowment for Democracy (NED) 230, 310, 332–33, 339, 375
National Institute of Allergy and Infectious Diseases (NIAID) 387, 388, 390, 391, 402
National Monetary Commission 61
National Security Act of 1947 242–43, 254
National Security Agency 242, 254, 287–88
National Turnpike 29
Nation of Islam 425
Native Americans 5, 7, 8, 64, 69, 77, 83, 95, 133, 225, 261, 330, 406, 414, 418–19
NATO 159, 201, 228, 244, 256–57, 264, 266, 315, 319, 320, 322–25, 334, 354, 355, 367, 374–76, 381, 394–95, 398, 400, 407, 410, 422
Nazis 152, 216, 233–35, 255, 399
NEED Act of 2011 215, 229, 313, 425, 429, 430
Neocons 303, 323–24, 331–32, 339, 361, 422
Neo Nazis 375
The Netherlands 151–52, 156, 237
New Deal 26, 212, 215–16, 227, 234, 261, 266, 386
New England 4, 5, 6, 18, 40, 48, 53
New Look 257
New Orleans 13, 45, 48–49, 88, 97, 126
The New Republic 339
New Testament 57
New World Order 309

New York City 54–55, 86, 125–26, 204, 208, 219, 254, 287, 293, 320, 339, 341, 386, 424
New York Herald 122–123, 125
The New York Morning News 64
The New York Times 135, 184, 186, 208, 255–56, 262, 308, 312, 316, 337, 339, 350, 353, 361, 368, 402
New-York Tribune 93, 121, 135
New Zealand 129, 151, 175, 177
Nez Perce Indians 116, 146, 149, 227
Nez Perce War of 1877 83, 116, 128, 146
Ngo Dinh Diem 288
Nicaragua 202, 303, 310–11, 355
Nicholas II of Russia 177, 200
Nightmare 251–275
"Night of the Long Knives" 233
Nixon, Richard 126, 255, 279
  near impeachment 286
Noah's Flood 72
Nord Stream 331, 395, 400, 404, 408
Norman Conquest 3
North American Free Trade Agreement (NAFTA) 316, 319, 338
Northern Pacific Railroad 82, 92, 114, 121, 146, 148, 219
North Korea 244, 381
North, Oliver 311
North Vietnam 258, 265, 270, 277–78
Northwest Ordinance 39
Northwest Territory 29
Norway 235, 237, 396, 408
notwendigkeit 405, 434
Nuclear Posture Review 355
nuclear weapons 242, 257, 262, 264, 313, 350, 352, 354–55, 373, 394, 410
Nuland, Victoria 331, 375, 395, 404

O

Obama administration 330, 355, 376
Obama, Barack 38, 330, 332, 336, 348–49, 352, 355–57, 361, 363, 365, 366–70, 375–78, 380, 382, 385–86, 398, 400–401, 403, 422

Occupy Wall Street 38, 360, 424
Odessa 375, 400
Oklahoma land rush 2
Old Ignace 78
Old Northwest 11, 14, 23, 39, 40, 44, 45, 118
Old Testament 45
Omar, Mullah 348
O'Neill, Paul 340, 362
"one pill can kill" 415
Ontario Bank 94
opening to China 279, 291
Open Society Foundation 333
Operation Belladonna 373
Operation Gladio 372
Operation Mockingbird 288
Operation Northwoods 263, 343
Operation Phoenix 138, 271, 277, 349, 353
Opium Wars 138, 162
Orange Free State 165–67
Orange Revolution 354, 374
Oregon Country 13, 35, 59, 64, 68, 76, 77, 118
Oregon Trail 76, 119
organized crime 37, 135, 209–210, 245, 269, 334, 338, 422
O-rings 311, 312
Orwell, George 198
OSHA 284
O'Sullivan, John L. 63
Oswald, Lee Harvey 268–69, 286
Ottoman Empire 65, 151, 160–63, 191, 194, 201

P

Palestine 194, 199, 201, 205, 245
Palin, Sarah 361, 366
Panama 133, 165–94
Panama Canal 133, 165
  Treaty 294
pandemic 332, 360, 386, 387, 388, 389, 390, 391, 403, 414, 422, 430
panics, financial 53, 57, 61, 62, 112, 172, 184
  Panic of 1819 54, 57

Panic of 1837 54, 57, 58, 60
Panic of 1857 54–55, 60–61
Panic of 1907 61
Paris Climate Accords 381
Paris Peace Talks 276, 277
Patriot Act 332, 344, 345, 351, 362, 368
Paul I 161
Paul, Ron 424
Paulson, Henry 360, 366
Paxlovid 390
Peabody, George 181
Pearl Harbor 237, 238, 250, 252, 256, 331, 341, 343–44, 362
Peilow, Carlton William 108, 129, 197, 218
Pelosi, Nancy 336, 360, 393, 424
Pentagon 321, 341–45, 353, 369, 389, 402–403
Pequot War of 1636–1637 6
perestroika 313
Perez, Jody Cahoon 416
Perle, Richard 299, 350
Pershing, General John 197
Persia 106, 158, 176, 194
Peter III 160
Peterson, Esther 292
Peter the Great 158, 176, 372
pharmacoterrorism 406
Phi Beta Kappa 279
Philippine Republic 137
Philippines 9, 119, 133, 137, 164, 238
Phillip II of Spain 153, 155–56
Pickett's charge 100
Pierce, Franklin 83, 87
pig iron 46
The Pilgrim's Progress (book) 44
Pinochet, General Augusto 281
Platte River 76
Poland 158, 159, 160, 176, 233–36, 239, 293, 322, 352, 364, 370, 401, 409–410
Politico 273, 378, 400
Polk, James K. 59, 64, 66, 68
Polson, Montana 414, 417
Pompeo, Mike 243, 382

Pope Pius V 155
Port Arthur 177
Powell, Colin 317, 349, 352, 363
Powers, Francis Gary, Jr. 259
powwows 148, 227, 418, 419, 432
The Precarious Balance (book) 152, 155, 171
Mr. Prince 41
Principles of Economics 56
Prins, Nomi 183, 190, 306, 356–58, 401
Privy Council 155
Prohibition 209–210
Project for the New American Century (PNAC) 331–32, 335, 341, 343, 400
Project Shamrock 288
Property Clause 23
Prophet Muhammad 405
proxy war 1, 140, 149, 156, 193, 228, 230, 277, 303, 308, 331, 332, 335, 339, 352, 362, 368, 369, 376, 381, 393, 395–98, 403, 408–410, 423, 428, 432
Prussia 33, 158–59, 160, 163
Psychological Strategy Board 255
Pueblo Indians 50
Puerto Rico 133
Pujo, Arsène 186
Puritans 3, 6
Putin, Vladimir 327, 354–55, 370

Q

Quebec 6, 9, 77
Quigley, Carroll 182, 203, 205, 236, 383

R

rattlesnake master 41
Ravalli, Anthony 81, 82, 146, 148
Reaganomics 305, 313
Reagan, Ronald 18, 208, 287, 292, 294, 298–300, 302, 308–309, 314
Reason to Live Native 415
Reconstruction 62, 125

Reconstruction Finance
    Corporation 214
"recover America for the British
    Empire" 337, 421
Redacted (newsletter) 389, 400
Red Cloud's War 120
redemption 14, 24, 26, 35–36, 58, 86,
    90, 134
Redford, Robert 221
Red Scare 252, 257
Red Street Journal 373, 399
Regan, Donald 304
Reichsmark 208, 233
remdesivir 390
"Remember the Maine" 137, 198, 263
rendition 320, 349
reparations 201, 202, 203, 207
Report on Manufactures 29
repossession agreements 329
Republican Party 93, 107, 114, 126,
    138, 179, 181, 184, 187, 271, 285,
    361, 379, 386
Republic of Mexico 107
reserve ratio 26, 93, 196
"Return to Normalcy" 204
Revolutionary War 2, 11, 14, 25, 28,
    36, 151, 284, 399
Rhodes, Cecil 24, 129, 131, 166, 168,
    171, 173, 174, 175, 1–76, 181–82,
    193, 197, 199, 203, 232, 236–37,
    363, 421
    Confession of Faith 168–170
    will 182
Rhodes Scholar elite 422
Rhodes scholarships 170
rhyolite quarries 425
Rice, Condoleezza 349, 351
"Right of Conquest" 11–12, 63–64
"Right of Discovery" 11–12, 63–64
Right Sector 374
Ritter, Scott 313, 333, 363, 364, 398,
    433
Roaring Twenties 202, 206–229
Rockefeller Commission 287
Rockefeller, David 236, 252, 254–55,
    267, 273, 280, 282–83, 292–93,
    298, 319, 323, 356, 423, 428

Rockefeller, John D. 133, 183
Rockefeller, John D., Jr. 183, 236, 252,
    308
Rockefeller, Nelson 252, 254, 256, 287,
    308, 336, 356, 386
"Rockefeller Republic" 276–308
Rocky Mountains 13, 71, 75, 77
Romania 108, 239, 354, 370, 401
Ronan, Mary 147, 150
Ronan, Peter 147, 177–78, 418, 427
Rooney, Mickey 224
Roosevelt, Franklin D. 145, 212, 227,
    234, 238, 250, 422
    Fireside Chat 213
Roosevelt, Theodore 139, 179, 204,
    337
Root, Elihu 204
Rose Revolution 354
de Rothschild, Lionel 86
Rothschild, Nathaniel 33, 86, 134, 166,
    167, 170, 173, 182, 199, 203, 232,
    337, 421
Rothschilds 24, 48, 57, 62, 86, 88, 91,
    106–108, 129, 131, 134, 151–52,
    166–67, 173, 181, 184, 189, 193,
    337–39, 382
The Round Table 182
Royal Institute of International Affairs,
    a.k.a. Chatham House 203–204,
    236
Ruhr (region in Germany) 207
Rule Britannia 151–171
Rumsey, James 45, 96
Rumsfeld, Donald 286, 298, 321, 333,
    349, 350, 366
Russia 1, 19, 33, 66, 76, 83, 85,
    106–108, 112, 132, 138, 140, 149,
    151, 156, 158–64, 171, 176–77,
    184, 191–94, 198, 200–203, 208,
    228, 230–33, 241, 243, 257, 259,
    264, 291, 293–94, 299, 307–308,
    315, 317–18, 320, 322, 324–27,
    331–32, 335, 339–40, 349, 352,
    354–56, 362, 364, 368–84,
    393–411, 422, 424, 428, 432
"Russiagate" 339, 361, 363, 379–81,
    384, 400, 422

Index

Russian Federation 326, 355, 374
Russo-Japanese War 164, 176, 177, 191

S

Sacajawea 75
Sadat, Anwwar 295
Saigon 265, 278
Salish Indians 7, 70, 72–82, 114–117, 124, 145, 147, 150, 177, 190, 224, 226–27, 260, 261, 274, 301, –302, 405, 412–13, 419, 427, 432
Salish Kootenai College 150, 177, 274, 302, 413, 419, 427
SALT I 280
SALT II 292
Sanders, Bernie 351, 366, 377–78, 392, 424
Sanders Capital 423
San Francisco 25, 36, 85, 106, 108, 246, 249
San Jacinto, Battle of 50
San Juan Hill 137, 138
Santa Fe, New Mexico 50
Sarkozy, Nicolas 426
Schacht, Hjalmar 208
ScheerPost 400, 401, 403, 431, 432
Schiff, Adam 384
Schiff, Jacob 129, 138, 181, 183, 200
Schlieffen Plan 193
Scholz, Olaf 409, 412
Schultz, Debbie Wasserman 378
Scotland 146, 153–54
scrip 25, 52, 113, 417
Seabees 224, 246–248
Second Bank of the United States 35, 54, 56–57, 66, 140
Secrets of the Temple (book) 297, 308
Sedition Act of 1918 193, 198
Seeley Lake Historical Society 221
Seeley Lake, Montana 2, 114, 129, 219, 220–21, 224, 433
Seminole Indians 11
Serbia 176, 193, 319, 332
Seton, Elizabeth Ann 96
Sevastopol 106, 162, 373, 375
Seven Years War 10, 12, 96, 159
"Shadow Men" 307, 382–86, 401

Shah of Iran 258, 293
Shangani River, Battle of 174
Shanksville, Pennsylvania 341, 342
Shawnee Indians 43, 143, 144
Sheridan, Philip H. 118
Sherman Anti-Trust Act of 1890 183
Sherman Silver Purchase Act of 1890 130, 134
Sherman's March to the Sea 100, 103, 104
Sherman, William Tecumseh 100, 103
Shoshone Indians 73–76
Sidney, Sir Philip 154
Singapore 177, 238
Sioux Indians 39, 73, 78, 115, 118–24, 127–28, 193, 230, 343
Sipido, Jean-Baptiste 175
Sitting Bull 122, 124
Skull and Bones 188, 314, 332
slavery 28, 37–38, 47, 49, 53, 66–67, 69, 83–85, 87, 93
Slidell, John 87, 88
smallpox 5, 7, 73–74, 78, 81–82
Smethurst, Joe 102–104
Smith, Jack 385
Snake River 71, 76
Snowden, Edward 289, 368
Social Security 145, 214, 216, 328
Soleimani, Qasem 381
Solid South 126, 271, 285
Solzhenitsyn, Alexander 326
Soros, George 333, 336, 375, 428
Souers, Sidney 253, 254
South Africa 129, 130, 163, 165–67, 170, 174–76, 182, 397
Southern Strategy 126, 271, 285, 379
South Korea 244
South Mountain 94–96, 98, 126, 425
  Battle of 98
South Vietnam 265, 277–78, 288
Soviet Union 156, 176, 208–209, 211, 228, 232, 233–40, 243–45, 250–52, 255, 257, 260, 262–64, 279–80, 288, 290, 292–94, 299–300, 303, 307, 313–15, 317, 319, 324–25, 327, 333, 341, 344, 347, 354–55, 364, 370, 372–74, 401, 403, 407

Soviet Union (cont.)
　collapse of 341, 354, 355, 370–74
Spain 12, 13, 20, 48, 66, 76–77, 83, 86–87, 107, 132, 133, 135, 137, 151, 153, 156, 158–59, 161, 163, 231, 360
Spanish Armada 153
Spaulding, Elbridge G. 91
Special Military Operation 376, 395
specie 20, 25–26, 29, 34,–36, 54, 58, 60–61, 68, 89–91
Specie Payment Resumption Act of 1875 130
Spenser, Edmund 154
"spirit mirror" 154, 171
"springtime in America" 298, 309–335
SS Minnow 408
stagflation 296
Stalingrad, Battle of 239
Standard Oil 133, 172, 182, 183, 207, 216, 233, 251
Starr, Kenneth 324
Star Wars 312, 354
Stead, W. T. 174, 181, 182
Steele, Ida Florence 51, 222
Stevens, Isaac 82
Stevensville, Montana 78, 116–17
St. Ignatius Mission 77, 145, 225–26, 229, 261, 433
Stillman, James 183, 184
St. Mary's Mission 77, 433
　Historical Foundation 77
stock market crash 196, 202
"Stop the Steal" 392
Strasser, Otto 250
Strategic Defense Initiative (SDI) 312
Strategic Hamlet program 265
Stuart, General Jeb 100
Sullivan, Jake 362, 427
Summers, Lawrence 326
Sund, Steven 393, 403
superdelegates 378
"survival of the fittest" 56
Sweden 158–61, 184, 396, 400
Switzer, Howard 431, 432

Syria 119, 201, 230, 315, 332, 352, 353, 367–68, 376, 38–82, 396, 399

T

Taft, Robert 256
Taft, William Howard 40, 179, 185, 186, 190
Taiwan 244, 294, 423–24
Taiwan Semiconductor Manufacturing Company (TSMC) 423–424
Talbott, Strobe 323
Taliban 346–48
tardis dyskinesia 279
Tavistock Institute 252
Tawney, Roger 85
Team B 299
Tehran 353
Teller, Edward 300
Tenet, George 319, 349
Tennessee Valley Authority 215
Tereshkova, Valentina 266
"termination" 260–262
Tesla, Nikola 133
Texas 13, 47, 49, 50–52, 62–64, 66–67, 69, 73, 88, 118, 141, 197, 222, 243, 268, 297, 299, 311, 314, 319, 332, 380, 401
　Republic of 50, 63
thermobaric bombs 342
Thiel, Peter 336
Thiokol, Inc. 312
Thirty Years War 156
Thompson, Hunter S. 342
thorazine 279
Thunderbirds 72
Tilden, Samuel 125
Tioga, Texas 51, 222
Tora Bora, Battle of 348
Tower Commission 311
Trail of Tears 11
Transjordan 194, 201
Trans-Pacific Partnership 381
Transvaal 165–67, 173, 175
Travelers' Rest 74, 433
Treaty of Brest-Litovsk 194
Treaty of Fort Laramie 76, 115, 119

Treaty of Fort Wayne 23
Treaty of Guadalupe Hidalgo 84
Treaty of Oregon 68, 84
Treaty of Paris 9, 96, 137
Treaty of Versailles 201–204, 228, 410
Treaty of Westphalia 156
Trilateral Commission 283, 292–93, 356
Triple Entente 151, 191
Triumpherate 254
Trotsky, Leon 138, 200, 320
Troubled Asset Relief Program (TARP) 360
Trudeau, Justin 417
Truman Doctrine 243, 244
Trump, Donald 149, 307, 313, 332, 335-36, 358, 361, 370, 376, 378–79, 381–87, 389, 392–93, 398, 400–401, 403, 427
   impeachments 383–85
Tubman, Harriet 85
Tulip Revolution 354
Turner, Ted 336, 428
Tweed, Boss 112
Twitter 339, 343, 384, 393, 403, 422
Two Eagle River School 302, 413
Tyler, John 59, 66, 187

U

U-2 spy plane 259
Uitlanders 174
Ukraine 1, 9, 19, 83, 140, 149, 156, 159, 193, 198, 228, 230, 251, 264, 308, 317, 325, 331–32, 334–35, 338–39, 341, 352, 354–55, 362, 364, 368, 369, 370–77, 379, 381–82, 384, 386, 393–95, 397–404, 408–410, 422, 428, 432
Underground Railroad 69, 85, 96
unemployment 134, 208, 216, 237, 284, 306, 360, 389
United Nations Charter 318, 395
United Nations Security Council 264, 348
University of Montana 77, 79, 220, 229, 249, 330, 433

The Untold History of the United States (book) 198, 205, 276, 362
Unz, Ron 389
U.S. Civil Service Commission 279, 291, 432
U.S. Constitution 9–11, 21, 31, 35, 155, 188–89, 215, 321, 371, 420–22
   Treaty Clause 23
U.S. Court of Claims 302
U.S. Customs and Border Protection 345
U.S. Department of Health and Human Services 391
"useless eaters" 421
U.S. Forest Service 220
U.S. governance 365–404
U.S. Indian Industrial School 127
U.S. Mints 130
   Philadelphia 24
   San Francisco 25
USS Cole 322
U.S. Secret Services 269, 276, 345
U.S. Steel 134, 172, 267
U.S. Treasury Department 2, 24, 59, 131, 149, 184, 185, 295, 298, 312, 328, 369, 399
usury 27–28, 31, 37, 53, 86, 297, 338, 371, 377, 397, 423

V

Van Buren, Martin 35, 58, 64, 66
Vance, Cyrus 292–93, 301
Vanderlip, Frank 186
Vanguard 283, 316, 361
Viet Cong 265, 270, 277–78
Viet Minh 258
Vietnamization 271, 277
Vietnam War 138, 270, 271, 272, 277, 282, 291–92, 296, 299
Volcker, Paul 283, 297, 297–298, 300
Volcker Recession 298, 423
vulture capitalism 334

W

Wallace, Ben 398, 400
Wallace, George 285

Wallace, Henry 401
Wall Street 38, 55, 125–26, 138, 183–84, 187, 195, 205–206, 228, 235, 251, 256, 273, 274, 304, 309, 357, 358, 359–60, 366, 377, 381–83, 385, 424
The Wall Street Journal 428
Walsingham, Sir Francis 154, 171
War and Peace Studies 237
Warburg, Paul 185, 186, 189
War Guilt Clause 410
War of 1812 2, 33, 34, 39, 66, 77, 87, 141, 151
War of the Spanish Succession 158
War on Poverty 283
War on Terror 230, 287, 320, 327, 344, 347, 349, 353, 357, 365, 366, 369, 386, 395, 407, 422
Warsaw Pact 244, 266, 314, 323, 324, 325
Washington, George 10, 14, 46, 95, 140, 372
The Washington Post 256, 268, 320, 331, 339, 352
Washita River, Battle of 120
waterboarding 350
Watergate 279, 285–86, 287, 307
Waterloo, Battle of 13
Watt, James 45
weapons of mass destruction (WMDs) 320, 350, 352
Webster, Daniel 53
Weimar Republic 201
Welles, Sumner 236
Wells, David A. 135
Wells, H.G. 203
Wellstone, Paul 351, 422
Wertheimer, Al 223–224
Western front 239
West Hollywood High School 224
Westinghouse, George 133
Whig Party 53, 55, 59
Whiskey Rebellion 32, 89
white colonies 129
White Helmets 368
Whitman, Walt 85

Whitworth, Charles 161
Wikileaks 355, 368, 378
Wikipedia 124, 150, 309, 334, 339, 364
wildcat banks 36
Wilhelm I 159, 163
Wilhelm II 152, 164, 177
William of Orange 157
Williamsburg, Virginia 2, 8, 12, 66, 246, 249, 267, 268, 271–72, 276, 278, 290, 308
Williamson, Marianne 424
Williams, Roger 6
Willis, Fani 385
Wilson, Woodrow 186, 187, 188, 191, 193, 194, 200, 204, 237, 270
    Fourteen Points 200–201
Winfrey, Oprah 336, 428
Winthrop, John 6
"woke" 339
Wolfowitz Doctrine 316–18, 331, 351, 395, 422
Wolfowitz, Paul 299, 315–16, 351
Working Men's Party 58
World Bank 30, 283, 296, 306, 316, 338, 364
World Health Organization (WHO) 338, 386–87, 390, 392, 403
World War I 337
World War II 337
World War III 376, 423, 424
Wray, Christopher 388
WTC-7 341, 342, 343
Wuhan Institute of Virology 387–88
Wyoming Massacre 11

Y

Yale University 188, 268, 272, 314
    Law School 268
Yanukovych, Viktor 370, 375
Yellowstone National Park 74
Yeltsin, Boris 323, 325–27, 374
Yom Kippur War (1973) 280, 295
Young Ignace 78–79
Yugoslav Bulldozer Revolution of 2000 354
Yugoslavia 230, 239–40, 319, 354

Z

Zarlenga, Stephen 429, 430
Zelensky, Volodymyr 375
Zimmerman Telegram 197
Zionism 199
Zuckerberg, Jeff 336
Zulu people 165